OPTIMIZATION WITH MULTIVALUED MAPPINGS

Theory, Applications, and Algorithms

Springer Series in Optimization and Its Applications

VOLUME 2

Managing Editor
Panos M. Pardalos (University of Florida)

Editor—Combinatorial Optimization
Ding-Zhu Du (University of Texas at Dallas)

Advisory Board
J. Birge (University of Chicago)
C.A. Floudas (Princeton University)
F. Giannessi (University of Pisa)
H.D. Sherali (Virginia Polytechnic and State University)
T. Terlaky (McMaster University)
Y. Ye (Stanford University)

Aims and Scope

Optimization has been expanding in all directions at an astonishing rate during the last few decades. New algorithmic and theoretical techniques have been developed, the diffusion into other disciplines has proceeded at a rapid pace, and our knowledge of all aspects of the field has grown even more profound. At the same time, one of the most striking trends in optimization is the constantly increasing emphasis on the interdisciplinary nature of the field. Optimization has been a basic tool in all areas of applied mathematics, engineering, medicine, economics and other sciences.

The series *Springer Optimization and Its Applications* publishes undergraduate and graduate textbooks, monographs and state-of-the-art expository works that focus on algorithms for solving optimization problems and also study applications involving such problems. Some of the topics covered include nonlinear optimization (convex and nonconvex), network flow problems, stochastic optimization, optimal control, discrete optimization, multi-objective programming, description of software packages, approximation techniques and heuristic approaches.

OPTIMIZATION WITH MULTIVALUED MAPPINGS

Theory, Applications, and Algorithms

Edited by

STEPHAN DEMPE
TU Bergakademie Freiberg, Germany

VYACHESLAV KALASHNIKOV
ITESM, Monterrey, Mexico

ISBN: 978-1-4419-4167-1
e-ISBN: 978-0-387-34221-4

Printed on acid-free paper.

AMS Subject Classifications: 90C26, 91A65

© 2006 Springer Science+Business Media, LLC
Softcover reprint of the hardcover 1st edition 2006

All rights reserved. This work may not be translated or copied in whole or in part without the written permission of the publisher (Springer Science+Business Media, LLC, 233 Spring Street, New York, NY 10013, USA), except for brief excerpts in connection with reviews or scholarly analysis. Use in connection with any form of information storage and retrieval, electronic adaptation, computer software, or by similar or dissimilar methodology now known or hereafter developed is forbidden.

The use in this publication of trade names, trademarks, service marks, and similar terms, even if they are not identified as such, is not to be taken as an expression of opinion as to whether or not they are subject to proprietary rights.

9 8 7 6 5 4 3 2 1

springer.com

This volume is dedicated to our families, our wives Jutta and Nataliya, our children Jana and Raymond, Vitaliy and Mariya

Contents

Preface

Optimization problems involving multivalued mappings in constraints or as the objective function are investigated intensely in the area of non-differentiable non-convex optimization. Such problems are well-known under the names bilevel programming problems [2, 4, 5], mathematical problems with equilibrium (complementarity) constraints (MPEC) [6, 9], equilibrium problems with equilibrium constraints (EPEC) [8, 10], set-valued optimization problems [3] and so on. Edited volumes on the field are [1, 7]. Since the publication of these volumes there has been a tremendous development in the field which includes the formulation of optimality conditions using different kinds of generalized derivatives for set-valued mappings (as e.g. the coderivative of Mordukhovich), the opening of new applications (as the calibration of water supply systems), or the elaboration of new solution algorithms (as for instance smoothing methods). We are grateful to the contributors of this volume that they agreed to publish their newest results in this volume. These results reflect most of the recent developments in the field.

The contributions are classified into three parts. Focus in the first part is on bilevel programming.

Different promising possibilities for the construction of optimality conditions are the topic of the paper "Optimality conditions for bilevel programming problems." Moreover, the relations between the different problems investigated in this volume are carefully considered.

The computation of best tolls to be payed by the users of a transportation network is one important application of bilevel programming. Especially in relation to recent economic developments in traffic systems this problem has attracted large interest. In the paper "Path-based formulations of a bilevel toll setting problem," M. Didi-Biha, P. Marcotte and G. Savard describe different formulations of this problem and develop efficient solution algorithms.

Applying Mordukhovich's coderivative to the normal cone mapping for the feasible set of the lower level problem, J. Dutta and S. Dempe derive necessary optimality conditions in "Bilevel programming with convex lower level problems."

Most papers in bilevel programming concentrate on continuous problems in both levels of hierarchy. D. Fanghänel investigates problems with discrete lower level problems in her paper "Optimality criteria for bilevel programming problems using the radial subdifferential." Important for her developments are the notions of the optimistic resp. the pessimistic optimal solution. Both notions reduce the bilevel problem to the minimization of a discontinuous function. To develop the optimality conditions she uses the analytic tool of a radial-directional derivative and the radial subdifferential.

In their paper "On approximate mixed Nash equilibria and average marginal functions for two-stage three-players games," L. Mallozzi and J. Morgan investigate bilevel optimization problems where the two lower level players react computing a Nash equilibrium depending on the leader's selection. Other than in D. Fanghänel's paper, they apply a regularization approach to the bilevel problem on order to get a continuous auxiliary function.

The second part collects papers investigating mathematical programs with equilibrium constraints.

Due to violation of (most of) the standard regularity conditions at all feasible points of mathematical programs with equilibrium constraints using classical approaches only necessary optimality conditions in form of the Fritz John conditions can be obtained. This leads to the formulation of weaker optimality conditions as A-, B-, C- and M-stationarity. This opens the way for the investigation of assumptions guaranteeing that one of these conditions can be shown to be necessary for optimality. In their paper "A direct proof for M-stationarity under MPEC-GCQ for mathematical programs with equilibrium constraints," M. L. Flegel and C. Kanzow give an interesting proof of such a result.

To solve and to investigate bilevel programming problems or mathematical programs with equilibrium constraints these are usually transformed into a (nondifferentiable) standard optimization problem. In "On the use of bilevel programming for solving a structural optimization problem with discrete variables," J. J. Júdice et al. use the opposite approach to solve an applied large-dimensional mixed integer programming problem. They transform this problem into a mathematical program with equilibrium constraints and solve the latter problem using their complementarity active-set algorithm. This then results in a promising algorithm for the original problem.

A further applied problem is investigated in the paper "On the control of an evolutionary equilibrium in micromagnetics" by M. Kočvara et al. They model the problem as an MPEC in infinite dimensions. The problem after discretization can be solved by applying the implicit programming technique since the solution of the lower level's evolutionary inequality in uniquely determined. The generalized differential calculus of B. Mordukhovich is used to compute the needed subgradients of the composite function.

An important issue of the investigations is the formulations of solution algorithms. Promising attempts of applying certain algorithms of nonlinear mathematical programming are quite recent. S. Leyffer investigates in his paper "Complementarity constraints as nonlinear equations: Theory and numerical experience" one such algorithm. The method is based on an exact smoothing approach to the complementarity conditions. Comprehensive numerical tests show the power of the method.

Again a challenging applied problem is investigated in the paper "A semi-infinite approach to design centering" by O. Stein. He formulated this problem as a general semi-infinite optimization problem. After certain reformulations, the intrinsic bilevel structure of the problem is detected.

The third part on multivalued set-valued optimization starts with the paper "Contraction mapping fixed point algorithms for solving multivalued mixed variational inequalities" by P. N. Anh and L. D. Muu. They use a fixed-point approach to solve multivalued variational inequalities and Banach's contraction mapping principle to find the convergence rate of the algorithm.

Mordukhovich's extremal principle is used in the paper "Optimality conditions for a d.c. set-valued problem via the extremal principle" by N. Gadhi to derive optimality conditions for a set-valued optimization problem with an objective function given as difference of two convex mappings. The necessary optimality conditions are given in form of set inclusions.

Second order necessary optimality conditions for set-valued optimization problems are derived in the last paper "First and second order optimality conditions in set optimization" in this volume by V. Kalashnikov et al. The main theoretical tool in this paper is an epiderivative for set-valued mappings.

We are very thankful to the referees of the enclosed papers who with their reports for the papers have essentially contributed to the high scientific quality of the enclosed papers. We are also thankful to Sebastian Lohse for carefully reading the text and to numerous colleagues for fruitful and helpful discussions on the topic during the preparation of this volume. Last but not least we thank John Martindale and Robert Saley from the publisher for their continuing support.

Freiberg and Monterrey,
February 2006

Stephan Dempe
Vyatcheslav V. Kalashnikov

References

1. G. Anandalingam and T.L. Friesz (eds.) (1992), Hierarchical Optimization, *Annals of Operations Research*, Vol. 34.
2. J.F. Bard (1998), *Practical Bilevel Optimization: Algorithms and Applications*, Kluwer Academic Publishers, Dordrecht.
3. G.Y. Chen and J. Jahn (eds.) (1998), *Mathematical Methods of Operations Research*, Vol. 48.
4. S. Dempe (2002), *Foundations of Bilevel Programming*, Kluwer Academic Publishers, Dordrecht.

5. S. Dempe (2003), Annotated Bibliography on Bilevel Programming and Mathematical Programs with Equilibrium Constraints, *Optimization*, 52: 333-359.
6. Z.-Q. Luo and J.-S. Pang and D. Ralph (1996), *Mathematical Programs with Equilibrium Constraints*, Cambridge University Press, Cambridge.
7. A. Migdalas and P.M. Pardalos and P. Värbrand (eds.) (1998), *Multilevel Optimization: Algorithms and Applications*, Kluwer Academic Publishers, Dordrecht.
8. B. S. Mordukhovich (2003), *Equilibrium problems with equilibrium constraints via multiobjective optimization*, Technical Report, Wayne State University, Detroit, USA.
9. J. Outrata and M. Kočvara and J. Zowe (1998), *Nonsmooth Approach to Optimization Problems with Equilibrium Constraints*, Kluwer Academic Publishers, Dordrecht.
10. J.-S. Pang and M. Fukushima (2005), Quasi-variational inequalities, generalized Nash equilibria, and multi-leader-follower games, *Comput. Manag. Sci.*, 2: 21–56.

Part I

Bilevel Programming

Optimality conditions for bilevel programming problems

Stephan Dempe[1], Vyatcheslav V. Kalashnikov[2] and Nataliya Kalashnykova[3]

[1] Department of Mathematics and Informatics, Technical University Bergakademie Freiberg, 09596 Freiberg, Germany dempe@tu-freiberg.de
[2] Departamento de Ingeniería Industrial y de Sistemas, Centro de Calidad ITESM, Campus Monterrey, Ave. Eugenio Garza Sada 2501 Sur Monterrey, N.L., México, 64849, on leave from the Central Economics and Mathematics Institute (CEMI) of the Russian Academy of Sciences, Nakhimovsky prospect 47, Moscow 117418, Russian Federation kalash@itesm.mx
[3] Facultad de Ciencias Físico-Matemáticas, Universidad Autónoma de Nuevo León, San Nicolás de los Garza, N.L., México, 66450 nkalash@fcfm.uanl.mx

Summary. Focus in the paper is on optimality conditions for bilevel programming problems. We start with a general condition using tangent cones of the feasible set of the bilevel programming problem to derive such conditions for the optimistic bilevel problem. More precise conditions are obtained if the tangent cone possesses an explicit description as it is possible in the case of linear lower level problems. If the optimal solution of the lower level problem is a PC^1-function, sufficient conditions for a global optimal solution of the optimistic bilevel problem can be formulated. In the second part of the paper relations of the bilevel programming problem to set-valued optimization problems and to mathematical programs with equilibrium constraints are given which can also be used to formulate optimality conditions for the original problem. Finally, a variational inequality approach is described which works well when the involved functions are monotone. It consists in a variational re-formulation of the optimality conditions and looking for a solution of the thus obtained variational inequality among the points satisfying the initial constraints. A penalty function technique is applied to get a sequence of approximate solutions converging to a solution of the original problem with monotone operators.

Key words: Bilevel Programming, Set-valued Optimization, Mathematical Programs with Equilibrium Constraints, Necessary and Sufficient Optimality Conditions, Variational Inequality, Penalty Function Techniques

1 The bilevel programming problem

Bilevel programming problems are hierarchical in the sense that two decision makers make their choices on different levels of hierarchy. While the first one,

S. Dempe and V. Kalashnikov (eds.), *Optimization with Multivalued Mappings*, pp. 3-28
©2006 Springer Science + Business Media, LLC

the so-called upper level decision maker or leader fixes his selections x first, the second one, the follower or lower level decision maker determines his solution y later in full knowledge of the leader's choice. Hence, the variables x play the role of parameters in the follower's problem. On the other hand, the leader has to anticipate the follower's selection since his revenue depends not only on his own selection but also on the follower's reaction.

To be more precise, let the follower make his decision by solving a parametric optimization problem

$$\Psi(x) := \underset{y}{\operatorname{argmin}} \{f(x,y) : g(x,y) \leq 0\}, \tag{1}$$

where $f, g_i : \mathbb{R}^n \times \mathbb{R}^m \to \mathbb{R}$, $i = 1, \ldots, p$ are smooth (at least twice continuously differentiable) functions, convex with respect to y for each fixed x.

Then, the leader's problem consists in minimizing the continuously differentiable function $F : \mathbb{R}^n \times \mathbb{R}^m \to \mathbb{R}$ subject to the constraints $y \in \Psi(x)$ and $x \in X$, where $X \subseteq \mathbb{R}^n$ is a closed set. This problem has been discussed in the monographs [1] and [5] and in the annotated bibliography [6].

Since the leader controls only the variable x , this problem is well-defined only in the case when the optimal solution of the lower level problem (1) is uniquely determined for all parameter values $x \in X$. If this is not the case the optimistic and pessimistic approaches have been considered in the literature, see e.g. [32]. Both approaches rest on the introduction of a new lower level problem.

The optimistic approach can be applied if the leader assumes that the follower will always take an optimal solution which is the best one from the leader's point of view, which leads to the problem

$$\min\{\varphi_o(x) : x \in X\}, \tag{2}$$

where

$$\varphi_o(x) := \min_y \{F(x,y) : y \in \Psi(x)\}. \tag{3}$$

Problem (2)-(3) is obviously equivalent to

$$\min_{x,y} \{F(x,y) : x \in X,\ y \in \Psi(x)\} \tag{4}$$

provided that the latter problem has an optimal solution. But note that this equivalence is true only for global minima [5]. It is easy to see that each locally optimal solution of problem (2)-(3) is also a locally optimal solution of problem (4), but the opposite implication is in general not true.

Example 1.1 Consider the simple linear bilevel programming problem

$$\min\{x : y \in \Psi(x), -1 \leq x \leq 1\},$$

where

$$\Psi(x) = \operatorname*{argmin}_y \{xy : 0 \le y \le 1\}$$

at the point $(x^0, y^0) = (0,0)$. Then, this point is a locally optimal solution to problem (4), i.e. there exists an open neighborhood $W_\varepsilon(0,0) = (-\varepsilon, \varepsilon) \times (-\varepsilon, \varepsilon)$ with $0 < \varepsilon < 1$ such that $x \ge 0$ for all $(x,y) \in W_\varepsilon(0,0)$ with $y \in \Psi(x)$ and $-1 \le x \le 1$. The simple reason for this is that there is no $-\varepsilon < y < \varepsilon$ with $y \in \Psi(x)$ for $x < 0$ since $\Psi(x) = \{1\}$ for $x < 0$. But if we consider the definition of a locally optimistic optimal solution by solving problem (2) then the point $(0,0)$ is not a locally optimistic optimal solution since $x^0 = 0$ is not a local minimum of the function $\varphi_o(x) = x$. △

The basic assumption for this approach is cooperation between the follower and the leader. If the follower cannot be assumed to cooperate with the leader, the latter applies the pessimistic approach

$$\min\{\varphi_p(x) : x \in X\}, \tag{5}$$

where

$$\varphi_p(x) := \max_y\{F(x,y) : y \in \Psi(x)\}. \tag{6}$$

Then, the following notions of optimality can be used:

Definition 1.1 *A point $(\overline{x}, \overline{y})$ is called a locally optimistic optimal solution of the bilevel programming problem if*

$$\overline{y} \in \Psi(\overline{x}), \overline{x} \in X, F(\overline{x}, \overline{y}) = \varphi_o(\overline{x})$$

and there is a number $\varepsilon > 0$ such that

$$\varphi_o(x) \ge \varphi_o(\overline{x}) \ \forall\, x \in X, \ \ \|x - \overline{x}\| < \varepsilon.$$

Definition 1.2 *A point $(\overline{x}, \overline{y})$ is called a locally pessimistic optimal solution of the bilevel programming problem if*

$$\overline{y} \in \Psi(\overline{x}), \overline{x} \in X, F(\overline{x}, \overline{y}) = \varphi_p(\overline{x})$$

and there is a number $\varepsilon > 0$ such that

$$\varphi_p(x) \ge \varphi_p(\overline{x}) \ \forall\, x \in X, \ \ \|x - \overline{x}\| < \varepsilon.$$

Using these definitions it is possible to determine assumptions guaranteeing the existence of locally optimal solutions [5].
(C) The set

$$\{(x,y) : x \in X, \ g(x,y) \le 0\}$$

is nonempty and bounded.
(MFCQ) The Mangasarian-Fromowitz constraint qualification is satisfied at a point $(\overline{x}, \overline{y})$ if there is a direction d such that

$$\nabla_y g_i(\overline{x}, \overline{y}) d < 0, \ \forall\, i \in \{j : g_j(\overline{x}, \overline{y}) = 0\}.$$

A point-to-set mapping $\Gamma : \mathbb{R}^p \to 2^{\mathbb{R}^q}$ maps points $w \in \mathbb{R}^p$ to sets $\Gamma(w) \subseteq \mathbb{R}^q$.

Definition 1.3 *A point-to-set mapping $\Gamma : \mathbb{R}^p \to 2^{\mathbb{R}^q}$ is said to be upper semicontinuous at a point $\overline{w} \in \mathbb{R}^p$ if for each open set $A \supseteq \Gamma(\overline{w})$ there is an open set $V \ni \overline{w}$ such that $\Gamma(w) \subseteq A$ for all $w \in V$. The point-to-set mapping Γ is lower semicontinuous at $\overline{w} \in \mathbb{R}^p$ provided that for each open set $A \subseteq \mathbb{R}^q$ with $\Gamma(\overline{w}) \cap A \neq \emptyset$ there is an open set $V \ni \overline{w}$ with $\Gamma(w) \cap A \neq \emptyset$ for all $w \in V$.*

Theorem 1.1 ([21],[33]) *A locally optimal optimistic solution of the bilevel programming problem exists provided the point-to-set mapping $\Psi(\cdot)$ is upper semicontinuous at all points $x \in X$ and assumption (C) is satisfied. A locally optimal pessimistic solution exists if upper semicontinuity of the mapping $\Psi(\cdot)$ is replaced by lower semicontinuity.*

It should be mentioned that the point-to-set mapping $\Psi(\cdot)$ is upper semicontinuous at a point $\overline{x} \in X$ if (C) and (MFCQ) are satisfied at all points $(\overline{x}, \overline{y})$ with $\overline{y} \in \Psi(\overline{x})$. In most cases, to guarantee lower semicontinuity of the point-to-set mapping $\Psi(\cdot)$, uniqueness of an optimal solution of problem (1) is needed.

2 Optimality conditions

To derive optimality conditions for the optimistic bilevel programming problem we have two possibilities. Either we apply the contingent or some other cone to the feasible set of the bilevel programming problem

$$M := \operatorname{Gph} \Psi \cap (X \times \mathbb{R}^m),$$

where $\operatorname{Gph} \Psi := \{(x,y)^\top : y \in \Psi(x)\}$ denotes the graph of the point-to-set mapping $\Psi(\cdot)$, or we use one of the known reformulations of the bilevel programming problem to get a one-level optimization problem and formulate optimality conditions for the latter problem. Focus in this paper is on possible advantages and difficulties related with the one or the other of these approaches. We start with the first one.

Definition 2.1 *The cone*

$$\begin{aligned} C_M(x,y) := \{ (u,v)^\top : \exists \{t_k\}_{k=1}^\infty \subset \mathbb{R}_+,\ \exists \{(u^k,v^k)^\top\}_{k=1}^\infty \subset \mathbb{R}^n \times \mathbb{R}^m \\ \text{with } (x,y)^\top + t_k (u^k,v^k)^\top \in \operatorname{Gph} \Psi\ \forall k,\ x + t_k u^k \in X, \\ \lim_{k\to\infty} t_k = 0,\ \lim_{k\to\infty} (u^k,v^k)^\top = (u,v)^\top \} \end{aligned}$$

is the contingent (or Bouligand) cone of M.

Theorem 2.1 *If the point $(\overline{x}, \overline{y})^\top \in \operatorname{Gph} \Psi$, $\overline{x} \in X$ is a locally optimal solution of the optimistic problem (4), then*

$$\nabla F(\overline{x},\overline{y})(d,r)^\top \geq 0$$

for all

$$(d,r)^\top \in C_M(x,y).$$

On the other hand, if $(\overline{x},\overline{y})^\top \in \operatorname{Gph}\Psi$, $\overline{x} \in X$ *and*

$$\nabla F(\overline{x},\overline{y})(d,r)^\top > 0$$

for all

$$(d,r)^\top \in C_M(x,y),$$

then the point $(\overline{x},\overline{y})^\top$ *is a locally optimal solution of (4).*

Proof. Let $(\overline{x},\overline{y})^\top \in \operatorname{Gph}\Psi$, $\overline{x} \in X$ be a locally optimal solution of problem (4). Assume that the proposition of the theorem is not satisfied. Then, there exists a direction $(d,r)^\top$ with

$$(d,r)^\top \in C_M(\overline{x},\overline{y})$$

and

$$\nabla F(\overline{x},\overline{y})(d,r)^\top < 0. \tag{7}$$

Then, by definition there are sequences $\{t_k\}_{k=1}^\infty \subset \mathbb{R}_+$, $\{(u^k,v^k)^\top\}_{k=1}^\infty \subset \mathbb{R}^n \times \mathbb{R}^m$ with $(\overline{x},\overline{y})^\top + t_k(u^k,v^k)^\top \in \operatorname{Gph}\Psi \ \forall k$, $\overline{x} + t_k u^k \in X$, $\lim_{k\to\infty} t_k = 0$, $\lim_{k\to\infty} (u^k,v^k)^\top = (d,r)^\top$. Hence, using the definition of the derivative we get

$$F(\overline{x} + t_k u^k, \overline{y} + t_k v^k) = F(\overline{x},\overline{y}) + t_k \nabla F(\overline{x},\overline{y})(u^k,v^k) + o(t_k)$$

for sufficiently large k, where $\lim_{k\to\infty} \frac{o(t_k)}{t_k} = 0$. Since

$$\lim_{k\to\infty} \left\{\nabla F(\overline{x},\overline{y})(u^k,v^k) + \frac{o(t_k)}{t_k}\right\} = \nabla F(\overline{x},\overline{y})(d,r) < 0$$

by the assumption this implies

$$\nabla F(\overline{x},\overline{y})(u^k,v^k) + \frac{o(t_k)}{t_k} < 0$$

for all sufficiently large k and, hence,

$$F(\overline{x} + t_k u^k, \overline{y} + t_k v^k) < F(\overline{x},\overline{y})$$

for large k. This leads to a contradiction to local optimality.

Now, let $\nabla F(\overline{x},\overline{y})(d,r)^\top > 0$ for all $(d,r)^\top \in C_M(\overline{x},\overline{y})$ and assume that there is a sequence $(x^k,y^k) \in M$ converging to $(\overline{x},\overline{y})^\top$ with $F(x^k,y^k) < F(\overline{x},\overline{y})$ for all k. Then,

$$\left(\frac{x^k - \overline{x}}{\|(x^k, y^k) - (\overline{x}, \overline{y})\|}, \frac{y^k - \overline{y}}{\|(x^k, y^k) - (\overline{x}, \overline{y})\|}\right)^\top$$

converges to some $(d, r)^\top \in C_M(\overline{x}, \overline{y})$. Using differential calculus, it is now easy to verify that

$$\nabla F(\overline{x}, \overline{y})(d, r)^\top \leq 0$$

contradicting our assumption. □

Applying this theorem the main difficulty is the computation of the contingent cone. This has been done e.g. in the paper [7].

2.1 The linear case

If bilevel programming problems with linear lower level problems are under consideration, an explicit description of this contingent cone is possible [8] under a certain regularity condition. For this, consider a linear parametric optimization problem

$$\max_y \{c^\top y : \ Ay = b, \ y \geq 0\} \tag{8}$$

with a (m, n)-matrix A and parameters in the right-hand side as well as in the objective function. Let $\Psi_L(b, c)$ denote the set of optimal solutions of (8). A special optimistic bilevel programming problem reads as

$$\min_{y,b,c} \{f(y) : Bb = \widetilde{b}, Cc = \widetilde{c}, y \in \Psi_L(b, c)\}. \tag{9}$$

Using linear programming duality problem (9) has a reformulation as

$$\begin{aligned} f(y) &\longrightarrow \min_{y,b,c,u} \\ Ay &= b \\ y &\geq 0 \\ A^\top u &\geq c \\ y^\top (A^\top u - c) &= 0 \\ Bb &= \widetilde{b} \\ Cc &= \widetilde{c}. \end{aligned} \tag{10}$$

It should be noted that the objective function in the upper level problem does not depend on the parameters of the lower level one. This makes a more precise definition of a locally optimal solution of problem (9) necessary:

Definition 2.2 *A point $\overline{y}$ is a locally optimal solution of problem (9) if there exists an open neighborhood U of $\overline{y}$ such that $f(\overline{y}) \leq f(y)$ for all y, b, c with $Bb = \widetilde{b}$, $Cc = \widetilde{c}$ and $y \in U \cap \Psi_L(b, c)$.*

The main result of this definition is the possibility to drop the explicit dependence of the solution of the problem (10) on c. This dependence rests on solvability of the dual problem and is guaranteed for index sets I in the set $\mathcal{I}(y)$ below.

Let the following index sets be determined at some point $\overline{y}$:

1. $I(\overline{y}) = \{i : \overline{y}_i = 0\}$,
2. $I(u,c) = \{i : (A^\top u - c)_i > 0\}$
3. $\mathcal{I}(\overline{y}) = \{I(u,c) : A^\top u \geq c, (A^\top u - c)_i = 0 \ \forall i \notin I(\overline{y}), \ Cc = \widetilde{c}\}$
4. $I^0(\overline{y}) = \bigcap_{I \in \mathcal{I}(\overline{y})} I$.

Using these definitions, problem (10) can be transformed into the following one by replacing the complementarity conditions:

$$
\begin{aligned}
f(y) &\longrightarrow \min_{y,b,I} \\
Ay &= b \\
y &\geq 0 \\
y_i &= 0 \quad \forall i \in I \\
Bb &= \tilde{b} \\
I &\in \mathcal{I}(y).
\end{aligned}
\tag{11}
$$

The tangent cone to the feasible set of the last problem is

$$
T(\overline{y}) := \bigcup_{I \in \mathcal{I}(\overline{y})} T_I(\overline{y}),
$$

where

$$
T_I(\overline{y}) = \{d | \ \exists r : \ Ad = r, \ Br = 0, \ d_i \geq 0, \ \forall i \in I(\overline{y}) \setminus I, \ d_i = 0, \ \forall i \in I\}
$$

for all $I \in \mathcal{I}(\overline{y})$. Note that $T(\overline{y})$ is the tangent cone to the feasible set of problem (9) with respect to Definition 2.2.

Theorem 2.2 *[Optimality conditions, [8]] If f is differentiable at $\overline{y}$, this point is a local optimum of (9) if and only if $\nabla f(\overline{y}) \cdot d \geq 0$ for all $d \in \operatorname{conv} T(\overline{y})$.*

For an efficient verification of the condition in Theorem 2.2 a compact formula for the convex hull of the tangent cone of the feasible set is crucial. For that consider the relaxed problem of (10)

$$
\begin{aligned}
f(y) &\longrightarrow \min_{y,b} \\
Ay &= b \\
y_i &\geq 0 \quad i = 1, \ldots, l \\
y_i &= 0 \quad i = l+1, \ldots, k \\
Bb &= \widetilde{b}
\end{aligned}
\tag{12}
$$

together with the tangent cone to its feasible set

$$T_R(\overline{y}) = \{d |\ \exists r : Ad = r,\ Br = 0,\ d_i \geq 0,\ i = 1, \ldots, l,\ d_i = 0,\ i = l+1, \ldots, k\}$$

(relative to y only) at the point $\overline{y}$. Here, it is assumed that $I(\overline{y}) = \{1, \ldots, k\}$ and $I^0(\overline{y}) = \{l+1, \ldots, k\}$.

Remark 2.1 ([8]) *We have* $j \in I(\overline{y}) \setminus I^0(\overline{y})$ *if and only if the system*

$$\begin{aligned} (A^\top u - c)_i &= 0 \quad \forall i \notin I(\overline{y}) \\ (A^\top u - c)_j &= 0 \\ (A^\top u - c)_i &\geq 0 \quad \forall i \in I(\overline{y}) \setminus \{j\} \\ Cc &= \tilde{c} \end{aligned}$$

has a solution.

In the following theorem we need an assumption: The point $\overline{y}$ is said to satisfy the *full rank condition*, if

$$\operatorname{span}(\{A_i :\ i = k+1, \ldots, n\}) = \mathbb{R}^m, \qquad \text{(FRC)}$$

where A_i denotes the ith column of the matrix A.

Theorem 2.3 ([8]) *Let (FRC) be satisfied at the point* $\overline{y}$. *Then,*

$$\operatorname{conv} T(\overline{y}) = \operatorname{cone} T(\overline{y}) = T_R(\overline{y}). \qquad (13)$$

This theorem together with Theorem 2.2 enables us to check local optimality for the problem (9) in polynomial time while, in general, this is an $\mathcal{NP}$-hard problem [31].

2.2 The regular case

It is clear that

$$C_M(\overline{x}, \overline{y}) \subseteq C_\Psi(\overline{x}, \overline{y}) \cap (C_X(\overline{x}) \times \mathbb{R}^m), \qquad (14)$$

where $C_\Psi(\overline{x}, \overline{y})$ denotes the contingent cone to the graph of $\Psi(\cdot)$:

$$\begin{aligned} C_\Psi(x, y) := \{ (u, v)^\top : \exists \{t_k\}_{k=1}^\infty \subset \mathbb{R}_+,\ \exists \{(u^k, v^k)^\top\}_{k=1}^\infty \subset \mathbb{R}^n \times \mathbb{R}^m \\ \text{with } (x, y)^\top + t_k (u^k, v^k)^\top \in \operatorname{Gph} \Psi\ \forall k, \\ \lim_{k \to \infty} t_k = 0,\ \lim_{k \to \infty} (u^k, v^k)^\top = (u, v)^\top \} \end{aligned}$$

and $C_X(\overline{x})$ is the contingent cone for the set X at $\overline{x}$. This implies that the sufficient conditions in Theorem 2.1 can be replaced by the assumption that $\nabla F(\overline{x}, \overline{y})(d, r)^\top > 0$ for all $(d, r)^\top \in C_\Psi(\overline{x}, \overline{y}) \cap (C_X(\overline{x}) \times \mathbb{R}^m)$. Conditions for the contingent cone of the solution set mapping of a parametric optimization problem can be found in the monograph [29] and in [30]. Moreover, $C_M(\overline{x}, \overline{y}) = C_\Psi(\overline{x}, \overline{y})$ if $X = \mathbb{R}^n$.

Theorem 2.4 *If $\Psi(x) = \{y(x)\}$ for some locally Lipschitz continuous, directionally differentiable function $y(\cdot)$, then $C_M(\overline{x}, \overline{y}) = C_\Psi(\overline{x}, \overline{y}) \cap (C_X(\overline{x}) \times \mathbb{R}^m)$.*

Here the directional derivative of a function $h : \mathbb{R}^q \to \mathbb{R}^s$ in direction $d \in \mathbb{R}^q$ at a point $w \in \mathbb{R}^q$ is given by

$$h(w; d) = \lim_{t \to 0+} t^{-1}[h(w + td) - h(w)].$$

Proof of Theorem 2.4: Obviously,

$$C_M(\overline{x}, \overline{y}) \subseteq C_\Psi(\overline{x}, \overline{y}) \cap (C_X(\overline{x}) \times \mathbb{R}^m).$$

Let $(d, r)^\top \in C_\Psi(\overline{x}, \overline{y}) \cap (C_X(\overline{x}) \times \mathbb{R}^m)$. Then, by the assumptions $(d, r)^\top \in C_\Psi(\overline{x}, \overline{y})$, i.e. $r = y'(\overline{x}; d)$ and the directional derivative of $y(\cdot)$ is also locally Lipschitz continuous with respect to perturbations of the direction d [10]. Now, take any sequences $\{u^k\}_{k=1}^\infty$ and $\{t_k\}_{k=1}^\infty$ converging to d respectively to zero from above with $\overline{x} + t_k u^k \in X$ for all k existing by definition of $T_X(\overline{x})$. Then, $(y(\overline{x} + t_k u^k) - y(\overline{x}))/t_k$ converges to $y'(\overline{x}; d)$, which completes the proof. □

To determine conditions guaranteeing the assumptions of the last theorem to be valid consider the lower level problem (1) under the assumptions (SSOC), (MFCQ), and (CRCQ):
(SSOC) The strong second-order sufficient optimality condition for problem (1) is satisfied at a point $(\overline{x}, \overline{y})$ with $g(\overline{x}, \overline{y}) \leq 0$ if:

1. The set

$$\Lambda(\overline{x}, \overline{y}) := \{\lambda : \lambda \geq 0, \lambda^\top g(\overline{x}, \overline{y}) = 0, \nabla_y L(\overline{x}, \overline{y}, \lambda) = 0\}$$

   is not empty and
2. for all $\overline{\lambda} \in \Lambda(\overline{x}, \overline{y})$ and for all $d \neq 0$ with

$$\nabla_y g_i(\overline{x}, \overline{y}) d = 0 \;\forall\, i : \overline{\lambda}_i > 0$$

   there is

$$d^\top \nabla^2_{yy} L(\overline{x}, \overline{y}, \overline{\lambda}) d > 0.$$

Here, $L(x, y, \lambda) = f(x, y) + \lambda^\top g(x, y)$ is the Lagrange function of problem (1).
(CRCQ) The constant rank constraint qualification is satisfied for the problem (1) at the point $(\overline{x}, \overline{y})$ with $g(\overline{x}, \overline{y}) \leq 0$ if there exists an open neighborhood V of $(\overline{x}, \overline{y})$ such that for each subset $J \subseteq \{i : g_i(\overline{x}, \overline{y}) = 0\}$ the set of gradients

$$\{\nabla_y g_i(\overline{x}, \overline{y}) : i \in J\}$$

has a constant rank on V.

Theorem 2.5 ([28], [38]) *Consider problem (1) at a point $(x, y) = (\overline{x}, \overline{y})$ with $\overline{y} \in \Psi(\overline{x})$ and let the assumptions (MFCQ) and (SSOC) be satisfied. Then* *there are an open neighborhood U of $\overline{x}$ and a uniquely determined function $y : U \to \mathbb{R}^m$ such that $y(x)$ is the unique (globally) optimal solution of problem (1) for all $x \in U$. Moreover, if the assumption (CRCQ) is also satisfied, then the function $y(\cdot)$ is locally Lipschitz continuous and directionally differentiable at $\overline{x}$.*

To compute the directional derivative of the solutions function $y(x)$ it is sufficient to compute the unique optimal solution of a quadratic optimization problem using an optimal solution of a linear programming problem as data [38].

Under the assumptions in Theorem 2.5, the bilevel programming problem (both in its optimistic (2) and pessimistic (5) formulations) is equivalent to the problem

$$\min\{G(x) := F(x, y(x)) : x \in X\}. \tag{15}$$

The necessary and sufficient optimality conditions resulting from Theorem 2.1 under the assumptions of Theorem 2.5 and convexity of the lower level problem can be found in [4]:

Theorem 2.6 ([4]) *Consider the bilevel programming problem and let the assumptions (SSOC), (MFCQ), (CRCQ) be valid at a point $(\overline{x}, \overline{y}) \in M$. Then,*

1. *if $(\overline{x}, \overline{y})$ is a locally optimal solution, we have*

$$G'(x; d) \geq 0 \; \forall \; d \in C_X(\overline{x}).$$

2. *if*

$$G'(x; d) > 0 \; \forall \; d \in C_X(\overline{x}),$$

the point $(\overline{x}, \overline{y})$ is a locally optimal solution.

2.3 Application of the protoderivative

Consider the bilevel programming problem in its optimistic formulation (4) and assume that the lower level problem is given in the simpler form

$$\Psi_K(x) := \operatorname*{argmin}_y \{f(x, y) : y \in K\}, \tag{16}$$

where $K \subseteq \mathbb{R}^m$ is a polyhedral set. Then,

$$\Psi_K(x) = \{y \in \mathbb{R}^m : 0 \in \nabla_y f(x, y) + N_K(y)\},$$

where $N_K(y)$ denotes the normal cone of convex analysis to the set K at y which is empty if $y \notin K$. Hence, assuming that $X = \mathbb{R}^n$ the problem (4) reduces to

$$\min_{x,y}\{F(x, y) : 0 \in \nabla_y f(x, y) + N_K(y)\}.$$

Then, if we assume that the regularity condition

$$\text{rank}(\nabla^2_{xy} f(x, y)) = m \quad \text{(full rank)} \tag{17}$$

is satisfied, from Theorem 7.1 in [11] we obtain that the solution set mapping Ψ_K is protodifferentiable. Using the formula for the protoderivative we obtain:

Theorem 2.7 ([7]) *Let $(\bar{x}, \bar{y})$ be a locally optimistic solution of the bilevel programming problem (4), where $\Psi_K(x)$ is given by (16). Assume that the solution set mapping Ψ_K is locally bounded and that the qualification condition (17) holds. Then one has*

$$\nabla F(\bar{x}, \bar{y})(u, v)^\top \geq 0$$

for all $(u, v) \in \mathbb{R}^n \times \mathbb{R}^m$ satisfying

$$0 \in \nabla^2_{xy} f(\bar{x}, \bar{y})u + \nabla^2_{xx} f(\bar{x}, \bar{y})v + N_{K_*}(v).$$

Here

$$K_* = \{d \in C_K(\bar{y}) : \nabla_y f(\bar{x}, \bar{y})d = 0\}$$

and

$$N_{K_*}(v) = \text{cone}\{a_i : i \in I(\bar{y})\} + \text{span}\{\nabla_y f(\bar{x}, \bar{y})\}.$$

provided that

$$K = \{y \in \mathbb{R}^m : a_i^\top y \leq b_i, \quad i = 1, \ldots, p\},$$

where $a_i \in \mathbb{R}^m$ for $i = 1, \ldots, p$ and $b \in \mathbb{R}$ for $i = 1, \ldots, p$. Here, $I(\bar{y})$ denotes the set of active indices at $\bar{y}$.

Optimality conditions for problem (4) using the coderivative of Mordukhovich can be found in the papers [7, 13, 44]. While in the paper [44] the coderivative is applied directly to the graph of the solution set mapping, the attempt in the papers [7, 13] applies the coderivative to the normal cone mapping to the feasible set mapping. We will not go into the details here but refer to the paper [13] in the same volume.

2.4 Global minima

The following sufficient condition for a global optimal solution applies in the case when $X = \mathbb{R}^n$.

Theorem 2.8 ([12]) *Consider the problem (1), (4), let the assumptions of Theorem 2.5 be satisfied at all feasible points $(x, y) \in M$. Let $\overline{x}$ be given and assume that*

$$G'(\widetilde{x}; d) > 0 \ \forall \ d \neq 0, \ \forall \ \widetilde{x} \text{ with } G(\widetilde{x}) = G(\overline{x})$$

Then, $\overline{x}$ is a global minimum of the function $G(x) = F(x, y(x))$.

Proof. First, we show that the point $\overline{x}$ is a strict local minimum of G. If $\overline{x}$ is not a strict local minimum then there exists a sequence $\{x^k\}$ converging to $\overline{x}$ such that $G(x^k) \le G(\overline{x})$. Put $d^k = \dfrac{x^k - \overline{x}}{\|x^k - \overline{x}\|}$. Then, $\{d^k\}_{k=1}^{\infty}$ is a bounded sequence and hence has a convergent subsequence $\{d^k\}$ converging to d^0 (say). If we denote this subsequence again by $\{d^k\}_{k=1}^{\infty}$ we have $x^k = \overline{x} + t_k d^k$ where $t_k = \|x^k - \overline{x}\|$. Hence,

$$G(\overline{x} + t_k d^k) - G(\overline{x}) \le 0.$$

This immediately leads to

$$t_k G'(\overline{x}; d^k) + o(t_k) \le 0.$$

Passing to the limit we obtain a contradiction to the assumption. Hence $\overline{x}$ is a strict local minimum.

Now assume that $\overline{x}$ is not a global minimum. Then, there exists x^0 with $G(x^0) < G(\overline{x})$. Consider the line $Z := \{x : x = \lambda x^0 + (1 - \lambda)\overline{x}, \lambda \in [0, 1]\}$. Then,

1. G is continuous on Z.
2. $x^0 \in Z$, $\overline{x} \in Z$.

Hence, there exist $\{x^1, \ldots, x^p\} \subseteq Z$ with $G(x^i) = G(\overline{x})$ for all i and $G(x) \neq G(\overline{x})$ for all other points in Z. By the assumption this implies that $G(x) \ge G(\overline{x})$ on Z (remember that Z is homeomorphic to a finite closed interval in $\mathbb{R}$ and that $g(\lambda) := G(\lambda x^0 + (1 - \lambda)\overline{x}) : \mathbb{R} \to \mathbb{R}$). But this contradicts $G(x^0) < G(\overline{x})$. □

2.5 Optimality conditions for pessimistic optimal solutions

The pessimistic bilevel programming problem is more difficult than the optimistic one. This may be the reason for attacking the optimistic problem (explicitly or not) in most of the references on bilevel programming problems. The investigations in the paper [15] in this volume indicate that this may not be true for discrete bilevel programming problems.

The approach for using the radial directional derivative for deriving necessary and sufficient optimality conditions for pessimistic optimal solutions has been earlier used for linear bilevel programming problems [5, 9].

Definition 2.3 *Let $U \subseteq \mathbb{R}^m$ be an open set, $\overline{x} \in U$ and $\alpha : U \to \mathbb{R}$. We say that α is radial-continuous at $\overline{x}$ in direction $r \in \mathbb{R}^m$, $\|r\| = 1$, if there exists a real number $\alpha_r(\overline{x})$ such that*

$$\lim_{t \downarrow 0} \alpha(\overline{x} + tr) = \alpha_r(\overline{x}).$$

If the radial limit $\alpha_r(\overline{x})$ *exists for all* $r \in \mathbb{R}^m, \|r\| = 1$, α *is called radial-continuous at* $\overline{x}$.

The function α *is radial-directionally differentiable at* $\overline{x}$, *if there exists a positively homogeneous function* $\mathrm{d}\alpha(\overline{x};\cdot) : \mathbb{R}^m \to \mathbb{R}$ *such that*

$$\alpha(\overline{x} + tr) - \alpha_r(\overline{x}) = t\mathrm{d}\alpha(\overline{x}; r) + o(\overline{x}, tr)$$

with $\lim\limits_{t\downarrow 0} \frac{o(\overline{x},tr)}{t} = 0$ *holds for all* $r \in \mathbb{R}^m, \|r\| = 1$, *and all* $t > 0$.

Obviously, the vector $\mathrm{d}\alpha(\overline{x};\cdot)$ *is uniquely defined and is called the radial-directional derivative of* α *at* $\overline{x}$.

It is not very difficult to show, that, for (mixed-discrete) linear bilevel programming problems, the functions $\varphi_o(\cdot)$ and $\varphi_p(\cdot)$ determined in (3) and (6) are radial-directionally differentiable [5].

A necessary optimality condition is given in the next theorem:

Theorem 2.9 ([9]) *Let* $\alpha : \mathbb{R}^m \to \mathbb{R}$ *be a radial-directionally differentiable function and* $\overline{x} \in \mathbb{R}^m$ *a fixed point. If there exists* $r \in \mathbb{R}^m$ *such that one of the following two conditions is satisfied then* $\overline{x}$ *is not a local optimum of the function* α:

- $\mathrm{d}\alpha(\overline{x}; r) < 0$ *and* $\alpha_r(\overline{x}) \le \alpha(\overline{x})$
- $\alpha_r(\overline{x}) < \alpha(\overline{x})$.

This optimality condition can be complemented by a sufficient one.

Theorem 2.10 ([9]) *Let* $\alpha : \mathbb{R}^m \to \mathbb{R}$ *be a radial-directionally differentiable function and* $\overline{x}$ *a fixed point which satisfies one of the following two conditions.*

- $\alpha(\overline{x}) < \alpha_r(\overline{x})\ \forall r \in \mathbb{R}^m$
- $\alpha(\overline{x}) \le \alpha_r(\overline{x})\ \forall r$ *and* $\mathrm{d}\alpha(\overline{x}; r) > 0\ \forall r : \alpha(\overline{x}) = \alpha_r(\overline{x}), \|r\| = 1$.

Then, α *achieves a local minimum at* $\overline{x}$.

3 Relations to set-valued optimization

Closely related to bilevel programming problems are also *set-valued optimization problems* e.g. of the kind

$$\text{“}\min_x\text{”}\{\mathcal{F}(x) : x \in X\}, \tag{18}$$

where $\mathcal{F} : X \to 2^{\mathbb{R}^p}$ is a point-to-set mapping sending $x \in X \subseteq \mathbb{R}^n$ to a subset of $\mathbb{R}^p$. To see this assume that $\mathcal{F}(x)$ corresponds to the set of all possible upper level objective function values

$$\mathcal{F}(x) := \bigcup_{y \in \Psi(x)} F(x, y).$$

Thus, the bilevel programming problem is transformed into (18) in the special case of $\mathcal{F}(x) \subseteq \mathbb{R}$.

An edited volume on the related set-valued optimization problems is [2], while [23] is a survey on that topic.

Definition 3.1 ([3]) *Let an order cone $C \subseteq \mathbb{R}^p$ with nonempty interior be given. A pair $(\overline{x}, \overline{z})$ with $\overline{x} \in X$, $\overline{z} \in \mathcal{F}(\overline{x})$ is called a weak minimizer of problem (18) if $\overline{z}$ is a weak minimal element of the set*

$$\mathcal{F}(X) := \bigcup_{x \in X} \mathcal{F}(x).$$

Here, $\overline{z} \in \mathcal{F}(X)$ is a weak minimal element of the set $\mathcal{F}(X)$ if

$$(\overline{z} + \operatorname{int} C) \cap \mathcal{F}(X) = \emptyset.$$

Let C be a polyhedral cone. Then, there exist a finite number of elements l^i, $i = 1, \ldots, p$ such that the dual cone C^* to C is

$$C^* = \{z : z^\top d \geq 0 \ \forall\ d \in C\} = \left\{ z : \exists\ \mu \in \mathbb{R}^p_+ \text{ with } z = \sum_{i=1}^p \mu_i l^i \right\}.$$

The following theorem is well-known:

Theorem 3.1 *If the set $\mathcal{F}(X)$ is convex then a point $\overline{z} \in \mathcal{F}(X)$ is a weak minimal element of $\mathcal{F}(X)$ if and only if $\overline{z}$ is an optimal solution of*

$$\min_z \left\{ \sum_{i=1}^p \mu_i l^{i\top} z : z \in \mathcal{F}(X) \right\}$$

for some $\mu \in \mathbb{R}^p_+$.

In the case of bilevel programming (i.e. $\mathcal{F}(X) \subseteq \mathbb{R}$) $p = 1, l_1 = 1, \mu_1 = 1$ can be selected and the problem in Theorem 3.1 reduces to

$$\min_z \{z : z \in \mathcal{F}(X)\}.$$

For this, the convexity assumption is not necessary but the set $\mathcal{F}(X)$ is only implicitly given. Possible necessary optimality conditions for this problem reduce to the ones discussed above.

The difficulty with the optimistic definition is the following: Assume that there are two decision makers, the first one is choosing $x^0 \in X$ and the second one selects $y^0 \in \mathcal{F}(x^0)$. Assume that the first decision maker has no control over the selection of the second one but that he intends to determine a solution $x^0 \in X$ such that for each selection $\widehat{y} \in \mathcal{F}(\widehat{x})$ and $\widehat{x}$ close to x^0 of the second one there exists $y^0 \in \mathcal{F}(x^0)$ which is preferable to $\widehat{y}$. In this case, since the selection of the second decision maker is out of control of the first one, the latter cannot evaluate the quality of her selection chosen according to the above definition.

To weaken this definition assume that the first decision maker is able to compute the sets $\mathcal{F}(x)$ for all $x \in X$. Then, he can try to compute a point $x^* \in X$ such that

$$\mathcal{F}(x) \subseteq \mathcal{F}(x^*) + C$$

for all $x \in X$ sufficiently close to x^*. In distinction to Definition 3.1 this reflects a pessimistic point of view in the sense that the first decision maker bounds the damage caused by the selection of the second one. Let

$$\mathcal{F}(x^1) \preccurlyeq_C \mathcal{F}(x^2) \Longleftrightarrow \mathcal{F}(x^2) \subseteq \mathcal{F}(x^1) + C.$$

Definition 3.2 *Let an order cone $C \subseteq \mathbb{R}^p$ be given. A point $\overline{x} \in X$ is called a pessimistic local minimizer of problem (18) if*

$$\mathcal{F}(\overline{x}) \preccurlyeq_C \mathcal{F}(x) \; \forall \; x \in X \cap \{z : \|z - \overline{x}\| < \varepsilon\}$$

for some $\varepsilon > 0$.

Theorem 3.2 *Let $\overline{x} \in X$ be not a pessimistic local minimizer and assume that C and $\mathcal{F}(x)$ are convex sets for all $x \in X$. Then there exist a vector $\hat{k} \in C^* \setminus \{0\}$ and a point $\hat{x} \in X$ such that*

$$\min\{\hat{k}^\top y : y \in \mathcal{F}(\hat{x})\} < \min\{\hat{k}^\top y : y \in \mathcal{F}(\overline{x})\}.$$

Proof: Let $\overline{x} \in X$ be not a pessimistic minimizer. Then, by definition there exists $\hat{x} \in X$ sufficiently close to $\overline{x}$ such that $\mathcal{F}(\overline{x}) \not\preccurlyeq_C \mathcal{F}(\hat{x})$. Then there necessarily exists $\hat{y} \in \mathcal{F}(\hat{x})$ with $\hat{y} \notin \mathcal{F}(\overline{x}) + C$. Since by our assumption both $\mathcal{F}(\overline{x})$ and C are convex there is a vector $\hat{k} \neq 0$ with

$$\min\{\hat{k}^\top y : y \in \mathcal{F}(\hat{x})\} \leq \hat{k}^\top \hat{y} < \min\{\hat{k}^\top y : y \in \mathcal{F}(\overline{x}) + C\}$$

by a strong separation theorem in convex analysis (see e.g. [42]). Now assume that $\hat{k} \notin C^*$. Then, since C is a cone, we get

$$\begin{aligned} & \min\{\hat{k}^\top y : y \in \mathcal{F}(\overline{x}) + C\} \\ = & \min\{\hat{k}^\top (y^1 + ty^2) : y^1 \in \mathcal{F}(\overline{x}), y^2 \in C\} \\ = & \min\{\hat{k}^\top y^1 : y^1 \in \mathcal{F}(\overline{x})\} + t \min\{\hat{k}^\top y^2 : y^2 \in C\} \end{aligned}$$

for all $t \geq 0$. But since $\hat{k} \notin C^*$, the last term tends to minus infinity for increasing t which cannot be true since it is bounded from below by $\hat{k}^\top \hat{y}$. This proves the theorem. □

This implies that, if for all $k \in C^*$

$$\min_{x,y}\{k^\top y : (x, y) \in \operatorname{Gph} \mathcal{F}, x \in X, \|x - \overline{x}\| \leq \varepsilon\} \geq \min_y\{k^\top y : y \in \mathcal{F}(\overline{x})\}$$

then $\overline{x}$ is a local pessimistic minimizer. The main difference of this result to Theorem 3.1 is that here this condition needs to be satisfied for all elements $k \in C^*$ whereas there must exist one element $k \in C^*$ with the respective condition in Theorem 3.1.

Applied to bilevel programming, the notions of both the optimistic and the pessimistic minimizer coincide.

4 Relation to mathematical programs with equilibrium conditions

Applying the Karush-Kuhn-Tucker conditions to the lower level problem (1) in (4) we derive the problem

$$\begin{aligned} F(x,y) &\to \min_{x,y,u} \\ \nabla_x L(x,y,u) &= 0 \\ g(x,y) &\leq 0 \\ u &\geq 0 \\ u^\top g(x,y) &= 0 \\ x &\in X \end{aligned} \tag{19}$$

provided that the lower level problem satisfies the (MFCQ) at all feasible points for all $x \in X$ and that it is a convex optimization problem for fixed $x \in X$. Problem (19) is called a mathematical program with equilibrium constraints (MPEC) [34, 37]. There has been many interesting results concerning optimality conditions for MPECs in the recent time, cf. e.g. [16, 17, 18, 19, 40, 43]. Here we are interested in conditions needed for applying such conditions.

Example 4.1 This example shows that the convexity assumption is crucial.
Consider the problem [35]

$$\min_{x,y}\{(x-2)+(y-1)^2 : y \in \Psi(x)\}$$

where $\Psi(x)$ is the set of optimal solutions of the following unconstrained optimization problem on the real axis:

$$-x\exp\{-(y+1)^2\} - \exp\{-(y-1)^2\} \to \min_y$$

Then, the necessary optimality conditions for the lower level problem are

$$x(y+1)\exp\{-(y+1)^2\} + (y-1)\exp\{-(y-1)^2\} = 0$$

which has three solutions for $0.344 \leq x \leq 2.903$. The global optimum of the lower level problem is uniquely determined for all $x \neq 1$ and it has a jump at the point $x = 1$. Here the global optimum of the lower level problem can be found at the points $y = \pm 0.957$. The point $(x^0; y^0) = (1; 0.957)$ is also the global optimum of the optimistic bilevel problem.

But if the lower level problem is replaced with its necessary optimality conditions and the necessary optimality conditions for the resulting problem are solved then three solutions: $(x, y) = (1.99; 0.895)$, $(x, y) = (2.19; 0.42)$, $(x, y) = (1.98; -0.98)$ are obtained. Surprisingly, the global optimal solution of the bilevel problem is not obtained with this approach. The reason for this is that the problem

$$\min\{(x-2)+(y-1)^2 : x(y+1)\exp\{-(y+1)^2\}+(y-1)\exp\{-(y-1)^2\}=0\}$$

has a much larger feasible set than the bilevel problem. And this feasible set has no jump at the point $(x, y) = (1; 0.957)$ but is equal to a certain connected curve in $\mathbb{R}^2$. And on this curve the objective function has no stationary point at the optimal solution of the bilevel problem. $\triangle$

The following example can be used to illustrate that the above equivalence between problems (2) and (19) is true only if global optima are searched for.

Theorem 4.1 ([5]) *Consider the optimistic bilevel programming problem (1), (2), (3) and assume that, for each fixed y, the lower level problem (1) is a convex optimization problem for which (MFCQ) is satisfied for each fixed x and all feasible points. Then, each locally optimal solution for the problem (1), (2), (3) corresponds to a locally optimal solution for problem (19).*

This theorem implies that it is possible to derive necessary optimality conditions for the bilevel programming problem by applying the known conditions for MPECs. This has been done e.g. in [5]. Using the recent conditions in the papers [16, 17, 18, 19, 43] interesting results can be obtained. It would be a challenging topic for future research to check if the assumptions used in these papers can successfully be interpreted for bilevel programming problems.

But, the application of these results to get strong necessary optimality conditions and also to get sufficient optimality conditions seems to be restricted. This can be seen in the following example.

Example 4.2 We consider a linear bilevel programming problem with an optimistic optimal solution $(\overline{x}, \overline{y})$. Assume that the linear independence constraint qualification is not satisfied at the lower level problem at $(\overline{x}, \overline{y})$ and that there are more than one Lagrange multiplier for the lower level problem at $\overline{y}$. Then, the situation is as depicted in fig. 1: We see a part of the feasible set of the upper level problem (which is the union of faces of the graph of the lower level feasible set $\{(x, y) : Ax \leq y\}$) in the right-hand side picture. There is a kink at the point $(\overline{x}, \overline{y})$. The point $(\overline{x}, \overline{y})$ belongs to the intersection of two faces of the graph of the lower level feasible set $\{(x, y) : Ax \leq y\}$. The optimal solution of the lower level problem is unique for all x, hence the lower level solution function can be inserted into the upper level objective function. In the left-hand side picture we see the graph of the set of Lagrange multipliers in the lower level problem which is assumed to reduce to a singleton for $x \neq \overline{x}$ and is multivalued for $x = \overline{x}$.

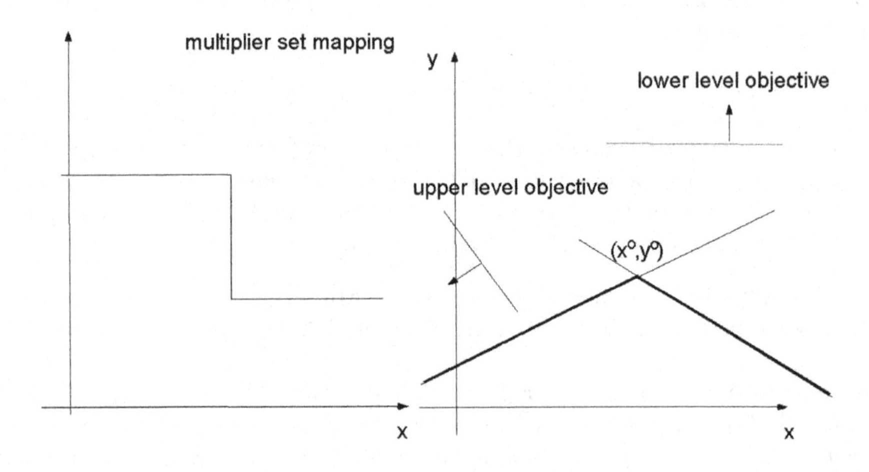

Fig. 1. Feasible set in the upper level problem and Lagrange multiplier mapping of the lower level problem

Now assume that we have used the MPEC corresponding to the bilevel programming problem for deriving necessary optimality conditions. For this we fix a feasible solution $(\overline{x}, \overline{y}, \overline{\lambda})$ of the MPEC, where $\overline{\lambda}$ denotes one Lagrange multiplier of the lower level problem at $\overline{y}$. Then, the necessary optimality conditions for the MPEC are satisfied. They show, that there does not exist a better feasible solution than $(\overline{x}, \overline{y}, \overline{\lambda})$ in a suitable small neighborhood of this point for the MPEC. This neighborhood of $(\overline{x}, \overline{y}, \overline{\lambda})$ restricts the λ-part to a neighborhood of $\overline{\lambda}$. Due to complementarity slackness this restriction implies for the bilevel programming problem that the feasible set of the upper level problem is restricted to one face of the graph of $\{(x, y) : Ax \leq y\}$, see fig. 2, where this face is the right-hand sided one. Hence, the necessary optimality conditions for the corresponding MPEC mean that the point under consideration is a stationary (here optimal) solution of the bilevel programming problem but only with respect to a part (and not with respect to an open neighborhood !) of the feasible set.

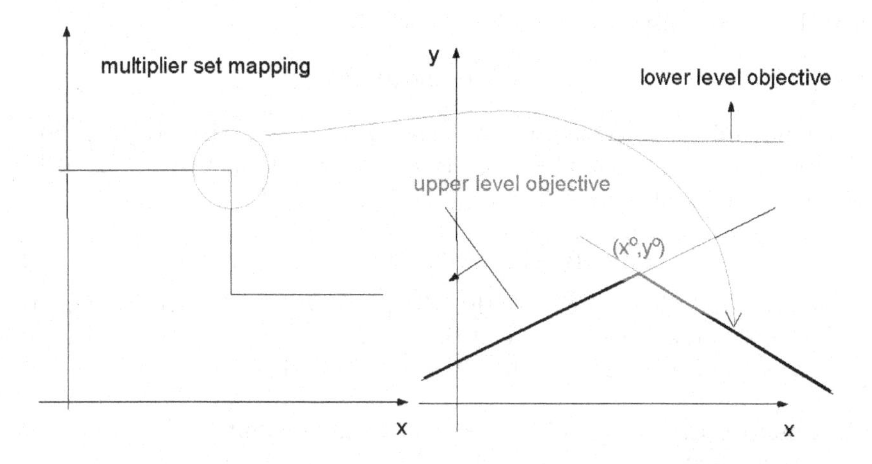

Fig. 2. Neighborhood of a Lagrange multiplier $\overline{\lambda}$ and corresponding part of the feasible set of the bilevel programming problem

5 Variational inequality approach

The problem of solving a mathematical program with variational inequalities or complementarity conditions as constraints arises quite frequently in the analysis of physical and socio-economic systems. According to a remark in the paper [20], the current state-of-the-art for solving such problems is heuristic. The latter paper [20] presents an exterior-point penalty method based on M.J. Smith's optimization formulation of the finite-dimensional variational inequality problem [41]. In the paper by J. Outrata [36], attention is also paid to this type of optimization problems.

An approach to solving the above-mentioned problem consists in a variational re-formulation of the optimization criterion and looking for a solution of the thus obtained variational inequality among the points satisfying the initial variational inequality constraints. This approach works well for the case when both operators involved are monotone and it is enlightened in the first part of the section. Namely, in subsection 5.2, we examine conditions under which the set of the feasible points is non-empty, and compare the conditions with those established previously [22]. Subsection 5.3 describes a penalty function method solving the bilevel problem after having reduced it to a single variational inequality with a penalty parameter.

5.1 Existence theorem

Let X be a non-empty, closed, convex subset of R^n and G a continuous mapping from X into R^n. Suppose that $\mathcal{F}$ is pseudo-monotone with respect to X, i.e.

$$(x-y)^{\mathrm{T}}\mathcal{F}(y) \geq 0 \quad \text{implies} \quad (x-y)^{\mathrm{T}}\mathcal{F}(x) \geq 0 \quad \forall x, y \in X, \tag{20}$$

and that there exists a vector $x^0 \in X$ such that

$$\mathcal{F}(x^0) \in \text{int } (0^+X)^*, \tag{21}$$

where int$(\cdot)$ denotes the interior of the set. Here 0^+X is the recession cone of the set X, i.e. the set of all directions $s \in R^n$ such that $X + s \subset X$; at last, C^* is the dual cone of $C \subset R^n$, i.e.

$$C^* = \{y \in R^n : y^{\mathrm{T}}x \geq 0 \qquad \forall x \in C\}. \tag{22}$$

Hence, condition (21) implies that the vector $\mathcal{F}(x^0)$ lies within the interior of the dual to the recession cone of the set X.

Under these assumptions, the following result obtains:

Proposition 5.1 ([24]) *The variational inequality problem: to find a vector* $z \in X$ *such that*

$$(x-z)^{\mathrm{T}}\mathcal{F}(z) \geq 0 \qquad \forall x \in X, \tag{23}$$

has a non-empty, compact, convex solution set.

Proof. It is well-known [27] that the pseudo-monotonicity (20) and continuity of the mapping G imply convexity of the solution set

$$Z = \{z \in X : (x-z)^{\mathrm{T}}\mathcal{F}(z) \geq 0 \qquad \forall x \in X\}, \tag{24}$$

of problem (23) provided that the latter is non-empty. Now we show the existence of at least one solution to this problem. In order to do that, we use the following fact [14]: if there exists a non-empty bounded subset D of X such that for every $x \in X \backslash D$ there is a $y \in D$ with

$$(x-y)^{\mathrm{T}}\mathcal{F}(x) > 0, \tag{25}$$

then problem (23) has a solution. Moreover, the solution set (24) is bounded because $Z \subset D$. Now, we construct the set D as follows:

$$D = \{\, x \in X : (x-x^0)^{\mathrm{T}}\mathcal{F}(x^0) \leq 0 \,\}. \tag{26}$$

The set D is clearly non-empty, since it contains the point x^0. Now we show that D is bounded, even if X is not such. On the contrary, suppose that a sequence $\{x^k\} \subseteq D$ is norm-divergent, i.e. $\|x^k - x^0\| \to +\infty$ when $k \to \infty$. Without lack of generality, assume that $x^k \neq x^0$, $k = 1, 2, \ldots$, and consider the inequality

$$\frac{(x^k - x^0)^{\mathrm{T}}\mathcal{F}(x^0)}{\|x^k - x^0\|} \leq 0, \qquad k = 1, 2, \ldots, \tag{27}$$

which follows from definition (26) of the set D. Again without affecting generality, accept that the normed sequence $(x^k - x^0)/\|x^k - x^0\|$ converges to a vector $s \in R^n$, $\|s\| = 1$. It is well-known (cf. [39], Theorem 8.2) that $s \in 0^+X$. From (27), we deduce the limit relation

$$s^{\mathrm{T}}\mathcal{F}(x^0) \leq 0. \tag{28}$$

Since $0^+X \neq \{0\}$ (as X is unbounded and convex), we have $0 \in \partial(0^+X)^*$, hence $\mathcal{F}(x^0) \neq 0$. Now it is easy to see that inequality (28) contradicts assumption (21). Indeed, the inclusion $\mathcal{F}(x^0) \in \text{int}\ (0^+X)^*$ implies that $s^{\mathrm{T}}\mathcal{F}(x^0) > 0$ for any $s \in 0^+X, s \neq 0$. The contradiction establishes the boundedness of the set D, and the statement of Proposition 5.1 therewith. Indeed, for a given $x \in X \backslash D$, one can pick $y = x^0 \in D$ with the inequality $(x - y)^{\mathrm{T}}\mathcal{F}(y) > 0$ taking place. The latter, jointly with the pseudo-monotonicity of $\mathcal{F}$, implies the required condition (25) and thus completes the proof. □

Remark 5.1 *The assertion of Proposition 5.1 has been obtained also in [22] under the same assumptions except for inclusion (21), which is obviously invariant with respect to an arbitrary translation of the set X followed by the corresponding transformation of the mapping G. Instead of (21), the authors [22] used another assumption $\mathcal{F}(x^0) \in \text{int}(X^*)$ which is clearly not translation-invariant.*

Now suppose that the solution set Z of problem (23) contains more than one element, and consider the following variational inequality problem: to find a vector $z^* \in Z$ such that

$$(z - z^*)^{\mathrm{T}}\mathcal{G}(z^*) \geq 0 \qquad \text{for all} \quad z \in Z. \tag{29}$$

Here, the mapping $\mathcal{G}\colon X \to R^n$ is continuous and strictly monotone over X; i.e.

$$(x - y)^{\mathrm{T}}\left[\mathcal{G}(x) - \mathcal{G}(y)\right] > 0 \qquad \forall x, y \in X, x \neq y. \tag{30}$$

In this case, the compactness and convexity of the set Z guaranties [14] the existence of a unique (due to the strict monotonicity of $\mathcal{G}$) solution z of the problem (29). We refer to problem (23), (24), (29) as the *bilevel variational inequality* (BVI). In the next subsection, we present a penalty function algorithm solving the BVI without explicit description of the set Z.

5.2 Penalty function method

Fix a positive parameter ε and consider the following parametric variational inequality problem: Find a vector $x_\varepsilon \in X$ such that

$$(x - x_\varepsilon)^{\mathrm{T}}[\mathcal{F}(x_\varepsilon) + \varepsilon\mathcal{G}(x_\varepsilon)] \geq 0 \qquad \text{for all} \quad x \in X. \tag{31}$$

If we assume that the mapping $\mathcal{F}$ is monotone over X, i.e.

$$(x - y)^{\mathrm{T}}[\mathcal{F}(x) - \mathcal{F}(y)] \geq 0 \qquad \forall x, y \in X, \tag{32}$$

and keep intact all the above assumptions regarding $\mathcal{F}, \mathcal{G}$ and Z, then the following result obtains:

Proposition 5.2 ([24]) *For each sufficiently small value $\varepsilon > 0$, problem (31) has a unique solution x_ε. Moreover, x_ε converge to the solution z^* of BVI (23), (24), (29) when $\varepsilon \to 0$.*

Proof. Since $\mathcal{F}$ is monotone and $\mathcal{G}$ is strictly monotone, the mapping $\Phi_\varepsilon = \mathcal{F} + \varepsilon\mathcal{G}$ is strictly monotone on X for any $\varepsilon > 0$. It is also clear that if x^0 satisfies (21) then the following inclusion holds

$$\Phi_\varepsilon(x^0) = \mathcal{F}(x^0) + \varepsilon\mathcal{G}(x^0) \in \text{int}\,(0^+X)^*, \tag{33}$$

if $\varepsilon > 0$ is small enough. Hence, Proposition 5.1 implies the first assertion of Proposition 5.2; namely, for every $\varepsilon > 0$ satisfying (33), the variational inequality (31) has a unique solution x_ε .

From the continuity of F and G, it follows that each (finite) limit point $\bar{x}$ of the generalized sequence $Q = \{x_\varepsilon\}$ of solutions to problem (31) solves variational inequality (23); that is, $\bar{x} \in Z$. Now we prove that the point $\bar{x}$ solves problem (29), too. In order to do that, we use the following relations valid for any $z \in Z$ due to (23), (29) and (31):

$$(z - x_\varepsilon)^{\mathrm{T}}[\mathcal{F}(z) - \mathcal{F}(x_\varepsilon)] \geq 0, \tag{34}$$

$$(z - x_\varepsilon)^{\mathrm{T}}\mathcal{F}(z) \leq 0, \tag{35}$$

$$(z - x_\varepsilon)^{\mathrm{T}}\mathcal{F}(x_\varepsilon) \geq -\varepsilon(z - x_\varepsilon)^{\mathrm{T}}\mathcal{G}(x_\varepsilon). \tag{36}$$

Subtracting (36) from (35) and using (34), we obtain the following series of inequalities

$$0 \leq (z - x_\varepsilon)^{\mathrm{T}}[\mathcal{F}(z) - \mathcal{F}(x_\varepsilon)] \leq \varepsilon(z - x_\varepsilon)^{\mathrm{T}}\mathcal{G}(x_\varepsilon). \tag{37}$$

From (37) we have $(z - x_\varepsilon)^{\mathrm{T}}\mathcal{G}(x_\varepsilon) \geq 0$ for all $\varepsilon > 0$ and $z \in Z$. Since $\mathcal{G}$ is continuous, the following limit relation holds: $(z-\bar{x})^{\mathrm{T}}\mathcal{G}(\bar{x}) \geq 0$ for each $z \in Z$, which means that $\bar{x}$ solves (29).

Thus we have proved that every limit point of the generalized sequence Q solves BVI (23), (24), (29). Hence, Q can have at most one limit point. To complete the proof of Proposition 5.2, it suffices to establish that the set Q is bounded, and consequently, the limit point exists. In order to do that, consider a norm-divergent sequence $\{x_{\varepsilon_k}\}$ of solutions to parametric problem (31) where $\varepsilon_k \to 0$ as $k \to \infty$. Without loss of generality, suppose that $x_{\varepsilon_k} \neq x^0$ for each k, and $\dfrac{(x_{\varepsilon_k} - x^0)}{\|x_{\varepsilon_k} - x^0\|} \to s \in R^n, \quad \|s\| = 1$; here x^0 is the vector from condition (21). Since $\|x_{\varepsilon_k} - x^0\| \to +\infty$, we get $s \in 0^+X$ (cf. [39]). As the mappings $\mathcal{F}$ and $\mathcal{G}$ are monotone, the following inequalities take place for all $k = 1, 2, \ldots$:

$$(x_{\varepsilon_k} - x^0)^{\mathrm{T}}[\mathcal{F}(x_{\varepsilon_k}) + \varepsilon_k\mathcal{G}(x_{\varepsilon_k})] \leq 0, \tag{38}$$

and hence,

$$(x_{\varepsilon_k} - x^0)^{\mathrm{T}}[\mathcal{F}(x^0) + \varepsilon_k \mathcal{G}(x^0)] \le 0. \tag{39}$$

Dividing inequality (39) by $\|x_{\varepsilon_k} - x^0\|$ we obtain

$$\frac{(x_{\varepsilon_k} - x^0)^{\mathrm{T}}}{\|x_{\varepsilon_k} - x^0\|} \cdot [\mathcal{F}(x^0) + \varepsilon_k \mathcal{G}(x^0)] \le 0, \qquad k = 1, 2, ..., \tag{40}$$

which implies (as $\varepsilon_k \to 0$) the limit inequality $s^{\mathrm{T}}\mathcal{F}(x^0) \le 0$. Since $s \ne 0$, the latter inequality contradicts assumption 21. This contradiction demonstrates the set Q to be bounded which completes the proof. □

Example 5.1 *Let $\Omega \subseteq R^m$, $\Lambda \subseteq R^n$ be subsets of finite-dimensional Euclidean spaces and $g : \Omega \times \Lambda \to R$, $f : \Omega \times \Lambda \to R^n$ be continuous mappings. Consider the following mathematical program with variational inequality constraint:*

$$\min_{(u,v) \in \Omega \times \Lambda} g(u, v), \tag{41}$$

subject to

$$f(u, v)^T (w - v) \ge 0, \qquad \forall w \in \Lambda. \tag{42}$$

If the function g is continuously differentiable, then problem (41)-(42) is obviously tantamount to BVI (23), (24), (29) with the gradient mapping $g'(z)$ used as $\mathcal{G}(z)$ and $\mathcal{F}(u, v) = [0; f(u, v)]$; here $z = (u, v) \in \Omega \times \Lambda$.

As an example, examine the case when

$$g(u, v) = (u - v - 1)^2 + (v - 2)^2; \quad f(u, v) = uv; \quad \Omega = \Lambda = R^1_+. \tag{43}$$

Then it is readily verified that $z^* = (1; 0)$ solves problem (41)-(42) and the parametrized mapping is given by

$$\Phi_\varepsilon(u, v) = \left[\varepsilon(2u - 2v - 2); uv + \varepsilon(-2u + 4v - 2)\right]. \tag{44}$$

Now solving the variational inequality: Find $(u_\varepsilon, v_\varepsilon) \in R^2_+$ such that

$$\Phi_\varepsilon(u_\varepsilon, v_\varepsilon)^T \left[(u, v) - (u_\varepsilon, v_\varepsilon)\right] \ge 0 \quad \forall (u, v) \in R^2_+, \tag{45}$$

we obtain

$$u_\varepsilon = v_\varepsilon + 1; \quad v_\varepsilon = -\frac{1}{2} - \varepsilon + \sqrt{(\frac{1}{2} + \varepsilon)^2 + 4\varepsilon}. \tag{46}$$

Clearly $(u_\varepsilon, v_\varepsilon) \to z^*$ when $\varepsilon \to 0$. △

Acknowledgment: The second and third authors' research was financially supported by the PAICyT project CA71033-05 at the UANL (Mexico), by the Russian Foundation for Humanity Research project RGNF 04-02-00172, and by the CONACyT-SEP project SEP-2004-C01-45786 (Mexico).

References

1. J.F. Bard (1998): *Practical Bilevel Optimization: Algorithms and Applications*, Dordrecht: Kluwer Academic Publishers.
2. G.Y. Chen and J. Jahn (eds.) (1998), *Mathematical Methods of Operations Research*, vol. 48.
3. G. Y. Chen and J. Jahn (1998). Optimality conditions for set-valued optimization problems. *Mathematical Methods of Operations Research*, 48: 187-200.
4. S. Dempe (1992). A necessary and a sufficient optimality condition for bilevel programming problems. *Optimization*, 25: 341–354.
5. S. Dempe (2002): *Foundations of Bilevel Programming*, Dordrecht: Kluwer Academic Publishers, 2002.
6. S. Dempe (2003). Annotated Bibliography on Bilevel Programming and Mathematical Programs with Equilibrium Constraints, *Optimization*, 52:, 333-359.
7. S. Dempe, J. Dutta, S. Lohse (2005) Optimality conditions for bilevel programming problems, Preprint, TU Bergakademie Freiberg.
8. S. Dempe, S. Lohse: Inverse linear programming, In: *Recent Advances in Optimization. Proceedings of the 12th French-German-Spanish Conference on Optimization held in Avignon, September 20-24, 2004*, Edited by A. Seeger. Lectures Notes in Economics and Mathematical Systems, Vol. 563, Springer-Verlag Berlin Heidelberg, 2006. pp. 19-28.
9. S. Dempe, T. Unger (1999). Generalized PC^1–functions, *Optimization*, 46:311-326.
10. V.F. Dem'yanov and A.M. Rubinov (1986): *Quasidifferential Calculus*, New York: Optimization Software.
11. A. Dontchev and R. T. Rockafellar (2001). Ample parametrization of variational inclusions, *SIAM Journal on Optimization*, 12: 170-187.
12. J. Dutta and S. Dempe (2004). *Global minima of locally Lipschitz functions*, KANGAL Research Report No. 2004/018, I.I.T. Kanpur, India, 2004.
13. J. Dutta, S. Dempe (2006). Bilevel programming with convex lower level problems, In: S. Dempe, V. Kalashnikov (eds.): *Optimization with Multivalued Mappings: Theory, Applications and Algorithms,* Springer Verlag, Berlin.
14. B.C. Eaves (1971). On the basic theorem of complementarity, *Mathematical Programming*, 1: 68-75.
15. D. Fanghänel (2006). Optimality criteria for bilevel programming problems using the radial subdifferential, In: S. Dempe, V. Kalashnikov (eds.): *Optimization with Multivalued Mappings: Theory, Applications and Algorithms,* Springer Verlag, Berlin.
16. M. L. Flegel and C. Kanzow (2003). A Fritz John approach to first order optimality conditions for mathematical programs with equilibrium constraints, *Optimization*, 52: 277-286.
17. M. L. Flegel and C. Kanzow. On the Guignard constraint qualification for mathematical programs with equilibrium constraints, *Optimization*, to appear.
18. M. L. Flegel and C. Kanzow. On M-stationarity for mathematical programs with equilibrium constraints, *Journal of Mathematical Analysis and Applications*, to appear.
19. M. L. Flegel and C. Kanzow (2005). Abadie-type constraint qualification for mathematical programs with equilibrium constraints, *Journal of Optimization Theory and Applications*, 124: 595–614.

20. P.T. Harker and S.-C. Choi (1991). A penalty function approach for mathematical programs with variational inequality constraints, *Information and Decision Technologies*, 17: 41-50.
21. P.T. Harker and J.-S. Pang (1988). Existence of optimal solutions to mathematical programs with equilibrium constraints, *Operations Research Letters*, 7: 61-64.
22. P.T. Harker and J.-S. Pang (1990). Finite-dimensional variational inequality and nonlinear complementarity problems: a survey of theory, algorithms and applications, *Mathematical Programming*, 48: 161-220.
23. J. Jahn (2000). *Set-valued optimization : a survey*, Preprint, Universität Erlangen-Nürnberg. Institut für Angewandte Mathematik.
24. G.Isac, V.A. Bulavsky and V.V. Kalashnikov (2002). *Complementarity, Equilibrium, Efficiency and Economics*, Kluwer Academic Publishers, Dordrecht/Boston/London, 468 p.
25. V.V. Kalashnikov and N.I. Kalashnikova (1996). Solving two-level variational inequality, *Journal of Global Optimization*, 8: 289-294.
26. S. Karamardian (1976). Complementarity over cones with monotone and pseudo-monotone maps, *Journal of Optimization Theory and Applications*, 18: 445-454.
27. S. Karamardian (1976). An existence theorem for the complementarity problem, *Journal of Optimization Theory and Applications*, 18: 455-464.
28. M. Kojima (1980). Strongly stable stationary solutions in nonlinear programs. In: S.M. Robinson (ed.): *Analysis and Computation of Fixed Points*, New York: Academic Press, pp. 93–138.
29. D. Klatte and B. Kummer (2002). *Nonsmooth Equations in Optimization; Regularity, Calculus, Methods and Applications*, Dordrecht: Kluwer Academic Publishers.
30. A. B. Levy and R. T. Rockafellar (1995). Sensitivity of solutions in nonlinear programming problems with nonunique multipliers. In: D.-Z. Du and L. Qi and R. S. Womersley (eds.): *Recent Advances in Nonsmooth Optimization*, World Scientific Publishing Co., Singapore et al., 215–223.
31. S. Lohse: Komplexitätsbeweis. Technical Report, TU Bergakademie Freiberg, 2005. see `http://www.mathe.tu-freiberg.de/~lohses/Dateien/komplex.pdf`
32. P. Loridan and J. Morgan (1996). Weak via strong Stackelberg problem: New results. *Journal of Global Optimization*, 8: 263-287.
33. R. Lucchetti and F. Mignanego and G. Pieri(1987). Existence theorem of equilibrium points in Stackelberg games with constraints. *Optimization*, 18: 857-866.
34. Z.-Q. Luo and J.-S. Pang and D. Ralph (1996). *Mathematical Programs with Equilibrium Constraints.* Cambridge University Press, Cambridge.
35. J. A. Mirrlees (1999). The theory of moral hazard and unobservable bevaviour: part I. *Review of Economic Studies*, 66: 3-21.
36. J.V. Outrata (1994). On optimization problems with variational inequality constraints, *SIAM Journal on Optimization*, 4: 340-357.
37. J. Outrata and M. Kočvara and J. Zowe (1998). *Nonsmooth Approach to Optimization Problems with Equilibrium Constraints.* Kluwer Academic Publishers, Dordrecht.
38. D. Ralph and S. Dempe (1995). Directional Derivatives of the Solution of a Parametric Nonlinear Program. *Mathematical Programming*, 70: 159-172.
39. R.T. Rockafellar (1970). *Convex Analysis*, Princeton University Press, Princeton, New Jersey.

40. H. Scheel and S. Scholtes (2000). Mathematical programs with equilibrium constraints: stationarity, optimality, and sensitivity. *Mathematics of Operations Research*, 25: 1-22.
41. M.J. Smith (1984). A descent algorithm for solving monotone variational inequalities and monotone complementarity problems, *Journal of Optimization Theory and Applications*, 44: 485-496.
42. H. Tuy (1998). *Convex Analysis and Global Optimization.* Kluwer Academic Publishers, Dordrecht.
43. J. J. Ye (2005). Necessary and sufficient optimality conditions for mathematical programs with equilibrium constraints, *Journal of Mathematical Analysis and Applications*, 307: 350-369.
44. R. Zhang (1994). Problems of hierarchical optimization in finite dimensions, *SIAM Journal on Optimization*, 4: 521-536.

Path-based formulations of a bilevel toll setting problem

Mohamed Didi-Biha[1], Patrice Marcotte[2] and Gilles Savard[3]

[1] Laboratoire d'Analyse non linéaire et Géométrie, Université d'Avignon et des Pays de Vaucluse, Avignon, France mohamed.didi-biha@univ-avignon.fr
[2] CRT, Département d'Informatique et de Recherche Opérationnelle, Université de Montréal, Montréal (QC), Canada marcotte@iro.umontreal.ca
[3] GERAD, Département de Mathématiques et de Génie Industriel, École Polytechnique de Montréal, Montréal (QC), Canada gilles.savard@polymtl.ca

Summary. A version of the toll setting problem consists in determining profit maximizing tolls on a subset of arcs of a transportation network, given that users travel on shortest paths. This yields a bilevel program for which we propose efficient algorithms based on path generation.

Key words. Pricing. Bilevel programming. Networks. Column generation. Combinatorial optimization.

1 Introduction

Bilevel programming offers a convenient framework for the modelling of pricing problems, as it allows to take explicitly into account user behaviour. One of the simplest instances was analyzed by Labbé et al. [8], who considered a toll optimization problem (TOP) defined over a congestion-free, multicommodity transportation network. In this setting, a highway authority (the "leader") sets tolls on a subset of arcs of the network, while the users (the "follower") assign themselves to shortest[4] paths linking their respective origin and destination nodes. The goal of the leader being to maximize toll revenue, it is not in its interest to set tolls at very high values, in which case the users will be discouraged from using the tolled subnetwork. The problem, which consists in striking the right balance between tolls that generate high revenues and tolls that attract customers, can be formulated as a combinatorial program that subsumes NP-hard problems, such as the Traveling Salesman Problem

[4] It is assumed that costs and travel times are expressed in a common unit, i.e., the monetary perception of one unit of travel time is uniform throughout the user population.

S. Dempe and V. Kalashnikov (eds.), *Optimization with Multivalued Mappings*, pp. 29-50
©2006 Springer Science + Business Media, LLC

(see Marcotte et al [11] for a reduction). Following the initial NP-hardness proof by Labbé et al., complexity and approximation results have also been obtained by Roch et al. [12] and Grigoriev et al. [5].

The aim of the present work is to assess the numerical performance of path-based reformulations of TOP, and to show their ability to solve to optimality medium-sized instances, and to near optimality large-scale instances. This stands in contrast with arc-based methods that have been proposed by Labbé et al. [8] and Brotcorne et al. [1]. Note that Bouhtou et al. [2] have recently proposed, together with arc-based methods, a path-based approach operating on a compact reformulation of the problem.

The structure of the paper is as follows: Section 2 introduces three Mixed Integer Programming (MIP) formulations for TOP; Section 3 introduces a path generation framework; Section 4 details a sequential implementation; Section 5 presents numerical results achieved on randomly generated test problems; Section 6 concludes with avenues for further research.

2 A bilevel formulation

In this section, we present three MIP formulations of TOP. The first, initially proposed by Labbé et al. [8], relies on the optimality conditions associated with an arc-commodity formulation. The second utilizes both arc and path variables, while the third is entirely path-based.

TOP can be analyzed as a leader-follower game that takes place on a multicommodity network $G = (K, N, A)$ defined by a set of origin-destination couples K, a node set N and an arc set A. The latter is partitioned into the subset A_1 of toll arcs and the complementary subset A_2 of toll-free arcs. We endow each arc $a \in A$ with a fixed travel delay c_a. Toll arcs $a \in A_1$ also involve a toll component t_a, to be determined, that is expressed in time units, for the sake of consistency. The demand side is represented by numbers n^k denoting the demand for travel between the origin node $o(k)$ and the destination $d(k)$ associated with commodity $k \in K$. With each commodity is associated a demand vector b^k whose components are, for every node i of the network:

$$b_i^k = \begin{cases} -n^k & \text{if } i = o(k), \\ n^k & \text{if } i = d(k), \\ 0 & \text{otherwise.} \end{cases}$$

Letting x_a^k denote the set of commodity flows and i^+ (respectively i^-) the set of arcs having i as their head node (respectively tail node), TOP can be formulated as a bilevel program involving bilinear objectives at both decision levels:

$$\text{TOP:} \quad \max_{t,x} \sum_{k \in K} \sum_{a \in A_1} t_a x_a^k$$

$$\text{subject to} \quad t_a \leq t_a^{\max} \qquad \forall a \in A_1$$

$$\forall k \in K \begin{cases} x^k \in \arg\min_{\bar{x}} \sum_{a \in A_1} (c_a + t_a)\bar{x}_a + \sum_{a \in A_2} c_a \bar{x}_a \\ \text{subject to} \sum_{a \in i^-} \bar{x}_a - \sum_{a \in i^+} \bar{x}_a = b_i^k \qquad \forall i \in N \\ \bar{x}_a \geq 0 \qquad \forall a \in A. \end{cases}$$

In the above formulation, the leader controls both the toll and flow variables. However the lower level 'argmin' constraint forces the leader to assign flows to shortest paths with respect to the current toll levels. In order to prevent the occurrence of trivial situations, the following conditions are assumed to hold throughout the paper:

1. There does not exist a profitable toll vector that induces a negative cost (delay) cycle in the network. This condition is clearly satisfied if all delays c_a are nonnegative.
2. For each commodity, there exists at least one path composed solely of toll-free arcs.

Under the above assumptions, the lower level optimal solution corresponds to a set of shortest paths, and the leader's profit is bounded from above.

A single-level reformulation of TOP is readily obtained by replacing the lower level program by its primal-dual optimality conditions. If one expresses the latter by the equality of the primal and dual objectives, we obtain the nonlinearly-constrained program

$$\text{MIP:} \quad \max_{t,x,\lambda} \sum_{k \in K} \sum_{a \in A_1} t_a x_a^k$$

$$\text{subject to} \quad t_a \leq t_a^{\max} \qquad \forall a \in A_1$$

$$\forall k \in K \begin{cases} \sum_{a \in i^-} x_a^k - \sum_{a \in i^+} x_a^k = b_i^k & \forall i \in N \\ \lambda_j^k - \lambda_i^k \leq c_a + t_a & \forall a = (i,j) \in A_1 \\ \lambda_j^k - \lambda_i^k \leq c_a & \forall a \in A_2 \\ \sum_{a \in A_1} (c_a + t_a) x_a^k + \sum_{a \in A_2} c_a x_a^k = (\lambda_{o(k)}^k - \lambda_{d(k)}^k) n^k & \\ x_a^k \geq 0 & \forall a \in A. \end{cases}$$

Now, for each commodity $k \in K$, one can substitute for the flow variables the *proportion* of the demand $d(k)$ assigned to arc a, and replace the node demand b_i^k by the unit demand $e_i^k = \mathrm{sgn}(b_i^k)$. Slightly abusing notation, we still denote the flow proportions by x_a^k. Since there exists an optimal extremal solution for the lower program (and the bilevel program as well) one may assume, without loss of generality, that the variables x_a^k are binary-valued, i.e., each commodity flow is assigned to a single path.

Next, we introduce unit commodity toll revenues t_a^k and replace the bilinear term $t_a x_a^k$ by the commodity toll t_a^k, which we force to take the common value t_a whenever the associated flow x_a^k assumes the value 'one'. These operations yield a mixed-integer program that involves relatively few integer variables, i.e., one per toll arc and per commodity.

$$\text{MIP I:} \quad \max_{t,x,\lambda} \sum_{k\in K} \sum_{a\in A_1} n_k t_a^k$$

$$\text{subject to} \quad t_a \le t_a^{\max} \qquad \forall a \in A_1$$

$$\forall k \in K \begin{cases} \sum_{a\in i^-} x_a^k - \sum_{a\in i^+} x_a^k = e_i^k & \forall i \in N \\ \lambda_j^k - \lambda_i^k \le c_a + t_a & \forall a = (i,j) \in A_1 \\ \lambda_j^k - \lambda_i^k \le c_a & \forall a \in A_2 \\ \sum_{a\in A_1} (c_a x_a^k + t_a^k) + \sum_{a\in A_2} c_a x_a^k = \lambda_{o(k)}^k - \lambda_{d(k)}^k & \\ -M_k x_a^k \le t_a^k \le M_k x_a^k & \forall a \in A_1 \\ -M(1 - x_a^k) \le t_a^k - t_a \le M(1 - x_a^k) & \forall a \in A_1 \\ x_a^k \in \{0,1\} & \forall a \in A_1 \\ x_a^k \ge 0 & \forall a \in A_2. \end{cases}$$

Note that, in the formulation MIP I, the parameter M_k can be set, for every commodity index k, to any value that exceeds the difference between the cost C_k^∞ of a shortest path that uses only arcs in A_2 and the cost C_k^0 of a shortest path with all tolls set at zero or, if $t_a^{\max}$ is bounded, to $t_a^{\max}$, simply. As for M, it can assume any value larger than the maximum of the M_k's. These assignments ensure that formulation MIP I is equivalent to the original bilevel program. In the case where tolls cannot assume negative values, i.e., subsidies are forbidden, these bounds have been refined by Dewez [4].

We now provide two *path-based* formulations for TOP. To this aim, we introduce the set P_k of paths from $o(k)$ to $d(k)$ and denote by I_a^k the set of elements of P_k that contain a, i.e.,

$$I_a^k = \{p \in P_k \mid a \in P\}, \quad \forall\, a \in A, \ \forall\, k \in K.$$

With each path $p \in P_k$, we associate the indicator variable z_p, which takes the value 1 if path p is used by commodity k, and takes the value 0 otherwise. From the identity

$$x_a^k = \sum_{p \in I_a^k} z_p, \quad \forall\, a \in A, \forall\, k \in K,$$

there comes the *arc-path formulation*

$$\text{MIP II:} \quad \max_{t,z,\lambda,s} \sum_{k \in K} \sum_{a \in A_1} n_k t_a^k$$

subject to

$$\forall k \in K \begin{cases} \lambda_i^k - \lambda_j^k = c_a + t_a - s_a^k & \forall a = (i,j) \in A_1 \\ \lambda_i^k - \lambda_j^k = c_a - s_a^k & \forall a = (i,j) \in A_2 \\ 0 \leq s_a^k \leq M(1 - \sum_{p \in I_a^k} z_p) & \forall a \in A \\ -M_k \sum_{p \in I_a^k} z_p \leq t_a^k \leq M_k \sum_{p \in I_a^k} z_p & \forall a \in A_1 \\ t_a^k \leq t_a \leq t_a^{\max} & \forall a \in A_1 \\ \sum_{p \in P_k} z_p = 1 & \\ z_p \in \{0,1\} & \forall p \in P_k, \end{cases}$$

where s is the vector of slack variables associated with the dual constraints. The first three constraints ensure that the selected paths are optimal with respect to the current toll vector. The fourth and fifth ones, together with the max operator, ensure that the commodity revenue t_a^k is equal to the true revenue t_a whenever arc a lies on the path actually used by commodity k, hence that the model is consistent. The set of values that can be assumed by the constants M and M_k is the same as that for MIP I.

This formulation involves $\sum_{k \in K} |P_k|$ binary variables. Although this number grows exponentially with the size of the network, it may very well be less than the number of variables involved in MIP I, whenever the number of 'reasonable' paths is small. In Section 3, we present a procedure that limits the number of paths to be considered, and consequently makes this approach practical for realistic instances of TOP.

The third formulation, MIP III, is entirely path-based. Let us first introduce T^k, the profit raised from commodity k, as well as L^k, the disutility (cost plus delay) associated with the shortest path p actually used by commodity k. Since at most one path is used for every commodity, we obtain

$$T^k = \sum_{a \in p \cap A_1} t_a$$

and

$$\begin{aligned} L^k &= \sum_{a \in p} (c_a + t_a) \\ &= T^k + \sum_{p \in P_k} z_p \sum_{a \in p} c_a. \end{aligned}$$

This leads to the path formulation MIP III, that involves a smaller number of variables than formulation MIP II.

$$\text{MIP III:} \qquad \max_{t,z,L} \sum_{k \in K} n_k T^k$$

$$\text{subject to} \quad t_a \leq t_a^{\max} \qquad \forall a \in A_1$$

$$\forall k \in K \begin{cases} T^k \leq \sum_{a \in p \cap A_1} t_a + M_k(1 - z_p) & \forall\, p \in P_k \\ \sum_{a \in p \cap A_1} t_a + \sum_{a \in p} c_a - M_k^p(1 - z_p) \leq L^k \leq \sum_{a \in p \cap A_1} t_a + \sum_{a \in p} c_a & \forall\, p \in P_k \\ L^k = T^k + \sum_{p \in P_k} z_p \sum_{a \in p} c_a & \\ \sum_{p \in P_k} z_p = 1 & \\ z_p \in \{0, 1\} & \forall\, p \in P_k. \end{cases}$$

In this formulation, a suitable value for M_k^p is given by:

$$M_k^p = \sum_{a \in p} c_a + \sum_{a \in p} t_a^{\max} - C_k^0.$$

3 A path generation algorithm

In this section we propose an algorithmic framework that relies on the following three observations:

- lower level solutions correspond to some shortest paths for each origin-destination pair;
- for a given lower level extremal solution (collection of shortest paths), one may efficiently recover a set of revenue-maximizing tolls that is compatible with this solution;
- one may expect to extract higher revenues from toll arcs belonging to paths having low rather than large initial delays. (This should come at no surprise.)

The algorithm generates a sequence of extremal lower level solutions, or *multipath* P, which corresponds to vectors of commodity paths, one per commodity, e.g.,

$$P = (p_1, p_2, \dots, p_{|K|}), \text{with } p_k \in P_k.$$

We denote by C the set of all multipaths:

$$C = \{(p_1, \dots, p_{|K|}) \mid p_k \in P_k;\ k = 1, \dots, |K|\}.$$

Given a multipath $P \in C$, we define $c(P)$ as the sum of delays on the arcs belonging to at least one of its paths, i.e.,

$$c(P) = \sum_{k=1}^{|K|} n_k \sum_{a \in p_k} c_a.$$

Without loss of generality, we assume that the elements of C are indexed in nondecreasing order of their respective total delays:

$$c(P^1) \leq c(P^2) \leq \dots \leq c(P^{|C|}).$$

The algorithm explores multipaths in increasing order of total delay, and stops as soon as no progress can be achieved. One iteration of the generic algorithmic scheme is composed of the following operations:

1. Generate the ith multipath P^i;
2. Update upper bound on total revenue;
3. Optimize toll schedule with respect to current multipath;
4. Update lower bound on total revenue;
5. If lower and upper bounds coincide, stop with an optimal solution to TOP.

The efficiency of the procedure rests on the quality of lower and upper bounds, which are crucial in limiting the scope of the enumeration process, and on the design of an efficient algorithmic procedure for generating multipaths. Those are considered in turn.

3.1 Upper bound

Let p_k^∞ (respectively p_k^0) denote the shortest path from $o(k)$ to $d(k)$ in the graph G obtained by setting all tolls to $+\infty$ (respectively 0) and let α_k^∞ (respectively α_k^0) denote the corresponding delay, i.e.,

$$\alpha_k^\infty = \sum_{a \in p_k^\infty} c_a$$

$$\alpha_k^0 = \sum_{a \in p_k^0} c_a.$$

For a given commodity index k in K, an upper bound on the revenue raised from this commodity k is given by the product of the demand n_k and the gap between α_k^∞ and α_k^0, i.e.,

$$UB(p_k) = n_k(\alpha_k^\infty - \alpha_k^0).$$

This bound can actually be tightened by making it dependent on the delay of the multipath under consideration:

$$UB(P) = \sum_{k \in K} n_k(\alpha_k^\infty - \sum_{a \in p_k} c_a).$$

3.2 Lower bound

If the lower level solution (multipath) is known a priori, all bilinear constraints become linear, and the resulting program is easy. Its solution yields a toll vector t that maximizes revenue while being compatible with the multipath. This operation is tentamount to solving an inverse problem. Loosely speaking, an inverse optimization problem occurs when one wishes to estimate the parameters of a primary optimization problem whose optimal solution is known a priori. In general, the set of such parameters is not unique, and one may therefore optimize a secondary objective over this set. In the context of toll optimization, one seeks tolls that maximize revenue (the secondary objective) while inducing a predetermined lower level solution, i.e., a toll-compatible multipath. The resulting toll schedule provides the best solution compatible with the multipath, and thus a valid lower bound on the problem's optimum value.

Let $LB(P)$ denote the optimal value of the inverse optimization problem associated with the multipath $P = \{p_k\}_{k \in K}$, i.e.,

$$LB(P) = \max_{t \in D} \sum_{k=1}^{|K|} n_k \sum_{a \in p_k \cap A_1} t_a$$

where

$$D = \left\{ t \mid \sum_{a \in p_k} c_a + \sum_{a \in p_k \cap A_1} t_a \leq \sum_{a \in \bar{p}_k} c_a + \sum_{a \in \bar{p}_k \cap A_1} t_a \quad \forall \bar{p}_k \in P_k \quad \forall k \in K \right\}.$$

While the number of constraints that define the set D is exponential, the associated separation problem can be solved in polynomial time. Indeed, a shortest path oracle can be used to check whether a given toll vector t lies in D, and exhibit, whenever $t \notin D$, a violated constraint. It follows from a result of Grötschel et al [6] that the inverse optimization problem is polynomially solvable. Alternatively, Labbé et al [8] have shown how to reduce the inverse problem to a transshipment problem over a suitably defined network, and for which polynomial algorithms are well known.

3.3 Computation of the ith shortest multipath

The computation of multipaths in increasing order of their fixed costs is an essential part of the sequential algorithm. In this subsection, we describe a procedure to compute the ith shortest multipath

$$P^i = (p_1^{i(1)}, p_2^{i(2)}, \ldots, p_{|K|}^{i(|K|)}),$$

where $i(k)$ denotes the ranking of the path associated with commodity k. In particular, the shortest multipath is

$$P^1 = (p_1^1, p_2^1, \ldots, p_{|K|}^1).$$

ith shortest multipath

Step 0 [Initialization]
Set $j \leftarrow 1$, and $LIST \leftarrow \emptyset$.
Compute p_k^1 and p_k^2 and set $j(k) \leftarrow 1$, $k = 1, \ldots, |K|$.
$P^1 = (p_1^1, p_2^1, \ldots, p_{|K|}^1)$.

Step 1 If $j = i$, stop: the ith multipath has been obtained.

Step 2 [updating of $LIST$]
For all $l \in \{1, \ldots, |K|\}$, let $P^{j,l} \leftarrow (p_1, \ldots, p_{|K|})$, where $p_k = p_k^{j(k)}$ if $k \neq l$, and $p_l = p_l^{j(l)+1}$. Set $LIST \leftarrow LIST \cup P^{j,l}$.
Set $j \leftarrow j + 1$.

Step 3
Remove the least costly multipath from $LIST$ distinct from $P^1, \ldots, P^{j-1}$, and output this element P^j as the jth shortest multipath.

Step 4
Let $j_0 \in \{1, \ldots, j-1\}$ and $l_0 \in \{1, \ldots, |K|\}$ such that $P^j = P^{j_0, l_0}$.
Compute $p_{l_0}^{j_0(l_0)+2}$ and set $j_0(l_0) \leftarrow j_0(l_0) + 1$. Return to step 1.□

The above algorithm requires the knowledge of commodity paths, sorted in increasing order of their costs. To this end, we adopt a procedure proposed by Lawler [10], which we describe, for the sake of completeness. Let the arcs of the directed graph be numbered $1, \ldots, m$. For a given path p, let $x_j = 1$ if arc j is contained in p, and $x_j = 0$ otherwise. Given an integer i, the procedure generates the first i shortest paths in sequence.

ith-shortest path algorithm

Step 0 [Initialization]
Compute a shortest path $x^{(1)} = (x_1^{(1)}, x_2^{(1)}, \ldots, x_m^{(1)})$, without fixing the values of any variables.
$LIST \leftarrow \{x^{(1)}\}$ and $j \leftarrow 1$.

Step 1 [Output the jth shortest path]
Remove the least costly solution from $LIST$ and output this solution, denoted by $x^{(j)} = (x_1^{(j)}, x_2^{(j)}, \ldots, x_m^{(j)})$, as the jth shortest path.

Step 2 If $j = i$, stop; the ith shortest path has been obtained.

Step 3 [Update of $LIST$]
Assume that the jth shortest path was obtained by fixing the following conditions

$$x_1 = x_2 = \cdots = x_q = 1,$$
$$x_{q+1} = x_{q+2} = \cdots = x_s = 0,$$

where a reordering has been assumed for notational purposes. Leaving these variables fixed as they are, create $m-s$ new shortest path problems that must satisfy the additional conditions

$$x_{s+1} = 1 - x_{s+1}^{(j)},$$
$$x_{s+1} = x_{s+1}^{(j)}, x_{s+2} = 1 - x_{s+2}^{(j)},$$
$$\vdots$$
$$x_{s+1} = x_{s+1}^{(j)}, x_{s+2} = x_{s+2}^{(j)}, \ldots, x_{m-1} = x_{m-1}^{(j)}, x_m = 1 - x_m^{(j)}.$$

Compute optimal solutions (i.e, the shortest path subject to conditions above) to each of these $m-s$ problems and place each of the $m-s$ solutions in LIST, together with a record of the variables which were fixed for each of them. Set $j = j + 1$. Return to Step 1. □

Remark that the first time Step 3 is executed, $q = s = 0$

3.4 Algorithm specification

We now formerly state the algorithm. Let LB^* be the current best profit, P^* the associated multipath, and UB^* the current upper bound. Note that, since the upper bound is non increasing, UB^* is actually the same as $UB(P)$ evaluated at the current multipath. Let N denote the number of distinct multipaths.

Multipath Algorithm

Step 0 [Initialization]
$LB^* \leftarrow -\infty$.
$i \leftarrow 1$.
Step 1 [Multipath generation and evaluation]
Generate the ith smallest element of C,
$P^i = (p_1^{i(1)}, p_2^{i(2)}, \ldots, p_{|K|}^{i(|K|)})$.
$UB^* \leftarrow \sum_{k \in K} n_k(\alpha_k^\infty - \sum_{a \in p_k^{i(k)}} c_a)$.
Compute $LB(P^i)$ by inverse optimization.
Set $LB^* \leftarrow \max\{LB^*, LB(P^i)\}$.
Step 2 [Stopping criterion]
If $UB^* \leq LB^*$ or $i = N$, stop. The optimal solution is the multipath P^* that has achieved the best lower bound.
$i \leftarrow i + 1$ and return to step 1. □

In order to prove the correctness of the algorithm, it suffices to remark that the upper bound UB^* does not increase at each iteration, so that an optimal multipath cannot be missed. Note that the algorithm may have to scan the entire list of multipaths, and that it may terminate with the local upper bound UB^* being *strictly* less than the lower bound LB^*. This can be observed on the single-commodity example illustrated in Figure 1, taken from Labbé et al. [9]. In this example, the first multipath P^1 (a single path in this case) generated is $\{(1-2-3-4-5)\}$. Its upper bound is $22 - 6 = 16$ while its lower bound (this easy to check), is equal to 15. The algorithm stops after generating $P^2 = \{(1-2-4-5)\}$, whose upper bound 11 is less than the lower bound of the first path. There does not exist a path that achieves an upper bound equal to the optimal value.

3.5 Redundant paths

A serious drawback of formulations MIP II and MIP III is that all paths between all origin-destination pairs must be enumerated a priori. Obviously,

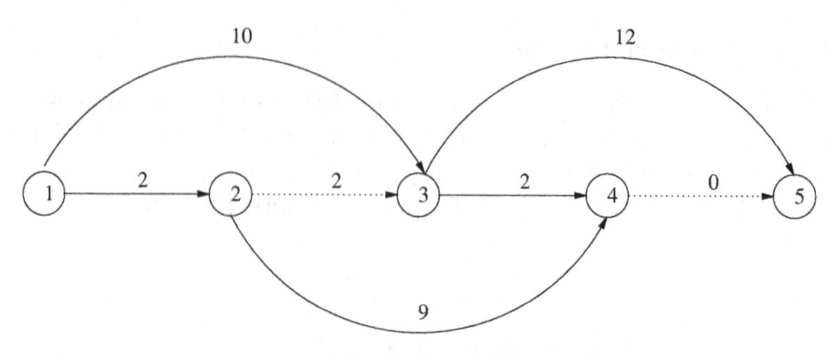

Fig. 1. The multipath algorithm terminates with $LB^* \neq UB^*$

many paths are suboptimal and irrelevant. For instance, one need not consider paths that contain toll-free subpath that are *not* shortest subpaths. Along this line of reasoning, Kraaij [7] constructed a Shortest Path Graph Model (SPGM), equivalent to TOP, where subpaths between toll arcs are replaced by single arcs with cost set to that of a shortest subpath; similarly, subpaths from the origins to the tail of toll arcs, and from the head of toll arcs to the destinations, are shrunk to single arcs. While this preprocessing does not affect the combinatorial nature of the problem, it may reduce computing times by a (roughly) constant factor. Note that, in some cases, the SPGM may contain *more* arcs than the original network, and the computational burden of setting up an SPGM may actually greatly exceed that of solving the resulting problem (see [2]).

While we have not adopted the SPGM formulation, we have implemented a technique for eliminating a subset of dominated paths, according to the criterion outlined in the following lemma. This allows to limit the number of problems created at step 3 of the ith-shortest path algorithm.

Lemma 3.1 (Bouhtou et al [2]) *Consider an instance of TOP where tolls are restricted to be nonnegative. Let p and p_* be two paths between an origin-destination pair $k \in K$, and let $p = (p^1, p^2)$ and $p_* = (p_*^1, p_*^2)$ denote their partition into toll and toll-free arcs, respectively. Assume that $p_*^1 \subseteq p^1$ and that*

$$\sum_{a \in p_*} c_a \leq \sum_{a \in p} c_a.$$

If p, together with a toll schedule t, is optimal for TOP, it follows that the couple (p_*, t) is also optimal for TOP.

Based on the above result, one may reduce the number of problems generated at step 3 of the ith shortest path algorithm by only considering the toll arcs and undominated paths. Suppose, without loss of generality, that the arcs of A_1 are numbered $1, \ldots, |A_1|$ and that the jth shortest path contains arcs

$1, \ldots, r$, $r \leq |A_1|$. Suppose, moreover, that the jth shortest path is the shortest path obtained by fixing the following variables

$$x_1 = x_2 = \ldots = x_p = 1,$$
$$x_{r+1} = x_{r+2} = \ldots = x_q = 0,$$

and the other toll variables (non fixed) verify

$$x_{p+1}^{(j)} = x_{p+2}^{(j)} = \ldots = x_r^{(j)} = 1,$$
$$x_{q+1}^{(j)} = x_{q+2}^{(j)} = \ldots = x_{|A_1|}^{(j)} = 0.$$

Leaving the fixed variables as they are, create $r-p$ new shortest path problems that must satisfy the additional conditions

$$x_{p+1} = 0,$$
$$x_{p+1} = 1, x_{p+2} = 0,$$
$$\vdots$$
$$x_{p+1} = x_{p+2} = \ldots = x_{r-1} = 1, x_r = 0.$$

4 A block sequential heuristic (BLOSH)

While, as we shall see in the next section, the three MIP reformulations and the exact multipath algorithm allow to tackle medium size problems, the NP-hard nature of TOP will ultimately limit the size of problems that can be solved to prove optimality. The main limitation is due both to the large number of commodities and their interactions. To circumvent the problem, we have implemented a windowing technique that consists in optimizing over a subset of commodities at a time, keeping fixed the paths associated with the (temporarily) fixed commodities. This results in a block sequential heuristic (BLOSH), reminiscent of the Gauss-Seidel approach, well known in optimization. At each iteration, the subproblems are solved using either one of the algorithms presented previously. For our purpose, we have used the MIP III formulation, which involves a small number of binary variables, and proved efficient for solving problems involving a small number of commodities. MIP III was solved using the commercial software CPLEX [3].

More precisely, let us consider a partition K into the set of *active* and *inactive* commodities (origin-destination pairs), i.e., $K = K^1 \cup K^2$. The restricted MIP III formulation then takes the form:

MIP III–R: $\max_{t,z,L} \sum_{k \in K} n_k T^k$

subject to

$$\forall k \in K^1 \begin{cases} T^k \leq \sum_{a \in p \cap A_1} t_a + M_k(1 - z_p) & \forall\, p \in P_k \\ \sum_{a \in p \cap A_1} t_a + \sum_{a \in p} c_a - M_k^p(1 - z_p) \leq L^k \leq \sum_{a \in p \cap A_1} t_a + \sum_{a \in p} c_a & \forall\, p \in P_k \\ L^k = T^k + \sum_{p \in P_k} z_p \sum_{a \in p} c_a & \\ \sum_{p \in P_k} z_p = 1 & \\ z_p \in \{0, 1\} & \forall\, p \in P_k \end{cases}$$

$$\forall k \in K^2 \begin{cases} T^k = \sum_{a \in p_k^{i(k)} \cap A_1} t_a & \\ L^k \leq \sum_{a \in p \cap A_1} t_a + \sum_{a \in p} c_a & \forall\, p \in P_k \\ L^k = T^k + \sum_{a \in p_k^{i(k)}} c_a. & \end{cases}$$

In practice, the number of active commodities, $|K^1|$ is small with respect to the number of inactive commodities $|K^2|$. In that case, MIP III-R involves a small number of binary variables and is efficient. A pseudocode for the algorithm is outlined below.

Algorithm BLOSH

Step 0 [initialization]
Compute a feasible paths solution and the associated optimal tax vector (e.g. based on the shortest multipath).
Let $\overline{k}$ be the number of active commodities.
$i \leftarrow 1$.

Step 1
$K^1 \leftarrow \{[(i-1) \mod |K|] + 1, [(i) \mod |K|] + 1, \ldots, [(i+\overline{k}-2) \mod |K|] + 1\}$
$K^2 \leftarrow K \setminus K^1$
Solve MIP III–R and let V^i be its optimal value.
If $V^i = V^{i+1}$, stop.

Step2
$i \leftarrow i + 1$.
Return to step 1.

5 Numerical results

The numerical tests have been performed on randomly generated networks, and give a good idea of the various algorithms' behaviour with respect to the size of the instances. They have been conducted in two steps. We first tested the multipath algorithm on medium-size instances, rapidly showing the limitations of this approach. We then assessed the efficiency of the three MIP formulations, using the commercial MIP solver CPLEX 6.0, versus that of the sequential heuristic. The tests were performed on a SUN ULTRA60 workstation.

The main parameters of the test problems were $|N|$ (number of nodes), $|A|$ (number of arcs), $|A_1|$ (number of toll arcs), $|K|$ (number of commodities). For each toll-free arc (respectively toll arc), an integer fixed cost c_a was uniformly chosen in the interval [2,20] (respectively [0,6]). The origin-destination pairs were uniformly chosen, and their demands set to uniform random variables on the interval [20,100]. Finally, to ensure connectivity of the underlying graph and the existence of toll-free paths for each origin-destination pair, a Hamiltonian circuit composed only of toll-free arcs was integrated within the network.

Tables 1,2 and 3 illustrate the (in)efficiency of the multipath algorithm. Each table presents the solution of 10 randomly generated problems. We observe that the multipath approach can solve small to medium scale problems, but fails on larger instances. Whenever the number of commodities increases, the approach rapidly shows its limits, mainly due to the large number of multipaths with similar length values. This disappointing performance is due to the existence of nearly identical multipaths. This resulted in upper bounds that decrease very slowly, as well as lower bounds that increase by steps, after having stalled for several iterations. This can be observed on Figures 2, 3 and 4.

The next tables illustrate the performance of CPLEX on the three MIP formulations, comparing with algorithm BLOSH, for various problem sizes. Each MIP problem was solved with the default parameters of CPLEX 6.0. Running times include the elimination of dominated multipaths.

Algorithm BLOSH was initiated with a shortest multipath. The first $\overline{k}$ active commodities (set to 20) were chosen as follows: we solved independent TOP problems, one for each commodity and reordered them from higher to lower revenue. These revenues were obtained by inverse optimization.

Random networks were generated for various values of the main parameters ($|N|$, $|A|$, $|A_1|$ and $|K|$). The results of our computational experiments are presented in Tables 4 to 9, where each line corresponds to one specific instance. Column headers show the instance number, as well as the running times and the number of Branch-and-Bound nodes explored by CPLEX, for the three MIP formulations. We also indicated, in the BLOSH column, the percentage of optimality reached by the algorithm, 100% indicating that an optimal solution was obtained.

Tables 4 and 5 provide the running times for the four algorithms. While no clear conclusion could come out concerning the average number of nodes, MIP III came out the winner, as it could process each node much faster. On these medium-sized problems, Algorithm BLOSH converged to an optimal solution on all but two instances, with running times slightly less, on the average, that those of MIP III. In Tables 6, 7, 8 and 9, we focused on the following issues:

- How close is the solution provided by BLOSH to an optimal solution?
- How many iterations are required for BLOSH to converge?

On the larger instances, it was not always possible to answer the first question, due to excessive running times. Indeed, MIP III could not reach an optimal solution, or *prove* that such solution was reached, within the imposed time limit set respectively at 10 000 seconds (Tables 6 and 7) and 15 000 seconds (Table 9). Three instances (marked with an asterisk in Table 7) were subsequentially allowed 40 000 of CPU time and yet failed to reach an optimum. Actually, the best solution achieved by MIPIII was improved for most instances reported in Tables 8 and 9, i.e., whenever the deviation from MIPIII's best value exceeded 100 in the corresponding entry of the percentage column.

Finally, note that it is not straightforward to compare our numerical results with those obtained by the MIP formulation of Bouhtou et al. [2]. Indeed:

- The nature of the problems generated in their paper is quite different from ours. First, the number of paths between OD pairs is less than 3, on average it is of the order of 30 undominated paths for our instances.
- The proportion of toll arcs is much higher in our experiments.

– Computers used are from different generations.

This being said, the ratio of improvement between MIP I and MIP III is comparable to the ratio observed in [2] between MIP I (AMIP according to their notation) and their path-based formulation PMIP, once the computational time associated with the generation of the SPGM has been taken into account. Note that, on the set of problems considered by these authors, most of the running time is spent in the preprocessing phase.

6 Conclusion

In this paper, we have proposed new approaches, based on path variables, for addressing an NP-hard problem having applications in the context of optimal pricing. The two main results of the paper were to assess the quality of MIP reformulations of TOP, and to show that the best formulation (MIP III) could be used as the core of a promising heuristic procedure (BLOSH). We are currently working along two lines of attack. First, we wish to embed the most efficient procedures within a decomposition framework. Second, sophisticated techniques (partial inverse optimization) are being investigated, with the aim of improving the upper bound of the multipath algorithm (which is typically of very bad initial quality) and of reducing the number of multipaths explored in the course of the algorithm.

References

1. Brotcorne, L, Labbé, M., Marcotte, P., Savard, G., "A bilevel model for toll optimization on a multicommodity transportation network", *Transportation Science*, 35, 345–358, 2001.
2. Bouhtou, M., van Hoesel, S., van der Kraaij, A., Lutton, J.-L., "Tariff optimization in networks", Research Memorandum 041, METEOR, Maastricht Research School of Economics of Technology and Organization, 2003.
3. CPLEX, ILOG CPLEX, v6.0, 2000.
4. Dewez, S., *On the toll setting problem.* PhD thesis, Université Libre de Bruxelles, Institut de Statistique et de Recherche Opérationnelle, 2004.
5. Grigoriev, A., van Hoesel, S., van der Kraaij, A., Uetz M., Bouhtou, M., "Pricing Network Edges to Cross a River", Research Memorandum 009, METEOR, Maastricht Research School of Economics of Technology and Organization, 2004.
6. Grötschel, M., Lovász, L., Schrijver, A., "The ellipsoid method and its consequences in combinatorial optimization", *Combinatorica*, 1, 169–197, 1981.
7. van der Kraaij, A., *Pricing in networks.* PhD thesis, Proefschrift Universiteit Maastricht, 2004.
8. Labbé, M., Marcotte, P., Savard, G., "A bilevel model of taxation and its applications to optimal highway pricing", *Management Science*, 44, 1608–1622, 1998.

9. Labbé, M., Marcotte, P., Savard, G., "On a class of bilevel programs", In: Nonlinear Optimization and Related Topics, Di Pillo and Giannessi eds., Kluwer Academic Publishers, 183-206, 1999.
10. Lawler, E.L., "A procedure to compute the K best solutions to discrete optimization problems and its application to the shortest path problem", *Management Science*, 18, 401-405, 1972.
11. Marcotte, P., Savard, G. and Semet, F. "A bilevel programming approach to the travelling salesman problem", *Operations Research Letters*, 32, 240-248, 2004.
12. Roch, S., Savard, G., Marcotte, P.,"Design and analysis of an algorithm for Stackelberg network pricing", *Networks*, 46, 57-67, 2005.

Instances	Nodes	time (s)	gap (%)
1	42	0.11	0.00
2	287	0.12	0.00
3	27	0.13	0.00
4	102	0.17	0.00
5	21	0.14	0.00
6	8	0.12	0.00
7	152	0.17	0.00
8	86	0.16	0.00
9	54	0.17	0.00
10	194	0.20	0.00

Table 1. Problems with 60 nodes, 200 arcs, 20 tolled arcs and 10 O-D-pairs

Instances	Nodes	time (s)	gap (%)
1	1061001	14022.78	18.04
2	1161001	14006.93	7.46
3	2313001	14006.94	8.91
4	495542	2559.90	0.00
5	1375001	14002.36	3.78
6	1270001	14021.93	21.00
7	1220001	14024.66	20.70
8	78019	1017.52	0.00
9	1501001	14012.94	3.92
10	1219001	14015.69	8.86

Table 2. Problems with 60 nodes, 200 arcs, 40 tolled arcs and 20 O-D-pairs

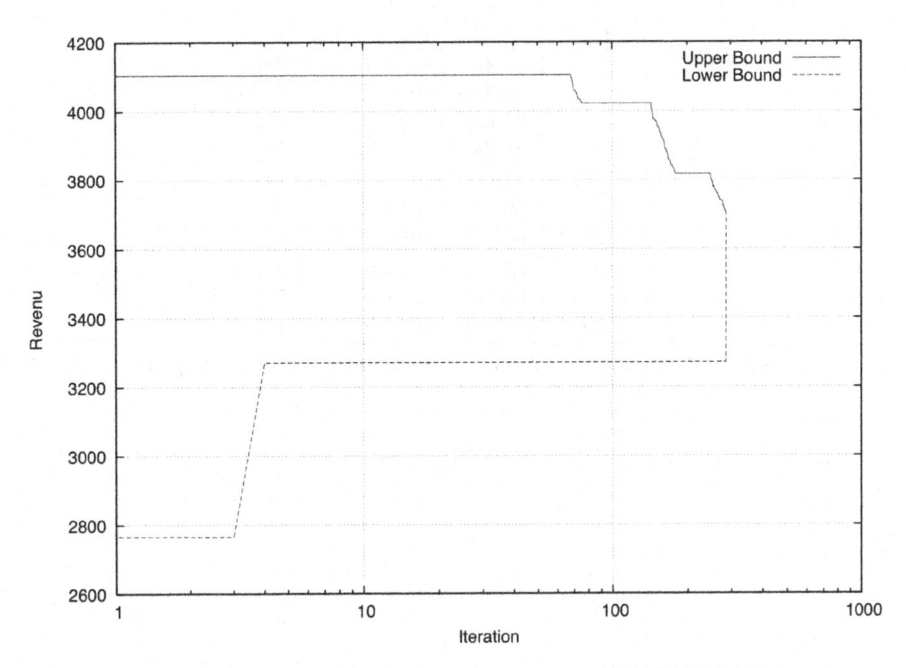

Fig. 2. Upper and lower bounds: instance 1 (60,200,20,10)

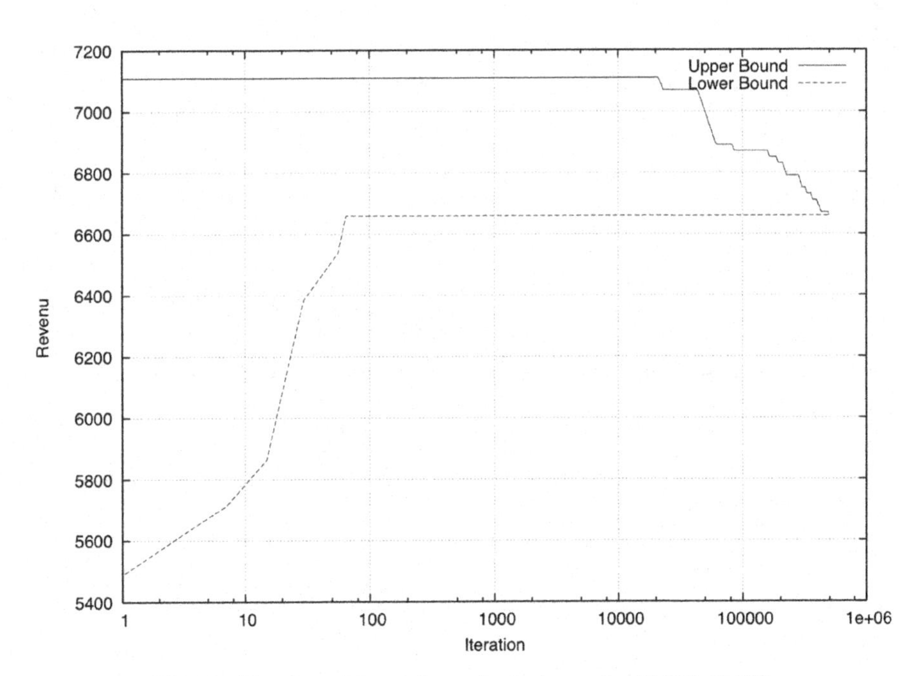

Fig. 3. Upper and lower bounds: instance 1 (60,200,40,20)

Instances	Nodes	time (s)	gap (%)
1	550001	14010.68	39.20
2	933001	14029.33	14.03
3	849001	14019.97	17.17
4	482001	14012.71	28.91
5	729001	14019.96	30.34
6	697001	14012.32	21.81
7	645001	14019.11	28.36
8	645001	14013.17	16.05
9	1016001	14015.23	25.54
10	1112001	14026.46	39.12

Table 3. 90 nodes, 300 arcs, 60 tolled arcs and 40 O-D-pairs

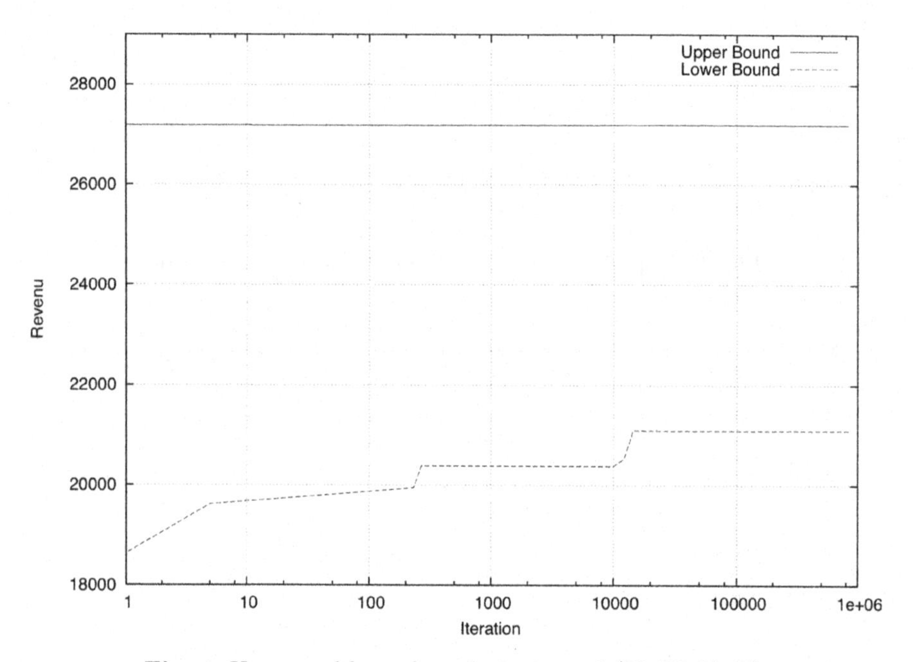

Fig. 4. Upper and lower bounds: instance 1 (60,300,60,40)

Instance	MIP I		MIP II		MIP III		BLOSH	
	Nodes	time (s)	Nodes	time (s)	Nodes	time (s)	%	time (s)
1	48	9.48	44	2.50	46	0.97	100.00	1.32
2	30	3.98	32	2.02	20	0.88	100.00	1.42
3	30	6.70	42	1.63	41	0.47	100.00	0.77
4	144	21.00	45	4.98	67	1.03	100.00	1.59
5	29	8.13	15	1.69	26	0.46	100.00	0.73
6	40	5.97	32	3.31	88	1.02	100.00	1.37
7	36	7.28	55	3.76	62	0.99	100.00	1.39
8	14	5.37	16	1.86	29	0.63	100.00	0.77
9	62	9.70	50	2.91	52	0.59	100.00	0.80
10	12	4.16	9	1.07	16	0.42	100.00	0.58

Table 4. 60 nodes, 20 O-D pairs, 200 arcs, 40 tolled arcs.

Instance	MIP I		MIP II		MIP III		BLOSH	
	Nodes	time (s)	Nodes	time (s)	Nodes	time (s)	%	time (s)
1	5010	5998.84	77500	19806.00	6881	34.11	99.76	12.51
2	496	147.98	232	114.89	1015	3.85	100.00	3.80
3	87	79.23	49	7.63	154	2.46	100.00	3.49
4	243	146.00	361	51.45	479	3.61	100.00	4.44
5	1076	830.20	1618	473.02	622	3.99	100.00	4.36
6	346	808.50	312	259.06	1264	8.61	100.00	7.74
7	425	237.03	1903	260.51	1829	6.87	100.00	5.27
8	145	67.19	295	47.39	856	7.34	100.00	8.04
9	736	2350.88	1129	904.61	3638	19.97	98.29	8.34
10	264	274.21	182	43.40	179	5.00	100.00	6.61

Table 5. 90 nodes, 40 O-D pairs, 300 arcs, 60 tolled arcs.

Instance	MIP III		BLOSH	
	Nodes	time (s)	%	time (s)
1	61897	460.00	97.08	25.37
2	2570196	29661.00	99.21	44.30
3	3542	38.00	98.81	11.12
4	202271	1027.00	97.76	20.52
5	38777	120.00	99.42	10.88
6	43588	239.00	98.20	16.05
7	378033	2609.00	98.75	33.67
8	133873	1002.00	98.39	26.09
9	3170	42.00	99.95	11.65
10	12549	181.00	99.24	31.57

Table 6. 120 nodes, 60 O-D pairs, 400 arcs, 80 tolled arcs.

Instance	MIP III		BLOSH	
	Nodes	time (s)	%	time (s)
1	800166	11304	97.15	122
2	852734	5739	97.46	66
3	3615113	39555	99.28	102
4	3059449	27559	100	91
5(*)	518756	10056	96.22	162
6	2129948	35160	99.39	127
7	182957	2868	100	133
8(*)	432250	10057	99.86	170
9(*)	1905100	10045	102.06	132
10	488533	5116	97.54	95

Table 7. 150 nodes, 80 O-D pairs, 600 arcs, 120 tolled arcs.

Instance	MIP III		BLOSH	
	Nodes	time (s)	%	time (s)
1	532884	10080	112.74	108
2	610662	10086	99.65	118
3	361924	10121	121.45	182
4	580082	10091	131.76	126
5	711571	10076	98.53	103
6	626277	10078	104.40	103
7	857079	10068	100.39	84
8	430176	10102	103.76	145
9	342673	10107	103.94	158
10	426247	10094	107.84	138

Table 8. 200 nodes, 100 O-D pairs, 800 arcs, 160 tolled arcs.

Instance	MIP III		BLOSH	
	Nodes	time (s)	%	time (s)
1	11591	15998	133.66	4991
2	20794	15181	123.98	4430
3	14927	15753	132.47	5139
4	47401	13597	114.17	2880
5	16987	14649	119.20	3962
6	36033	13690	117.84	2992
7	37476	13823	116.80	3006
8	24560	14778	109.41	4040
9	24274	15144	122.73	4450
10	16280	15195	118.39	4425

Table 9. 500 nodes, 200 O-D pairs, 5000 arcs, 1000 tolled arcs.

Bilevel programming with convex lower level problems

Joydeep Dutta[1] and Stephan Dempe[2]

[1] Department of Mathematics, Indian Institute of Technology, Kanpur-208106, India jdutta@iitk.ac.in
[2] Department of Mathematics and Informatics, Technical University Bergakademie Freiberg, 09596 Freiberg, Germany dempe@tu-freiberg.de

Summary. In this article we develop certain necessary optimality condition for bilevel programming problems with convex lower-level problem. The results are abstract in nature and depend on an important construction in nonsmooth analysis called the coderivative of a set-valued map.

Key words: Optimistic bilevel programming problem, coderivative of set-valued map, necessary optimality conditions.

1 Introduction

Bilevel programming is among the frontier areas of modern optimization theory. Apart from its importance in application it is also a theoretically challenging field. The first such challenge comes when one wants to write Karush-Kuhn-Tucker type optimality conditions for bilevel programming problems. The major drawback is that most standard constraint qualifications are never satisfied for a bilevel programming problem. Thus it is interesting to devise methods in which one may be able to develop in a natural way constraint qualifications associated with bilevel problems and thus proceed towards obtaining Karush-Kuhn-Tucker type optimality conditions. The recent literature in optimization has seen quiet a few attempts to obtain optimality conditions for bilevel programming problems. See for example Ye and Zhu [27],[28],[29], Ye and Ye [26], Dempe [9],[10], Loridan and Morgan [15], Bard [3], [4],[5] and the references there in. In 1984 J. F. Bard [3] made an attempt to develop optimality conditions for bilevel programming problems though it was later observed to have some error. The recent monograph by Dempe [9] is one helpful source to study optimality conditions for bilevel programming problems.

In this article we consider the special type of bilevel programming which has a convex programming problem as its lower-level problem. Using the recent advances made in the understanding of the solutions sets of variational

S. Dempe and V. Kalashnikov (eds.), *Optimization with Multivalued Mappings*, pp. 51-71
©2006 Springer Science + Business Media, LLC

systems (see for example Dontchev and Rockafellar [11],[12] and Levy and Mordukhovich [14]) we will develop some new necessary conditions for bilevel programming problems with convex lower-level problems. In section 2 we begin with the basic formulation of a bilevel programming problem and motivate the type of problems we intend to study in this article. Then we outline the variational tools that are required to represent the necessary optimality conditions. In section 4 we present the main results, i.e the necessary optimality conditions for each of the problem formulations that we have described earlier.

2 Motivation

Consider the following bilevel programming problem (P)

$$\min_x F(x,y) \quad \text{subject to} \quad y \in S(x), x \in X,$$

where $F : \mathbb{R}^n \times \mathbb{R}^m \to \mathbb{R}$, $X \subseteq \mathbb{R}^n$ and $S(x)$ is the solution set of the following problem (LLP)

$$\min_y f(x,y) \quad \text{subject to} \quad y \in K(x),$$

where $f : \mathbb{R}^n \times \mathbb{R}^m \to \mathbb{R}$ and $K(x) \subseteq \mathbb{R}^m$ is a set depending on x. The problem (LLP) is called the lower-level problem and the problem (P) is called the upper-level problem. For simplicity we consider only the case where $X = \mathbb{R}^n$ and where for each x the function $f(x,\cdot)$ is convex and the set $K(x)$ is convex. Thus in our setting the problem (P) will be as follows

$$\min_x F(x,y) \quad \text{subject to} \quad y \in S(x)$$

where $S(x)$ as before denotes the solution set of the lower-level problem (LLP)

$$\min_y f(x,y) \quad \text{subject to} \quad y \in K(x).$$

From now on we will assume that the problem (LLP) always has a solution. However the term min in the upper level problem is slightly ambiguous since one is not sure whether the lower-level problem has an unique solution or not. If $f(x,\cdot)$ is strictly convex in y for each x then $S(x)$ is a singleton for each x or rather S is a single-valued function. If however $f(x,\cdot)$ is only assumed to be convex one cannot always guarantee the single-valuedness of $S(x)$. It is important to observe that the main complication in bilevel programming arises when the lower-level problem does not have a unique solution, i.e. $S(x)$ is not a singleton for some x. Thus for that particular case the objective function of the upper-level problem would look like

$$\bigcup_{y \in S(x)} F(x,y) = F(x, S(x)).$$

The bilevel programming problem (P) is then rather a set-valued optimization problem, see e.g. G.Y. Chen and J. Jahn [7]. To treat this situation within bilevel programming problems at least two approaches have been reported in the literature, namely the *optimistic solution* and the *pessimistic solution*. We will consider here the optimistic solution approach for reasons which will be clear as we progress further. For details on the pessimistic solution approach see for example Dempe [9]. One of the reasons are the less restrictive assumptions needed to guarantee the existence of an optimal solution in the optimistic case. To introduce the optimistic case consider the function

$$\varphi_0(x) = \inf_y \{F(x,y) : y \in S(x)\}.$$

We remark that $\varphi_0(x)$ denotes the infimal function value of the upper level objective over the solution set of the lower-level problem parameterized by x and that we do not demand that it is attained. For each $x \in \mathbb{R}^n$, this function gives the lowest bound for possible objective function values of the upper level objective function on the set of optimal solutions of (LLP). Then, the optimistic bilevel problem reads as

$$\min_x \varphi_0(x). \tag{1}$$

Definition 2.1 *A point $\overline{x}$ is called a (global) optimistic solution of the problem (P) if $\varphi_0(x) \geq \varphi_0(\overline{x})$ for all x.*

An optimal solution of this problem exists whenever the function $\varphi_0(x)$ is lower semicontinuous and some boundedness assumptions are satisfied.

Theorem 2.1 *Consider problem (P) with continuous functions F, f and a continuous point-to-set mapping $K : \mathbb{R}^n \rightrightarrows \mathbb{R}^m$. Then, if gph K is bounded, problem (P) has a solution.*

Here $\operatorname{gph} K = \{(x,y) : y \in S(x)\}$ denotes the graph of the mapping K. To guarantee continuity of the point-to-set mapping some regularity condition (as Slater's conditions for all x) is needed. The main reason for this result is that the assumptions imply upper semicontinuity of the point-to-set mapping $S : \mathbb{R}^n \rightrightarrows \mathbb{R}^m$ which in turn implies lower semicontinuity of the function $\varphi_0(\cdot)$, see Bank et al. [2].

Let us now consider the problem (P1) given as

$$\min_{x,y} F(x,y) \quad \text{subject to} \quad (x,y) \in \operatorname{gph} S.$$

If local optimal solutions are under consideration it is easy to find examples showing that local optimal solutions of problem (P1) need not to correspond to local optimal solutions of (P). But, for each local optimal solution $\overline{x}$ of (1), some point $(\overline{x}, \overline{y})$ with $\overline{y} \in S(\overline{x})$ is a local optimal solution of (P1).

Proposition 2.1 *Let $\bar{x}$ be a local optimistic solution to the bilevel programming problem (P) whose solution set mapping S is upper-semicontinuous as a set-valued map. Then $(\bar{x}, \bar{y})$ with $\bar{y} \in S(\bar{x})$ and $\varphi_0(\bar{x}) = F(\bar{x}, \bar{y})$ is also a solution of (P1).*

Remark 2.1 We note that we have used the implicit assumption in the proposition that the lower-level problem (LLP) has an optimal solution for $x = \bar{x}$. Let us recall that we have already made this assumption in the beginning of this section.

Proof. Let $\bar{x}$ be a local optimistic solution to (P) and assume that there exists $\bar{y}$ with the properties as formulated in the statement. Then we first have $\bar{y} \in S(\bar{x})$ and

$$F(\bar{x}, \bar{y}) \leq F(\bar{x}, y), \quad \forall y \in S(\bar{x}).$$

By assumption $\varphi_0(\bar{x}) = F(\bar{x}, \bar{y})$. Further we also have

$$\varphi_0(\bar{x}) \leq \varphi_0(x), \quad \forall x \in \mathbb{R}^n \tag{2}$$

sufficiently close to $\bar{x}$. By definition of $\varphi_0(x)$ one has $\varphi_0(x) \leq F(x, y)$ for all $y \in S(x)$. Using (2) we immediately have

$$F(\bar{x}, \bar{y}) = \varphi_0(\bar{x}) \leq \varphi_0(x) \leq F(x, y), \ \forall y \in S(x) \text{ and } x \text{ sufficiently close to } \bar{x}.$$

Let V be an open neighborhood of $S(\bar{x})$, i.e. $S(\bar{x}) \subset V$. Since S is upper-semicontinuous as a set-valued map we have that there exists an open neighborhood U of $\bar{x}$ such that for all $x \in U$ one has $S(x) \subset V$(For a definition of upper-semicontinuous set-valued map see for example Berge [6]). Thus we can find a $\delta > 0$ such that $B_\delta(\bar{x}) \subset V$ and for all $x \in B_\delta(\bar{x})$ we have $\varphi(x) \geq \varphi(\bar{x})$. Here $B_\delta(\bar{x})$ denotes a ball centered at $\bar{x}$ and of radius δ. Thus arguing in a similar manner as before one has

$$F(\bar{x}, \bar{y}) \leq F(x, y) \quad \forall y \in S(x) \quad \text{and} \quad x \in B_\delta(\bar{x}).$$

However for all $x \in B_\delta(\bar{x})$ we have $S(x) \subset V$. This shows that

$$F(\bar{x}, \bar{y}) \leq F(x, y) \quad \forall (x, y) \in (B_\delta(\bar{x}) \times V) \cap \operatorname{gph} S.$$

Thus $(\bar{x}, \bar{y})$ is a local optimal solution for (P1). □

Remark 2.2 It is important to note that assumption of upper-semicontinuity on the solution set-mapping is not a strong one since it can arise under natural assumptions. Assume that feasible set $K(x)$ of the lower-level problem (LLP) is described by convex inequality constraints, i.e.

$$K(x) = \{y \in \mathbb{R}^m : g_i(x, y) \leq 0, i = 1, \ldots, p\},$$

where for each x the function $g(x,\cdot)$ is convex in y. Assume now that the set $\{(x,y) \in \mathbb{R}^n \times \mathbb{R}^m : g_i(x,y) \le 0, i = 1,\ldots,p\}$ is non-empty and compact and the lower-level problem (LLP) satisfies the Slater's constraint qualification then we can conclude that the solution set mapping S is upper-semicontinuous as a set-valued map, cf. Bank et al. [2].

Let us note that if we now consider a global optimistic solution then there is no necessity for any additional assumption on the solution set mapping. Thus we have the following proposition.

Proposition 2.2 *Let $\bar{x}$ be a global optimistic solution to the bilevel programming problem (P). Then $(\bar{x},\bar{y})$ with $\bar{y} \in S(\bar{x})$ and $\varphi_0(\bar{x}) = F(\bar{x},\bar{y})$ is also a global solution of (P1).*

Proof. Let $\bar{x}$ be a global optimistic solution to (P) and assume that there exists $\bar{y}$ with the properties as formulated in the statement. Then we first have $\bar{y} \in S(\bar{x})$ and

$$F(\bar{x},\bar{y}) \le F(\bar{x},y), \quad \forall y \in S(\bar{x}).$$

By assumption $\varphi_0(\bar{x}) = F(\bar{x},\bar{y})$. Further we also have

$$\varphi_0(\bar{x}) \le \varphi_0(x), \quad \forall x \in \mathbb{R}^n. \tag{3}$$

By definition of $\varphi_0(x)$ one has $\varphi_0(x) \le F(x,y)$ for all $y \in S(x)$. Using (3) we immediately have

$$F(\bar{x},\bar{y}) = \varphi_0(\bar{x}) \le \varphi_0(x) \le F(x,y), \quad \forall y \in S(x).$$

Hence the result. □

The opposite implication is also valid for global optima.

Proposition 2.3 *Let $(\bar{x},\bar{y})$ be a global optimal solution of problem (P1). Then, $\bar{x}$ is a global optimal solution of problem (P).*

Proof. Assume that $\bar{x}$ is not a global optimal solution of problem (1) then there is $\widetilde{x}$ with $\varphi_0(\widetilde{x}) < \varphi_0(\bar{x})$ and, by definition of the function $\varphi_0(\cdot)$ there is $\widetilde{y} \in S(\widetilde{x})$ with $\varphi_0(\widetilde{x}) \le F(\widetilde{x},\widetilde{y}) < \varphi_0(\bar{x})$. Now, $\bar{y} \in S(\bar{x})$ and, hence,

$$F(\bar{x},\bar{y}) = \varphi_0(\bar{x}) > F(\widetilde{x},\widetilde{y}).$$

Then $F(\widetilde{x},\widetilde{y}) < F(\bar{x},\bar{y})$ which contradicts optimality of $(\bar{x},\bar{y})$. □

The last two propositions enable us to reformulate the bilevel problem in its optimistic version to the problem (P1). Note that this excludes the case when the function φ_0 is determined as the infimal objective function value of the lower-level problem (LLP) (which is then not assumed to have an optimal solution), implying that the function φ_0 may have a minimum even in the case when the problem (P1) has no solution.

We have already stated that we will restrict ourselves to the case where the lower-level problem is a convex minimization problem. Let for the moment the set $K(x)$ in (LLP) be expressed in terms of convex inequalities:

$$K(x) = \{y : g_i(x, y) \leq 0, i = 1, \ldots, p\},$$

where $g_i(x, \cdot)$ are convex in y for each x and are sufficiently smooth i.e of class C^2. Finding the minimum of a regular convex problem is equivalent to solving the Karush-Kuhn-Tucker type conditions associated with the problem. Thus a bilevel programming problem can be posed as single level problem with the lower-level problem being replaced with its Karush-Kuhn-Tucker system which now become additional constraints to the problem (P1). Thus the problem (P1) can be reformulated as

$$\begin{aligned} & \min_{x,y,\lambda} F(x, y) \\ \text{subject to } & \nabla_y f(x, y) + \sum_{i=1}^{p} \lambda_i \nabla_y g_i(x, y) = 0 \\ & \lambda_i g_i(x, y) = 0 \quad i = 1, \ldots, p. \\ & g_i(x, y) \leq 0, \ \lambda_i \geq 0 \quad i = 1, \ldots, p. \end{aligned} \tag{4}$$

This is the so-called Karush-Kuhn-Tucker (KKT) formulation of a bilevel programming problem with a convex lower-level problem. Problem (4) is a special kind of the so-called Mathematical Program with Equilibrium Constraints (MPEC). It is well-known that many standard constraint qualifications like the Mangasarian-Fromowitz constraint qualification and the Abadie constraint qualification fail due to the presence of the complementary slackness condition of the lower-level problem which is now posed as an equality constraint. The reader is referred to the paper Ye [23] where possible regularity conditions are identified. The challenge therefore is to devise natural qualification conditions which can lead to KKT type optimality conditions for a bilevel programming problem. Here we suggest one new approach through which this may be possible. Similar investigations have been done in Ye [23] under the assumptions that the problem functions are either Gâteaux differentiable or locally Lipschitz continuous using the Michel-Penot subdifferential.

It should be mentioned that the KKT reformulation is equivalent to the problem (P1) only in the case when the lower-level problem is a convex regular one and global optimal solutions of the upper level problem are investigated (cf. Propositions 2.2 and 2.3). Without convexity, problem (4) has a larger feasible set than (P1) and an optimal solution of (4) need not to correspond to a feasible solution of problem (P1). Even more, an optimal solution of problem (P1) need also not be an optimal solution of (4), see Mirrlees [16].

What concerns optimality conditions the main difficulty in using the reformulation (4) of the bilevel programming problem is the addition of new variables. If these Lagrange multipliers of the lower-level problem are not uniquely determined, the optimality conditions of the MPEC depend on the selection of the multiplier but the conditions for the bilevel problem must not. This can easily been seen e.g. in the case when the lower-level problem is a convex one for which the Mangasarian-Fromowitz constraint qualification together with the strong sufficient optimality condition of second order and the constant rank constraint qualification are satisfied at a point $(\overline{x}, \overline{y})$. Then, the optimal solution of the lower-level problem is strongly stable in the sense of Kojima [13], Lipschitz continuous and directionally differentiable, see Ralph and Dempe [20] and the bilevel programming problem can be reformulated as

$$\min\{F(x, y(x)) : x \in \mathbb{R}^n\}.$$

Necessary optimality conditions for this problem reduce to nonexistence of directions of descent for the function $x \mapsto F(x, y(x))$, cf. Dempe [10]. If this problem is reformulated as (4) and a Lagrange multiplier is fixed it is possible that there is no direction of decent in the problem (4). But what we have done is to compute the directional derivative of the function $x \mapsto F(x, y(x))$ only in directions which correspond to the selected Lagrange multiplier, i.e. directions for which a certain linear optimization problem has a solution, see Ralph and Dempe [20]. But there is no need that the directional derivative of the function $F(x, y(x))$ into other directions (corresponding to other Lagrange multipliers) does not give a descent.

With other words, if optimality conditions for an MPEC are investigated, a feasible solution of this problem is fixed and optimality conditions are derived as in Pang and Fukushima [19], Scheel and Scholtes [22]. Considering the optimality conditions in primal space (i.e. formulating them as nonexistence of descent directions in the contingent cone) we see some combinatorial structure since the contingent cone is not convex. This approach has been applied to the KKT reformulation of a bilevel programming problem in Ye and Ye [26]. But to obtain a more useful condition for selecting a locally optimal solution we have to investigate the resulting systems for all Lagrange multipliers of the lower-level problem, or at least for all the vertices of the set of Lagrange multipliers, if some condition as the constant rank constraint qualification in the differentiable case is satisfied. Hence, this approach needs to be complemented by e.g. a method for an efficient computation of all Lagrange multipliers.

Hence we believe that other approaches are more promising. These are on the one hand approaches using the normal cone (or the contingent cone) to the graph of the solution set mapping of the lower-level problem and on the other hand approaches using the reformulation of the bilevel programming problem using the optimal value function of the lower-level problem. The latter approach has been used e.g. in the papers Babahadda and Gadhi [1], Ye [23].

Here we investigate the possibility to derive necessary optimality conditions using the normal cone.

3 Basic tools

Let $(\bar{x}, \bar{y})$ be a local (or global) solution of (P1) and let us assume that F is smooth. Then one has

$$0 \in \nabla F(\bar{x}, \bar{y}) + N_{\text{gph}S}(\bar{x}, \bar{y}).$$

In the above expression $N_{\text{gph}\, S}(\bar{x}, \bar{y})$ denotes the *Mordukhovich normal cone* or the *basic normal cone* to the graph of the set-valued map S at $(\bar{x}, \bar{y})$. For more details on Mordukhovich normal cone and the derivation of the above necessary optimality condition see for example Mordukhovich [17] and Rockafellar and Wets [21]. It is moreover important that the Mordukhovich normal cone is in general a closed and non-convex object. The basic normal cone to a convex set coincides with the usual normal cone of convex analysis. In order to obtain a KKT type optimality condition our main task is now to compute the basic normal cone to the graph of the solution set mapping at the point $(\bar{x}, \bar{y})$. Thus the qualification conditions that are required to compute the normal cone are indeed the natural qualification conditions for the bilevel programming problem. However let us note that it is in fact a formidable task to compute the normal cone to the graph of the solution set mapping. This is mainly due to the fact that even if the lower-level problem is convex the graph $\text{gph}\, S$ of the solution set mapping S need not be convex. The following simple example demonstrates this fact.

Example 3.1 Let the lower-level problem be given as

$$S(x) = \underset{y}{\text{argmin}}\, \{f(x, y) = -xy : 0 \leq y \leq 1\}.$$

Observe here that $K(x) = [0, 1]$ for all $x \in \mathbb{R}$. Observe that the problem is a convex problem in y. Also note that the solution set mapping S in this particular case is given as

$$S(x) = \begin{cases} \{0\} & : x < 0 \\ [0, 1] & : x = 0 \\ \{1\} & : x > 0. \end{cases}$$

It is now simple to observe that the $\text{gph}\, S$ is a non-convex set. △

Professor Rockafellar suggested that an interesting approach to bilevel programming may be obtained by having a minimax problem or rather a primal-dual problem in the lower-level instead of just a convex minimization problem. This can in fact be motivated from the KKT representation of a bilevel programming problem with convex lower-level problems. Observe that the KKT problem brings in an additional variable $\lambda \in \mathbb{R}^p_+$ which is actually the Lagrange multiplier associated with the problem as well as the dual variable associated with the Lagrangian dual of the convex lower-level problem. Hence both the primal variable y and the dual variable λ of the convex lower-level problem are present in the KKT formulation of the bilevel programming problem with convex lower-level problems. Thus one may as well define a lower-level problem which has both the primal and dual variable and that naturally suggests us to consider the lower-level problem as a minmax problem. Thus we can have a new formulation of the bilevel programming problem (P2) with a minimax lower-level problem as follows

$$\min_{(x,y,\lambda)} F(x,y,\lambda) \quad \text{subject to} \quad (y,\lambda) \in S(x),$$

where $F : \mathbb{R}^n \times \mathbb{R}^m \times \mathbb{R}^p \to \mathbb{R}$ and the set-valued map $S : \mathbb{R}^n \rightrightarrows \mathbb{R}^m \times \mathbb{R}^p$ is a solution set of the following problem (LLP2)

$$\text{minimaximize}\, L(x,y,\lambda) \quad \text{subject to} \quad (y,\lambda) \in Y \times W,$$

where $Y \subset \mathbb{R}^m$ and $W \subset \mathbb{R}^p$ are non-empty convex sets and $L(x,y,\lambda)$ is convex with respect to y for each $(x,\lambda) \in \mathbb{R}^n \times W$ and is concave in λ for each $(x,y) \in \mathbb{R}^m \times Y$. Thus we can write

$$S(x) = \{(y,\lambda) : (y,\lambda) \quad \text{solves} \quad (LLP2)\}.$$

By a solution $(y,\lambda) \in S(x)$ we mean

$$y \in \operatorname*{argmin}_{y \in Y} L(x,y,\lambda) \text{ and } \lambda \in \operatorname*{argmax}_{\lambda \in W} L(x,y,\lambda).$$

Let us end this section by defining the nonsmooth tools that would be required for the proofs of the optimality conditions. We first begin with the definition of the normal cone to a set C at a given point in C. Let C be a non-empty subset of $\mathbb{R}^n$ and let $\bar{x} \in C$. A vector v is called a *regular normal* to C at $\bar{x}$ if

$$\langle v, x - \bar{x} \rangle \leq o(\|x - \bar{x}\|),$$

where $\dfrac{o(\|x-\bar{x}\|)}{\|x-\bar{x}\|} \to 0$ as $\|x - \bar{x}\| \to 0$. The set of all regular normals form a convex cone denoted by $\hat{N}_C(\bar{x})$. This is also known as the *Fréchet normal cone* in the literature.

A vector $v \in \mathbb{R}^n$ is said to be a normal or a basic to C at $\bar{x}$ if there exist a sequence $\{v_k\}$, with $v_k \to v$ and a sequence $\{x_k\}$, $x_k \in C$ with $x_k \to \bar{x}$ and $v_k \in \hat{N}_C(x_k)$. The set of all normals forms a closed (but not necessarily convex) cone denoted as $N_C(\bar{x})$. The basic normal cone has also been referred

to as the Mordukhovich normal cone in the literature. For more details on the basic normal cone in the finite dimensional setting see for example Mordukhovich [17] or Rockafellar and Wets [21]. It is important to note that if the interior of C is nonempty and $\bar{x} \in \text{int}C$ then $N_C(\bar{x}) = \{0\}$.
Let $S : \mathbb{R}^n \rightrightarrows \mathbb{R}^m$ be a set-valued map and let $(x, y) \in \text{gph}\, S$. Then the coderivative at $(\bar{x}, \bar{y})$ is a set-valued map $D^*S(\bar{x}|\bar{y}) : \mathbb{R}^m \rightrightarrows \mathbb{R}^n$ given as

$$D^*S(\bar{x}|\bar{y})(w) = \{v \in \mathbb{R}^n : (v, -w) \in N_{\text{gph}\, S}(\bar{x}, \bar{y})\}.$$

For more details on the properties of the coderivative see for example Mordukhovich [17] and Rockafellar and Wets [21]. Further given a function $f : \mathbb{R}^n \to \mathbb{R} \cup \{+\infty\}$ and a point $\bar{x}$ where f is finite the *subdifferential* or the *basic subdifferential* at $\bar{x}$ is given as

$$\partial f(\bar{x}) = \{\xi \in \mathbb{R}^m : (\xi, -1) \in N_{\text{epi}\, f}(\bar{x}, f(\bar{x}))\},$$

where epi f denotes the epigraph of the function f. The asymptotic subdifferential of f at $\bar{x}$ is given as

$$\partial^\infty f(\bar{x}) = \{\xi \in \mathbb{R}^m : (\xi, 0) \in N_{\text{epi}\, f}(\bar{x}, f(\bar{x}))\}.$$

We will now present Theorem 2.1 in Levy and Mordukhovich [14] in the form of two lemmas whose application would lead to the necessary optimality conditions.

Lemma 3.1 *Consider the set-valued map $S : \mathbb{R}^n \rightrightarrows \mathbb{R}^m$ given as follows*

$$S(x) = \{y \in \mathbb{R}^m : 0 \in G(x, y) + M(x, y)\}, \tag{5}$$

where $G : \mathbb{R}^n \times \mathbb{R}^m \to \mathbb{R}^d$ is a smooth vector-valued function and $M : \mathbb{R}^n \times \mathbb{R}^m \rightrightarrows \mathbb{R}^d$ is a set-valued map with closed graph. Let $(\bar{x}, \bar{y}) \in$ gph S and let the following qualification condition hold

$$v \in \mathbb{R}^d \quad \text{with} \quad 0 \in \nabla G(\bar{x}, \bar{y})^T v + D^*M((\bar{x}, \bar{y})| - G(\bar{x}, \bar{y}))(v) \Longrightarrow v = 0.$$

Then one has

$$D^*S(\bar{x}|\bar{y})(y^*) \subseteq \{x^* : \exists v^* \in \mathbb{R}^d, (x^*, -y^*) \in \nabla G(\bar{x}, \bar{y})^T v^* + D^*M((\bar{x}, \bar{y})| - G(\bar{x}, \bar{y}))(v^*)\}.$$

Lemma 3.2 *Consider the set-valued map $S : \mathbb{R}^n \rightrightarrows \mathbb{R}^m$ given in formula (5) where $G : \mathbb{R}^n \times \mathbb{R}^m \to \mathbb{R}^d$ is a smooth vector-valued function and $M : \mathbb{R}^n \times \mathbb{R}^m \rightrightarrows \mathbb{R}^d$ is a set-valued map with closed graph. Further assume that M only depends on y i.e $M(x, y) = M(y)$. Assume that the matrix $\nabla_x G(\bar{x}, \bar{y})$ has full rank. Then one has*

$$D^*S(\bar{x}|\bar{y})(y^*) = \{x^* : \exists v^* \in \mathbb{R}^d, x^* = \nabla_x G(\bar{x}, \bar{y})v^*, \\ -y^* = \nabla_y G(\bar{x}, \bar{y})^T v^* + D^*M(\bar{y}| - G(\bar{x}, \bar{y}))(v^*)\}.$$

4 Main Results

In this section we shall present necessary optimality conditions for the two classes of bilevel programming problems which we have discussed in the previous sections. First we shall derive necessary optimality conditions for the problem format defined by (1) in which the lower-level problem is a convex minimization problem. Then we shall consider the case when $K(x) = K$ for all x and then move on to the case where the lower-level problem is given in the form a primal-dual problem. Then we will present a more refined optimality condition using the second-order subdifferential of the indicator function which would appear to be a very novel feature. Before we begin let us define the following set-valued map which can also be called as the normal cone map

$$N_K(x,y) = \begin{cases} N_{K(x)}(y) & : y \in K(x) \\ \emptyset & : y \notin K(x) \end{cases}$$

Theorem 4.1 *Consider the problem (P1) given as*

$$\min_{x,y} F(x,y) \quad \text{subject to} \quad (x,y) \in gph\ S,$$

where $F : \mathbb{R}^n \times \mathbb{R}^m \to \mathbb{R}$ is a smooth function and $S : \mathbb{R}^n \rightrightarrows \mathbb{R}^m$ is a set-valued map denoting the solution set of the problem (LLP) i.e.

$$S(x) = \operatorname*{argmin}_y \{f(x,y) : y \in K(x)\},$$

where $f(x,\cdot)$ is a smooth convex function in y for each x and $K(x)$ is a closed convex set for each x. Let $(\bar{x},\bar{y})$ be a local (or global) solution of (P1). Further assume that $\nabla_y f : \mathbb{R}^n \times \mathbb{R}^m \to \mathbb{R}^m$ is continuously differentiable. Set $\bar{p} = \nabla_y f(\bar{x},\bar{y})$. Assume also that the following qualification condition holds at $(\bar{x},\bar{y})$:

$$v \in \mathbb{R}^m \quad \text{with} \quad 0 \in \nabla(\nabla_y f(\bar{x},\bar{y}))^T v + D^* N_K((\bar{x},\bar{y})| - \bar{p})(v) \Longrightarrow v = 0.$$

Then there exists $v^* \in \mathbb{R}^m$ such that

$$0 \in \nabla F(\bar{x},\bar{y}) + \nabla(\nabla_y f(\bar{x},\bar{y}))^T v^* + D^* N_K((\bar{x},\bar{y})| - \bar{p})(v^*).$$

Proof. They key to the proof of this result is Lemma 3.1. To begin with note that since (LLP) is a convex minimization problem in y for each given x we can write $S(x)$ equivalently as

$$S(x) = \{y \in \mathbb{R}^m : 0 \in \nabla_y f(x,y) + N_{K(x)}(y)\}.$$

It is not much difficult to show that the normal cone map has a closed graph. Since $(\bar{x},\bar{y})$ is a local (or global) solution of the problem (P1) then we have

$$-\nabla F(\bar{x},\bar{y}) \in N_{\text{gph}\, S}(\bar{x},\bar{y}).$$

Now by using the definition of the coderivative and then applying Lemma 3.1 we have that there exists $v^* \in \mathbb{R}^m$ such that

$$-\nabla F(\bar{x},\bar{y}) \in \nabla(\nabla_y f(\bar{x},\bar{y}))^T v^* + D^* N_K((\bar{x},\bar{y})| - \bar{p})(v^*).$$

This proves the result □

We will now apply the above result to bilevel programming

Corollary 4.1 *Let us consider the bilevel programming problem (P)*

$$\min_x F(x,y) \quad \textit{subject to} \quad y \in S(x),$$

where $F : \mathbb{R}^n \times \mathbb{R}^n \to \mathbb{R}$ is a smooth function and $S(x)$ is the solution set of the following problem (LLP)

$$\min_y f(x,y) \quad \textit{subject to} \quad y \in K(x),$$

where $f : \mathbb{R}^n \times \mathbb{R}^m \to \mathbb{R}$ is a smooth strictly convex function in y for each x and $K(x)$ is a compact convex set for each x. Let $(\bar{x},\bar{y})$ be a local solution of (P). Further assume that $\nabla_y f : \mathbb{R}^n \times \mathbb{R}^m \to \mathbb{R}^m$ is continuously differentiable. Set $\bar{p} = \nabla_y f(\bar{x},\bar{y})$. Assume further that the following qualification condition holds at $(\bar{x},\bar{y})$:

$$v \in \mathbb{R}^m \quad \textit{with} \quad 0 \in \nabla(\nabla_y f(\bar{x},\bar{y}))^T v + D^* N_K((\bar{x},\bar{y})| - \bar{p})(v) \Longrightarrow v = 0.$$

Then there exists $v^* \in \mathbb{R}^m$ such that

$$0 \in \nabla F(\bar{x},\bar{y}) + \nabla(\nabla_y f(\bar{x},\bar{y}))^T v^* + D^* N_K((\bar{x},\bar{y})| - \bar{p})(v^*).$$

Proof. By the hypothesis of the theorem for each x the problem (LLP) has a unique solution. Hence the solution of the problem (P) is also a solution of (P1). The rest of the proof follows as in Theorem 4.1. □

Corollary 4.2 *Let us consider the bilevel programming problem (P)*

$$\min_x F(x,y) \quad \textit{subject to} \quad y \in S(x),$$

where $F : \mathbb{R}^n \times \mathbb{R}^n \to \mathbb{R}$ is a smooth function and $S(x)$ is the solution set of the following problem (LLP)

$$\min_y f(x,y) \quad \textit{subject to} \quad y \in K(x),$$

where $f : \mathbb{R}^n \times \mathbb{R}^m \to \mathbb{R}$ is a smooth convex function in y for each x and $K(x)$ is a convex set for each x. Further assume that the solution set mapping S is upper-semicontinuous as a set-valued map. Let $\overline{x}$ be a local optimistic solution of (P) and assume that $\overline{y} \in S(\overline{x})$ with $F(\overline{x},\overline{y}) = \varphi_0(\overline{x})$ exists. Further assume *that* $\nabla_y f : \mathbb{R}^n \times \mathbb{R}^m \to \mathbb{R}^m$ *is continuously differentiable. Set* $\bar{p} = \nabla_y f(\bar{x}, \bar{y})$. *Assume further that the following qualification condition hold at* $(\bar{x}, \bar{y})$:

$$v \in \mathbb{R}^m \quad \textit{with} \quad 0 \in \nabla(\nabla_y f(\bar{x}, \bar{y}))^T v + D^* N_K((\bar{x}, \bar{y})| - \bar{p})(v) \Longrightarrow v = 0.$$

Then there exists $v^* \in \mathbb{R}^m$ *such that*

$$0 \in \nabla F(\bar{x}, \bar{y}) + \nabla(\nabla_y f(\bar{x}, \bar{y}))^T v^* + D^* N_K((\bar{x}, \bar{y})| - \bar{p})(v^*).$$

Proof. Our assumptions imply that $(\overline{x}, \overline{y})$ is a local solution of the problem (P1) due to Proposition 2.1. The rest of the proof follows as in Theorem 4.1. □

Remark 4.1 An interesting feature in the optimality conditions presented in the above theorem is the presence of second-order partial derivatives in the expression of first order optimality conditions. This is essentially due to presence of the matrix $\nabla(\nabla_y f(\bar{x}, \bar{y}))$. The presence of second-order partial derivatives in the first conditions is a hallmark of bilevel programming. Further note that one can have analogous results for global optimistic solution using Proposition 2.2 and without any additional assumption on the nature of the solution set mapping S.

We will now turn to the case when $K(x) = K$ for all $x \in \mathbb{R}^n$. In such a case we have a much simplified qualification condition which amounts to checking whether a matrix is of full rank.

Theorem 4.2 *Consider the problem (P1) given as*

$$\min_{x,y} F(x, y) \quad \textit{subject to} \quad (x, y) \in \operatorname{gph} S,$$

where $F : \mathbb{R}^n \times \mathbb{R}^m \to \mathbb{R}$ *is a smooth function and* $S : \mathbb{R}^n \rightrightarrows \mathbb{R}^m$ *is a set-valued map denoting the solution set of the problem (LLP) i.e.*

$$S(x) = \operatorname*{argmin}_y \{f(x, y) : y \in K(x)\},$$

where $f(x, \cdot)$ *is a smooth convex function in* y *for each* x *and* $K(x) = K$ *for all* x *where* K *is a fixed closed and convex set. Let us also assume that the function* f *is twice continuously differentiable. Let* $(\bar{x}, \bar{y}) \in \operatorname{gph} S$ *be a solution of problem (P1). Set* $\bar{p} = \nabla_y f(\bar{x}, \bar{y})$. *Further assume that the matrix* $\nabla_x(\nabla_y f(\bar{x}, \bar{y})) = \nabla^2_{xy} f(\bar{x}, \bar{y})$ *has full rank, i.e.*

$$\textit{rank}\left(\nabla^2_{xy} f(\bar{x}, \bar{y})\right) = m.$$

Then there exists $v^* \in \mathbb{R}^m$ *such that the following conditions hold*

i) $0 = \nabla_x F(\bar{x}, \bar{y}) + \nabla^2_{xy} f(\bar{x}, \bar{y}) v^*$
ii) $0 \in \nabla_y F(\bar{x}, \bar{y}) + \nabla^2_{yy} f(\bar{x}, \bar{y}) v^* + D^* N_K(\bar{y}| - \bar{p})(v^*)$.

Proof. In this particular case when $K(x) = K$ then one can write $N_K(x, y) = N_K(y)$. Further the solution set mapping S can also be equivalently written as

$$S(x) = \{y \in \mathbb{R}^m : 0 \in \nabla_y f(x, y) + N_K(y)\}.$$

Since $(\bar{x}, \bar{y})$ solves (P1) we have

$$-\nabla F(\bar{x}, \bar{y}) \in N_{\text{gph } S}(\bar{x}, \bar{y}).$$

This shows that

$$-(\nabla_x F(\bar{x}, \bar{y}), \nabla_y F(\bar{x}, \bar{y})) \in N_{\text{gph } S}(\bar{x}, \bar{y}).$$

Hence by definition of the coderivative we have

$$-\nabla_x F(\bar{x}, \bar{y}) \in D^* S(\bar{x}|\bar{y})(\nabla_y F(\bar{x}, \bar{y})).$$

Now by using Lemma 3.2 we see that there exists $v^* \in \mathbb{R}^m$ such that

$$-\nabla_x F(\bar{x}, \bar{y}) = \nabla^2_{xy} f(\bar{x}, \bar{y}) v^*$$

and

$$-\nabla_y F(\bar{x}, \bar{y}) \in \nabla^2_{yy} f(\bar{x}, \bar{y})^T v^* + D^* N_K(\bar{y}| - \bar{p})(v^*).$$

Hence the result. □

Remark 4.2 The qualification condition that we have used in the above theorem is called the *ample parametrization condition* in Dontchev and Rockafellar [11]. However in Dontchev and Rockafellar [11] the proto-derivative of the solution set mapping S is computed. The proto-derivative is the tangent cone to the graph of S at $(\bar{x}, \bar{y})$. Thus the approach due to Dontchev and Rockafellar [11] can be used in the dual setting given in terms of the tangent cone. However as we have noted the approach through coderivatives is essential in surpassing the computation (a difficult one that too) that is required to compute the normal cone to the graph of S at $(\bar{x}, \bar{y})$. Thus the results in Levy and Mordukhovich [14] will play a very fundamental role in the study of mathematical programming with equilibrium constraints (MPEC) and also bilevel programming with convex lower-level problems.

It is now easy to observe that the above theorem can be used to deduce optimality conditions for a bilevel programming problem with a convex lower-level problem with $K(x) = K$ for all $x \in \mathbb{R}^n$ if the lower-level problem has a unique solution or we consider an optimistic solution of the bilevel programming problem. However we are not going to explicitly state the results here since this can be done as in the corollaries following Theorem 4.1.

One of the main drawback of the optimality conditions derived above for problem (P) and (P1) is the presence of the coderivative of the normal cone mapping. Thus the optimality conditions are more abstract in nature. The computation of the coderivative of the normal cone map seems to be very difficult. However by using an approach due to Outrata [18] by using some different qualification condition we can derive an optimality condition in which the explicit presence of the coderivative of the normal cone map is not there though as we will see that it will be implicity present. We now present the following result.

Theorem 4.3 *Consider the problem (P1) given as*

$$\min_{x,y} F(x,y) \quad \textit{subject to} \quad (x,y) \in \operatorname{gph} S,$$

where $F : \mathbb{R}^n \times \mathbb{R}^m \to \mathbb{R}$ *is a smooth function and* $S : \mathbb{R}^n \rightrightarrows \mathbb{R}^m$ *is a set-valued map denoting the solution set of the problem (LLP) i.e.*

$$S(x) = \operatorname*{argmin}_{y} \{f(x,y) : y \in K(x)\},$$

where $f(x,\cdot)$ *is a smooth convex function in* y *for each* x *and* $K(x) = K$ *for all* x *where* K *is a fixed closed and convex set. Let us also assume that the function* f *is twice continuously differentiable. Let* $(\bar{x},\bar{y}) \in \operatorname{gph} S$ *be a local solution of problem (P1). Further assume that the following qualification condition holds at* $(\bar{x},\bar{y})$ *:*

$$\begin{gathered} (w,z) \in N_{\operatorname{gph} N_k}(\bar{y}, -\nabla_y f(\bar{x},\bar{y})) \quad \textit{with} \\ (\nabla^2_{xy} f(\bar{x},\bar{y}))^T z = 0, \quad w - (\nabla^2_{yy} f(\bar{x},\bar{y}))^T z = 0 \Longrightarrow w = 0, z = 0. \end{gathered}$$

Then there exists a pair $(\bar{w},\bar{z}) \in N_{\operatorname{gph} N_k}(\bar{y}, -\nabla_y f(\bar{x},\bar{y}))$ *such that*

i) $\nabla_x F(\bar{x},\bar{y}) = (\nabla^2_{xy} f(\bar{x},\bar{y}))^T \bar{z}$.
ii) $-\nabla_y F(\bar{x},\bar{y}) = \bar{w} - (\nabla^2_{yy} f(\bar{x},\bar{y}))^T \bar{z}$.

Proof. Observe that according the hypothesis of the theorem the problem (P1) is equivalent to the following problem (P4)

$$\min_{x,y} F(x,y) \quad \text{subject to} \quad 0 \in \nabla_y f(x,y) + N_K(y).$$

Now by applying Theorem 3.1 in Outrata [18] we reach our desired conclusion. □

Observe that the qualification condition in Theorem 4.3 guarantees that $(\nabla^2_{xy} f(\bar{x},\bar{y}))^T$ has full rank which is similar to the qualification condition appearing in Theorem 4.2. However there is also an extra qualification condition since we now have two Lagrange multipliers instead of one. Though the coderivative does not appear explicitly in the representation of the optimality condition but the condition $(\bar{w}, \bar{z}) \in N_{\mathrm{gph}\, N_k}(\bar{y}, -\nabla_y f(\bar{x}, \bar{y}))$ tells us that

$$\bar{w} \in D^*(\bar{y}| - \nabla_y f(\bar{x}, \bar{y}))(-\bar{z}).$$

Thus the conditions obtained in Theorem 4.2 are same as that of Theorem 4.3. However the approach due to Outrata [18] seems to have an additional advantage. This apparent advantage is that we can use Outrata's approach even when in the problem x is lying in a proper closed set X of $\mathbb{R}^m$. In such a situation the problem (P1) gets slightly modified and looks as follows

$$\min_{x,y} F(x,y), \quad \text{subject to} \quad (x,y) \in \mathrm{gph}\, S \quad x \in X.$$

Also note that since $N_K(y) = \emptyset$, when $y \notin K$ it is clear that $y \in K$ is implied by $0 \in \nabla_y f(x,y) + N_K(y)$. Thus $S(x)$ can also be written as

$$\begin{aligned} S(x) &= \{y \in \mathbb{R}^m : 0 \in \nabla_y f(x,y) + N_K(y)\} \\ &= \{y \in K : 0 \in \nabla_y f(x,y) + N_K(y)\}. \end{aligned}$$

Hence $S(x) \subset K$. So when we write the expression for $S(x)$ there is no need to explicitly write that $y \in K$. Thus when $x \in X$ the modified version of the problem (P1) is equivalent to

$$\min_{x,y} F(x,y) \quad \text{subject to} \quad 0 \in \nabla_y f(x,y) + N_K(y), \quad (x,y) \in X \times \mathbb{R}^m$$

Hence, if $(\bar{x}, \bar{y})$ is a solution of the modified (P1) then it also solves the above problem. Thus in this scenario the qualification condition in Theorem 3.1 in Outrata reduces to the following. Consider $(w,z) \in N_{\mathrm{gph}\, N_k}(\bar{y}, -\nabla_y f(\bar{x}, \bar{y}))$. Then

$$((\nabla^2_{xy} f(\bar{x}, \bar{y}))^T z, w - (\nabla^2_{yy} f(\bar{x}, \bar{y}))^T z) \in N_{X \times \mathbb{R}^m}(\bar{x}, \bar{y}) \Longrightarrow w = 0, z = 0.$$

Then by applying Theorem 3.1 in Outrata [18] we arrive at the conclusion that there exists a pair $(\bar{w}, \bar{z}) \in N_{\mathrm{gph}\, N_k}(\bar{y}, -\nabla_y f(\bar{x}, \bar{y}))$ and $(\gamma, 0) \in N_{X \times \mathbb{R}^m}(\bar{x}, \bar{y})$ such that

i) $-\nabla_x F(\bar{x}, \bar{y}) = -(\nabla^2_{xy} f(\bar{x}, \bar{y}))^T \bar{z} + \gamma$.
ii) $-\nabla_y F(\bar{x}, \bar{y}) = \bar{w} - (\nabla^2_{yy} f(\bar{x}, \bar{y}))^T \bar{z}$.

Note that $N_{X \times \mathbb{R}^m}(\bar{x}, \bar{y}) = N_X(\bar{x}) \times N_{\mathbb{R}^m}(\bar{y})$ (see for example Rockafellar and Wets [21]). And since $N_{\mathbb{R}^m}(\bar{y}) = \{0\}$, it is clear that

$$N_{X \times \mathbb{R}^m}(\bar{x}, \bar{y}) = \{(\gamma, 0) : \gamma \in N_X(\bar{x})\}.$$

Further if $X = \mathbb{R}^n$ which is the case in Theorem 4.3 one has $N_{\mathbb{R}^n \times \mathbb{R}^m}(\bar{x}, \bar{y}) = \{(0,0)\}$.
Let us now turn our attention of how to calculate the normal cone to the

graph of the normal cone mapping associated with a given set K in the lower-level problem. However if K has some special form then one can have an explicit expression for $N_{\text{gph}\, N_K}(\bar{y}, \bar{z})$. For example if $K = \mathbb{R}^m_+$ then such an explicit expression for $N_{\text{gph}\, N_K}(\bar{y}, \bar{z})$ is given by Proposition 3.7 in Ye [24]. The result in Proposition 3.7 in Ye [24] depends on Proposition 2.7 in Ye [25]. In Proposition 2.7 of Ye [25] the normal cone to the graph of the normal cone mapping $N_{\mathbb{R}^m_+}$ is calculated. Let $C \subset \mathbb{R}^n$ be a closed set. Then $v \in \mathbb{R}^n$ is said to be proximal normal to C at $\bar{x} \in C$ if there exists $\sigma > 0$ such that

$$\langle v, x - \bar{x} \rangle \leq \sigma \|x - \bar{x}\|^2$$

The set of all proximal normals forms a cone called the proximal normal cone which is denoted by $N^P_C(\bar{x})$. It is also important to note that if C is a closed set then a normal vector can be realized as a limit of proximal normal vectors. More precisely if C is closed and $v \in N_C(\bar{x})$ then there exist sequences $v_k \to v$ and $x_k \to \bar{x}$ with $v_k \in N^P_C(\bar{x})$. It is clear from the definition of the proximal normal cone that

$$N^P_C(\bar{x}) \subseteq \hat{N}_C(\bar{x}) \subseteq N_C(\bar{x}).$$

For more details on the proximal normal cone see for example Clarke, Ledyaev, Stern and Wolenski [8].
We will now consider the simple case when $(x, y) \in \mathbb{R}^2$ and we shall consider the set $K(x) = K = [0, 1]$ as the feasible set of the lower-level problem. Our aim is to precisely calculate $N_{\text{gph}\, N_K}(\bar{y}, \bar{z})$. Observe that

$$N_K(y) = \begin{cases} (-\infty, 0] & : y = 0 \\ \{0\} & : 0 < y < 1 \\ [0, +\infty) & : y = 1 \end{cases}$$

It is easy to sketch the graph of the normal cone map N_K where $K = [0, 1]$. The proximal normal cone to $\text{gph}\, N_K$ is given as follows.

$$N^P_{\text{gph}\, N_K}(\bar{y}, \bar{z}) = \begin{cases} (-\infty, 0] \times [0, +\infty) & : \bar{y} = 0, \bar{z} = 0 \\ \{0\} \times \mathbb{R} & : 0 < \bar{y} < 1, \bar{z} = 0 \\ \mathbb{R} \times \{0\} & : \bar{y} = 1, \bar{z} > 0 \\ \mathbb{R} \times \{0\} & : \bar{y} = 0, \bar{z} < 0 \\ [0, +\infty) \times (-\infty, 0] & : \bar{y} = 1, \bar{z} = 0 \end{cases}$$

Using the fact that the basic normal cone can be obtained as a limit of the proximal normal cone we obtain the following

$$N_{\text{gph}\, N_K}(0, 0) = \{(w, v) \in \mathbb{R}^2 : w < 0, v > 0\} \cup \{(w, v) \in \mathbb{R}^2 : v = 0\} \\ \cup \{(w, v) \in \mathbb{R}^2 : w = 0\}$$

and

$$N_{\mathrm{gph}\,N_K}(1,0) = \{(w,v) \in \mathbb{R}^2 : w > 0, v < 0\} \cup \{(w,v) \in \mathbb{R}^2 : v = 0\} \cup \{(w,v) \in \mathbb{R}^2 : w = 0\}.$$

For all other points the basic normal cone coincides with the proximal normal cone.

We have shown earlier that using the approach of Outrata [18] we are able to develop optimality conditions for the problem (P1) when $x \in X$ and X is a closed subset of $\mathbb{R}^n$. We would like to remark that by using the conditions (a), (b) and (c) in Theorem 3.2 of Ye and Ye [26] we can arrive at the same conditions as we have obtained using Outrata's approach. However the most interesting condition in Theorem 3.2 of [26] is (b). In our case this corresponds to the assumption that for each fixed $x \in X$ the function $y \mapsto f(x,y)$ is strongly convex. This may actually appear in practical situations.

We will now turn our attention to study the bilevel programming problem (P2) whose lower-level problem (LLP2) is of a primal-dual nature i.e. a minimax problem.

Theorem 4.4 *Let us consider the problem (P3) given as follows*

$$\min_{x,y,\lambda} F(x,y,\lambda) \quad \textit{subject to} \quad (x,y,\lambda) \in gph\ S,$$

where $F : \mathbb{R}^n \times \mathbb{R}^m \times \mathbb{R}^p \to \mathbb{R}$ *is a smooth function and the set-valued map* $S : \mathbb{R}^n \rightrightarrows \mathbb{R}^m \times \mathbb{R}^p$ *is a solution set of the following problem (LLP2)*

$$minimaximize L(x,y,\lambda) \quad \textit{subject to} \quad (y,\lambda) \in Y \times W,$$

where $Y \subset \mathbb{R}^m$ *and* $W \in \mathbb{R}^p$ *are non-empty convex sets and* $L(x,y,\lambda)$ *is convex with respect to* y *for each* $(x,\lambda) \in \mathbb{R}^n \times W$ *and is concave in* λ *for each* $(x,y) \in \mathbb{R}^n \times Y$*. Further assume that* $L(x,y,\lambda)$ *is a twice continuously differentiable function. Let* $(\bar{x},\bar{y},\bar{\lambda})$ *be a solution to (P3). Set*

$$\bar{p} = (\nabla_y L(\bar{x},\bar{y},\bar{\lambda}), -\nabla_\lambda L(\bar{x},\bar{y},\bar{\lambda})).$$

Further assume that the following qualification condition holds :

$$rank\left[\nabla^2_{xy} L(\bar{x},\bar{y},\bar{\lambda}) | \nabla^2_{x\lambda} L(\bar{x},\bar{y},\bar{\lambda})\right] = m + p$$

Then there exists $v^* \in \mathbb{R}^{m+p}$ *such that*

i) $0 \in \nabla_x F(x,y,\lambda) + \left[\nabla^2_{xy} L(\bar{x},\bar{y},\bar{\lambda}) | \nabla^2_{x\lambda} L(\bar{x},\bar{y},\bar{\lambda})\right] v^*$

ii) $0 \in \nabla_{(y,\lambda)} F(\bar{x},\bar{y},\bar{\lambda}) + \left[\nabla^2_{yy} L(\bar{x},\bar{y},\bar{\lambda}) | \nabla^2_{y\lambda} L(\bar{x},\bar{y},\bar{\lambda})\right] v^* + D^* N_{Y \times W}((\bar{y},\bar{\lambda}) | -\bar{p})(v^*)$.

Proof. Since $(\bar{x},\bar{y},\bar{\lambda})$ is a solution of (P3) then $(\bar{x},\bar{y},\bar{\lambda}) \in \mathrm{gph}\, S$. Hence from Proposition 1.4 in Dontchev and Rockafellar [12] we have that

$$-\nabla_y L(\bar{x},\bar{y},\bar{\lambda}) \in N_Y(\bar{y}) \quad \text{and} \quad \nabla_\lambda L(\bar{x},\bar{y},\bar{\lambda}) \in N_W(\bar{\lambda}).$$

This is equivalent to the fact that $(\bar{y}, \bar{\lambda})$ is solving the following variational inequality over $Y \times W$ namely

$$0 \in G(\bar{x}, y, \lambda) + N_{Y \times W}(y, \lambda),$$

where

$$G(\bar{x}, y, \lambda) = (\nabla_y L(\bar{x}, y, \lambda), -\nabla_\lambda L(\bar{x}, y, \lambda)).$$

The result then follows by direct application of Lemma 3.2. □

It is important to note that all the above optimality conditions are expressed in terms of the coderivative of the normal cone map. We can however provide a slightly different reformulation of the optimality conditions by using the second-order subdifferential of the indicator function. To do this let us observe that if the lower-level problem (LLP) in (P) is convex then the solution set mapping S can be equivalently written as follows

$$S(x) = \{y \in \mathbb{R}^m : 0 \in \nabla_y f(x, y) + \partial_y \delta_K(x, y)\},$$

where $\delta_K(x, y) = \delta_{K(x)}(y)$ denotes the indicator function for the set $K(y)$. For any function $f : \mathbb{R}^n \to \mathbb{R} \cup \{+\infty\}$ which is finite at $\bar{x}$ the second-order subdifferential of f at $(\bar{x}, \bar{y})$ is the coderivative of the subdifferential map i.e.

$$\partial^2 f(\bar{x}|\bar{y})(u) = D^*(\partial f)(\bar{x}|\bar{y})(u).$$

Theorem 4.5 *Consider the problem (P1) and let $(\bar{x}, \bar{y})$ be a local solution of the problem. Consider that f is a twice continuously differentiable function. Assume that the following qualification condition holds:*

$$(u, 0) \in \partial^\infty \delta_K(\bar{x}, \bar{y}) \Longrightarrow u = 0.$$

Additionally assume that the following qualification condition also holds

$$0 \in \nabla^2 f(\bar{x}, \bar{y})^T (0, v_2) + \bigcup_{w \in \partial \delta_K(\bar{x}, \bar{y}), \operatorname{proj}_2 w = -\nabla_y f(\bar{x}, \bar{y})} \partial^2 \delta_K((\bar{x}, \bar{y})|w)(0, v_2)$$
$$\Longrightarrow v_2 = 0,$$

where proj_2 *denotes the projection on $\mathbb{R}^m$. Then there exists $v_2^* \in \mathbb{R}^m$ and $\bar{w} \in \partial \delta_K(\bar{x}, \bar{y})$ with* $\operatorname{proj}_2 \bar{w} = -\nabla_y f(\bar{x}, \bar{y})$ *such that*

$$0 \in \nabla F(\bar{x}, \bar{y}) + \nabla^2 f(\bar{x}, \bar{y})^T (0, v_2^*) + \partial^2 \delta_K((\bar{x}, \bar{y})|\bar{w})(0, v_2^*).$$

Proof. The key to the proof of this result is the Corollary 2.2 in Levy and Mordukhovich [14]. As per the Corollary 2.2 in Levy and Mordukhovich [14] the function δ_K should satisfy the following properties. First $\operatorname{gph} \partial_y \delta_K$ is closed. This is true since $\operatorname{gph} \partial_y \delta_K = \operatorname{gph} N_K$ and we know that $\operatorname{gph} N_K$ is closed. Second δ_k should be subdifferentially continuous at $(\bar{x}, \bar{y})$ for any $v \in \operatorname{gph} N_K$. This is true since the indicator function is subdifferentially continuous (see Rockafellar and Wets [21] pages 610-612). Now the result follows by a direct application of Corollary 2.2 in [14] □

Acknowledgments
We are grateful to Jiri Outrata and an anonymous referee whose constructive suggestions have vastly improved the presentation of the paper.

References

1. H. Babahadda and N. Gadhi: *Necessary optimality conditions for bilevel optimization problems using convexificators*, Technical Report, Department of Mathematics, Dhar El Mehrez, Sidi Mohamed Ben Abdellah University, Fes, Morocco, 2005.
2. B. Bank and J. Guddat and D. Klatte and B. Kummer and K. Tammer, *Non-Linear Parametric Optimization*, Akademie-Verlag, Berlin, 1982.
3. J. F. Bard, Optimality conditions for bilevel programming, *Naval Research Logistics Quaterly*, Vol 31, 1984, pp 13-26.
4. J. F. Bard, Convex two level optimization, *Mathematical Programming*, Vol 40, 1988, 15-27.
5. J. F. Bard, On some properties of bilevel programming, *Journal of Optimization Theory and Applications*, Vol 68, 1991, pp 371-378.
6. C. Berge, *Espaces topologiques: Fonctions multivoques*, (French) Collection Universitaire de Mathématiques, Vol. III Dunod, Paris, 1959. (In English) *Topological spaces : Including a treatment of multi-valued functions*, vector spaces and convexity. Translated from the French original by E. M. Patterson. Reprint of the 1963 translation. Dover Publications, Inc., Mineola, NY, 1997.
7. , G. Y. Chen and J. Jahn (eds.), *Mathematical Methods of Operations Research* Vol. 48, 1998.
8. F. H. Clarke, Yu. Ledyaev, R. Stern and P. Wolenski, *Nonsmooth Analysis and Control Theory*, Springer, 1998.
9. S. Dempe, *Foundations of Bilevel Programming*, Kluwer Academic Publishers, 2002.
10. S. Dempe, A necessary and sufficient optimality condition for bilevel prgramming problems, *Optimization*, Vol 25, 1992, pp 341-354.
11. A. L. Dontchev and R. T. Rockafellar, Ample paramerization of variational inclusions, *SIAM Journal on Optimization*, Vol 12, 2001, pp 170-187.
12. A.L. Dontchev and R. T. Rockafellar, Primal-dual solution perturbations in convex optimization, *Set-Valued Analysis*, Vol 9, 2001, pp 49-65.
13. M. Kojima, Strongly stable stationary solutions in nonlinear programs, In: S.M. Robinson (ed.): Analysis and Computation of Fixed Points, Academic Press, New York, 1980,, pp. 93–138.
14. A. B. Levy and B.S.Mordukhovich, Coderivatives in parametric optimization, *Mathematical Programming*, Vol 99, 2004, pp 311-327.
15. P. Loridan and J. Morgan, Weak via strong Stackelberg problem : New Results. *Journal of Global Optimization*, Vol 8, 1996, pp 263-287.
16. J. A. Mirrlees, The theory of moral hazard and unobservable bevaviour: part I, *Review of Economic Studies*, Vol. 66, 1999, pp. 3-21.
17. B. S. Mordukhovich, Generalized differential calculus for nonsmooth and set-valued mappings, *Journal of Mathematical Analysis and Applications*, Vol 183, 1994, pp 250-288.

18. J. Outrata, A generalized mathematical program with equilibrium constraints, *SIAM Journal on Control and Optimization*, Vol 38, 2000, pp 1623-1638.
19. J.-S. Pang and M. Fukushima, Complementarity constraint qualifications and simplified B-stationarity conditions for mathematical programs with equilibrium constraints, *Computational Optimization and Applications*, Vol. 13, 1999, pp. 111-136.
20. D. Ralph and S. Dempe, Directional Derivatives of the Solution of a Parametric Nonlinear Program, *Mathematical Programming*, Vol. 70, 1995, pp. 159-172.
21. R. T. Rockafellar and R. J. B. Wets, *Variational Analysis*, Springer-Verlag, 1998 (Second corrected printing 2004).
22. H. Scheel and S. Scholtes, *Mathematical programs with equilibrium constraints: stationarity, optimality, and sensitivity*, Mathematics of Operations Research, Vol. 25, 2000, pp. 1-22.
23. J. J. Ye, Nondifferentiable multiplier rules for optimization and bilevel optimization problems, *SIAM Journal on Optimization*, Vol. 15, 2004, pp. 252-274.
24. J. J. Ye, Constraint qualifications and necessary optimality conditions for optimization problems with variational inequality constraints, *SIAM Journal on Optimization*, Vol 10, 2000, pp 943-962.
25. J. J. Ye, Optimality conditions for optimization problems with complementarity constraints, *SIAM Journal on Optimization* Vol9, 1999, pp 374-387.
26. J. J. Ye and X. Y. Ye, Necessary optimality conditions for optimization problems with variational inequality constraints, *Mathematics of Operations Research*, Vol 22, 1997, pp 977-997.
27. J. J. Ye and D. L. Zhu, Optimality conditions for bilevel programming problems, *Optimization* Vol 33, 1995, pp 9-27.
28. J. J. Ye and D.L. Zhu, A note on optimality conditions for bilevel programming,*Optimization 39*,1997,pp 361-366.
29. J. J. Ye, D. L. Zhu and Q. Z. Zhu, Exact penalization and necessary optimality conditions for generalized bilevel programming problem, *SIAM Journal on Optimization*,Vol 4, 1994, pp 481-507.

Optimality criteria for bilevel programming problems using the radial subdifferential

D. Fanghänel

Technical University Bergakademie Freiberg, Akademiestraße 6, 09596 Freiberg, Germany difang@math.tu-freiberg.de

Summary. The discrete bilevel programming problems considered in this paper have discrete parametric lower level problems with linear constraints and a strongly convex objective function. Using both the optimistic and the pessimistic approach this problem is reduced to the minimization of auxiliary nondifferentiable and generally discontinuous functions. To develop necessary and sufficient optimality conditions for the bilevel problem the radial-directional derivative and the radial subdifferential of these auxiliary functions are used.

Key words: Bilevel programming, necessary and sufficient optimality conditions, discrete parametric optimization, minimization of discontinuous functions, radial-directional derivative.

1 Introduction

Bilevel programming problems are hierarchical optimization problems where the constraints of one problem (the so-called upper level problem) are defined in part by a second parametric optimization problem (the lower level problem) [1, 2]. These problems occur in a large variety of practical situations [3]. Many approaches are known to attack continuous bilevel programming problems. But, the number of references for bilevel programming problems with discrete variables is rather limited. Focus in the paper [15] is on existence of optimal solutions for problems which have discrete variables in the upper resp. the lower level problems. Solution algorithms have been developed in [5, 8, 9, 16]. The position of constraints in the upper resp. in the lower level problems is critical. The implications of and gains obtained from shifting a 0-1 variable from the lower to the upper level problems have been investigated in [4].

Focus in this paper is on optimality conditions for bilevel programming problems with discrete variables in the lower level problem. Verification of optimality conditions for continuous linear problems is $\mathcal{NP}$–hard [14] even if

S. Dempe and V. Kalashnikov (eds.), *Optimization with Multivalued Mappings*, pp. 73-95
©2006 Springer Science + Business Media, LLC

the optimal solution of the lower level problem is unique for all upper level variable values.

If the lower level problems may have nonunique optimal solutions, useful concepts are the optimistic and the pessimistic approaches. Both concepts lead to the minimization of a discontinuous auxiliary function φ. In the case of a linear bilevel programming problem, this function is a generalized PC^1–function and the formulation of optimality conditions can be based on the radial-directional derivative [2, 6].

In this paper a similar approach is investigated for discrete bilevel programming problems.

The outline of the paper is as follows. In Section 2 the investigated bilevel program is formulated and some introductory examples are given. Structural properties of the solution set mapping of the lower level problem are investigated in Section 3. In Sections 4 and 5 focus is on properties of the auxiliary function φ. Optimality conditions using the radial-directional derivative of the function φ are developed in Section 6, and in Section 7 the same is done by the help of the radial subdifferential of the function φ.

Throughout this paper the gradient of a function is the row vector of the partial derivatives. Further we will use the abbreviation $\{z^k\}$ for a sequence $\{z^k\}_{k=1}^{\infty}$ if this will not cause any confusion. The abbreviations $\mathrm{cl}\,A$, $\mathrm{int}\,A$ and $\mathrm{relint}\,A$ will be used for the closure, the interior an the relative interior of a set A.

2 A bilevel problem with discrete lower level

In this paper we consider the following bilevel programming problem

$$\begin{cases} \min\{g(x,y) : y \in Y,\ x \in \Psi_D(y)\} \\ \\ \Psi_D(y) = \underset{x}{\operatorname{argmin}}\,\{f(x,y) :\ x \in S_D\} \end{cases} \tag{1}$$

with the following requirements:

1. $Y \subseteq \mathbb{R}^n$ is convex, closed and $\mathrm{int}\,Y \neq \emptyset$.
2. $f(x,y) = F(x) - y^\top x$ with $F : \mathbb{R}^n \to \mathbb{R}$ being differentiable and strongly convex [10] with modulus $\theta > 0$, i.e. for all $x, x^0 \in \mathbb{R}^n$ it holds
$$F(x) \geq F(x^0) + \nabla F(x^0)(x - x^0) + \theta\|x - x^0\|^2.$$
3. $g(x,y)$ is continuously differentiable with respect to y.
4. The set $S_D \subseteq \mathbb{R}^n$ is required to be nonempty and discrete, i.e. there exists some $\omega > 0$ with $\|x - x'\| \geq \omega$ for all $x, x' \in S_D$, $x \neq x'$.
S_D denotes the set of all feasible solutions of the lower level problem.

Thus, the problem under consideration is continuous in the upper level and discrete with some special structure in the lower level.

In general the solution of the lower level is not unique. This causes some uncertainty in the definition of the upper level objective function [2]. Thus, instead of $g(x,y)$, we will investigate the following functions

$$\varphi_o(y) = \min_{x\in\Psi_D(y)} g(x,y), \tag{2}$$

$$\varphi_p(y) = \max_{x\in\Psi_D(y)} g(x,y). \tag{3}$$

The function $\varphi_o(y)$ is called optimistic solution function and $\varphi_p(y)$ pessimistic solution function. While most of the papers on bilevel programming with possible nonunique lower level solutions investigate (implicitly) the optimistic approach (see e.g. [1] and the references therein), focus for instance in the paper [11] is on the pessimistic approach and both approaches have been compared in [12]. A local optimal solution of the optimististic/pessimistic solution function is a local optimistic/pessimistic solution of (1).

In this paper we investigate necessary and sufficient conditions under which some point $y^0 \in Y$ is a local optimistic/pessimistic solution of (1).

We will use the notation $\varphi(y)$ if the statement holds for both $\varphi_o(y)$ and $\varphi_p(y)$.

For our considerations the so-called regions of stability are very important. They are defined as follows.

Definition 2.1 *Let $x^0 \in S_D$. Then the set*

$$\begin{aligned} R(x^0) &= \{y \in \mathbb{R}^n : f(x^0,y) \le f(x,y) \text{ for all } x \in S_D\} \\ &= \{y \in \mathbb{R}^n : x^0 \in \Psi_D(y)\} \end{aligned}$$

is called region of stability for the point x^0.

Thus the set $R(x^0)$ denotes the set of all parameters for which the point x^0 is optimal.

To make the subject more clear consider the following example.

Example 2.1

$$\begin{aligned} &\min\{\sin(xy) : y \in [0,5],\ x \in \Psi_D(y)\} \\ &\Psi_D(y) = \operatorname*{argmin}_x \left\{\frac{1}{2}x^2 - xy : 0 \le x \le 5,\ x \in \mathbb{Z}\right\} \end{aligned}$$

Since the upper level objective function is continuous on the regions of stability the latter ones can be seen in figure 1. Formally the regions of stability are

$$R(0) = (-\infty, 0.5],\ R(1) = [0.5, 1.5],\ R(2) = [1.5, 2.5],\ R(3) = [2.5, 3.5],$$
$$R(4) = [3.5, 4.5] \text{ and } R(5) = [4.5, \infty).$$

Using the definitions of the optimistic and pessimistic solution functions at the intersection points of the regions of stability, we get

$$\varphi_o(y) = \begin{cases} 0 & y \leq 0.5 \\ \sin(y) & 0.5 < y < 1.5 \\ \sin(2y) & 1.5 \leq y \leq 2.5 \\ \sin(3y) & 2.5 < y \leq 3.5 \\ \sin(4y) & 3.5 < y \leq 4.5 \\ \sin(5y) & y > 4.5 \end{cases} \qquad \varphi_p(y) = \begin{cases} 0 & y < 0.5 \\ \sin(y) & 0.5 \leq y \leq 1.5 \\ \sin(2y) & 1.5 < y < 2.5 \\ \sin(3y) & 2.5 \leq y < 3.5 \\ \sin(4y) & 3.5 \leq y < 4.5 \\ \sin(5y) & y \geq 4.5 \end{cases} .$$

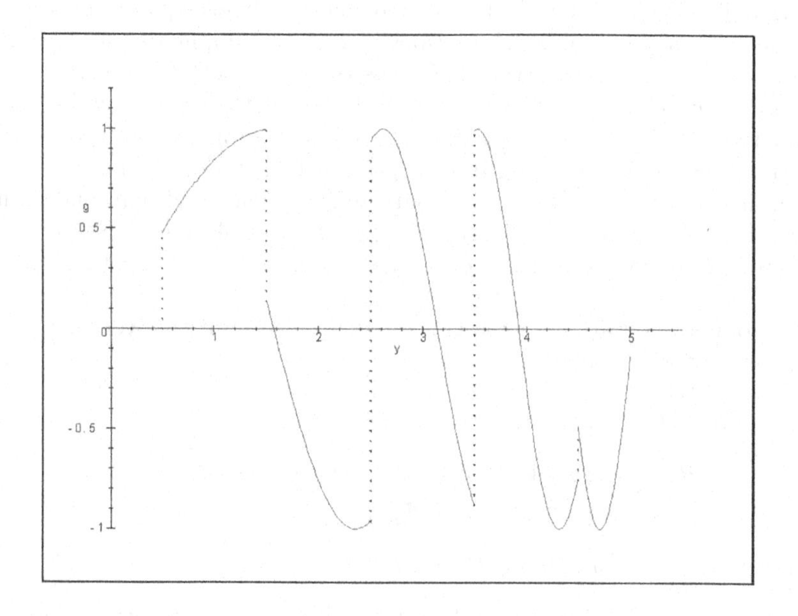

Fig. 1. Solution function φ for example 1

As it can be seen in figure 1 the local optimal solutions of φ_o are

$$y \in [0, 0.5], \ y = \frac{3\pi}{4}, \ y = 3.5, \ y = \frac{11\pi}{8}, \ y = \frac{3\pi}{2}$$

and

$$y \in [0, 0.5), \ y = \frac{3\pi}{4}, \ y = \frac{11\pi}{8}, \ y = \frac{3\pi}{2}$$

are the local optimal solutions of φ_p. △

In Example 2.1 the optimistic and the pessimistic solution functions are not continuous but rather selections of finitely many continuously differentiable functions.

3 Some remarks on the sets $\Psi_D(y)$ and $R(x)$

In this section we want to derive some properties of the sets $\Psi_D(y)$ and $R(x)$ which we will need later.

Lemma 3.1 *For each $x^0 \in S_D$ the set $R(x^0)$ is a closed convex set with $\nabla F(x^0)^\top$ in its interior.*

Proof. Let $x^0 \in S_D$. Then for all $y \in R(x^0)$ it holds $f(x^0, y) \le f(x, y)$ for all $x \in S_D$ and therefore

$$(x - x^0)^\top y \le F(x) - F(x^0) \quad \forall x \in S_D.$$

Thus, $R(x^0)$ corresponds to the intersection of (maybe infinitely many) half-spaces. This implies that $R(x^0)$ is convex and closed.

Now we want to show that $\nabla F(x^0)^\top \in \operatorname{int} R(x^0)$. Since $F : \mathbb{R}^n \to \mathbb{R}$ is strongly convex there exists some $\theta > 0$ with $F(x) \ge F(x^0) + \nabla F(x^0)(x - x^0) + \theta\|x - x^0\|^2$ for all $x \in \mathbb{R}^n$.

Consider $y = \nabla F(x^0)^\top + \alpha h$ with $h \in \mathbb{R}^n$, $\|h\| = 1$ and $\alpha \in [0, \theta\omega]$. Then, for all $x \in S_D$, $x \neq x^0$, the following sequence of inequalities is valid by $\|x - x^0\| \ge \omega$ for $x \neq x^0$:

$$\begin{aligned} F(x) &\ge F(x^0) + \nabla F(x^0)(x - x^0) + \theta\|x - x^0\|^2 \\ &= F(x^0) + y^\top(x - x^0) - \alpha h^\top(x - x^0) + \theta\|x - x^0\|^2 \\ &\ge F(x^0) + y^\top(x - x^0) - \alpha\|x - x^0\| + \theta\|x - x^0\|^2 \\ &\ge F(x^0) + y^\top(x - x^0) + (\theta\omega - \alpha)\|x - x^0\| \\ &\ge F(x^0) + y^\top(x - x^0). \end{aligned}$$

Thus we obtain $(\nabla F(x^0)^\top + \alpha h) \in R(x^0)$ for all $\alpha \in [0, \theta\omega]$, i.e. the assumption holds. □

Lemma 3.2 *1. For each $y \in \mathbb{R}^n$ the set $\Psi_D(y)$ has finite cardinality.*
2. If $y^0 \in \operatorname{int} R(x^0)$ for some $x^0 \in S_D$, then $\Psi_D(y^0) = \{x^0\}$.
3. Let some point $y^0 \in \mathbb{R}^n$ be given. Then there exists a positive real number $\varepsilon > 0$ such that $\Psi_D(y) \subseteq \Psi_D(y^0)$ for all $y \in U_\varepsilon(y^0) = \{y : \|y - y^0\| < \varepsilon\}$.

Proof. 1. If $S_D = \emptyset$ the assumption holds obviously. Assume that $S_D \neq \emptyset$ and take a point $x^0 \in S_D$. Let an arbitrary $y \in \mathbb{R}^n$ be given. Then for all $x \in \Psi_D(y)$ it holds

$$F(x) - y^\top x \le F(x^0) - y^\top x^0$$

implying

$$F(x^0) + \nabla F(x^0)(x - x^0) + \theta\|x - x^0\|^2 \le F(x^0) + y^\top(x - x^0)$$

for some $\theta > 0$ since F is strongly convex. Thus,

$$\theta\|x - x^0\|^2 \leq (y^\top - \nabla F(x^0))(x - x^0) \leq \|y - \nabla F(x^0)^\top\|\|x - x^0\|$$
$$\|x - x^0\| \leq \frac{1}{\theta}\|y - \nabla F(x^0)^\top\|.$$

Therefore $\Psi_D(y)$ has finite cardinality.

2. The inclusion $y^0 \in \operatorname{int} R(x^0)$ implies $\{x^0\} \subseteq \Psi_D(y^0)$ by definition. To prove the opposite direction assume that there exists a point $x \in \Psi_D(y^0)$, $x \neq x^0$. Then,

$$F(x) - {y^0}^\top x = F(x^0) - {y^0}^\top x^0$$
$$F(x) - F(x^0) = {y^0}^\top (x - x^0) > \nabla F(x^0)(x - x^0)$$

since F is strongly convex. Due to $y^0 \in \operatorname{int} R(x^0)$ there exists some $\varepsilon > 0$ such that
$y := y^0 + \varepsilon(y^0 - \nabla F(x^0)^\top) \in R(x^0)$. Now we obtain

$$\begin{aligned} f(x^0, y) &= F(x^0) - y^\top x^0 = F(x) - y^\top x^0 - {y^0}^\top (x - x^0) \\ &= f(x, y) + (y - y^0)^\top (x - x^0) \\ &= f(x, y) + \varepsilon({y^0}^\top - \nabla F(x^0))(x - x^0) > f(x, y) \end{aligned}$$

which is a contradiction to $y \in R(x^0)$.

3. Assume that the assertion does not hold. Then there exist sequences

$$\{y^k\}_{k=1}^\infty \text{ with } y^k \to y^0, k \to \infty, \text{ and } \{x^k\}_{k=1}^\infty \text{ with } x^k \in \Psi_D(y^k)$$

but $x^k \notin \Psi_D(y^0)$ for all k.
Thus, for fixed $x^0 \in S_D$, it holds

$$\begin{aligned} F(x^k) - {y^k}^\top x^k &\leq F(x^0) - {y^k}^\top x^0 \\ F(x^0) + \nabla F(x^0)(x^k - x^0) + \theta\|x^k - x^0\|^2 &\leq F(x^0) + {y^k}^\top (x^k - x^0) \\ \|x^k - x^0\| &\leq \frac{\|y^k - \nabla F(x^0)^\top\|}{\theta}. \end{aligned}$$

This yields

$$\|x^k - x^0\| \leq \underbrace{\frac{\|y^k - y^0\|}{\theta}}_{\to 0} + \frac{\|y^0 - \nabla F(x^0)^\top\|}{\theta},$$

i.e. $\{x^k\}$ is bounded and has finitely many elements. Therefore we can assume that all x^k are equal, i.e. $\exists x \in S_D$ with $x \in \Psi_D(y^k)\ \forall k$ but $x \notin \Psi_D(y^0)$.
That means $y^k \in R(x)\ \forall k$ but $y^0 \notin R(x)$. This is a contradiction to Lemma 3.1.

□

4 Basic properties of $\varphi(y)$

In this section we want to show first that for each $y^0 \in \mathbb{R}^n$ there exists some $\varepsilon > 0$ such that in the neighborhood $U_\varepsilon(y^0)$ the optimistic/pessimistic solution function is a selection of finitely many continuously differentiable functions. Further, for this special $\varepsilon > 0$ we will investigate the support set

$$Y_x(y^0) := \{y \in U_\varepsilon(y^0) \cap R(x) : g(x,y) = \varphi(y)\}$$

and its contingent cone

$$T_x(y^0) := \{r : \exists \{y^s\} \subseteq Y_x(y^0) \ \exists \{t_s\} \subseteq \mathbb{R}_+ \text{ with } y^s \to y^0, t_s \downarrow 0, \lim_{s\to\infty} \frac{y^s - y^0}{t_s} = r\}.$$

That means, $Y_x(y^0)$ is the set of all $y \in U_\varepsilon(y^0)$ for which both $x \in \Psi_D(y)$ and $g(x,y) = \varphi(y)$ hold for a fixed point $x \in S_D$. Properties of these sets are essential for the investigation of generalized PC^1–functions (in short: GPC^1–functions) in the paper [6] leading to optimality conditions for linear bilevel programming problems in [2]. The following two theorems show that the objective functions in the two auxiliary problems (2) and (3) have many properties of GPC^1–functions, but they are not GPC^1–functions as it is shown by Example 4.1 below.

Theorem 4.1 *For the function φ and each $y^0 \in \mathbb{R}^n$ it holds:*

1. *There exists an open neighborhood $U_\varepsilon(y^0)$ of y^0 and a finite number of points $x \in \Psi_D(y^0)$ with*

$$\varphi(y) \in \{g(x,y)\}_{x\in\Psi_D(y^0)} \quad \forall y \in U_\varepsilon(y^0).$$

2. $\operatorname{int} Y_x(y^0) = U_\varepsilon(y^0) \cap \operatorname{int} R(x)$ *and* $Y_x(y^0) \subseteq \operatorname{cl}\ \operatorname{int} Y_x(y^0)$ *for* $x, y^0 \in \mathbb{R}^n$.
3. $T_x(y^0) \subseteq \operatorname{cl}\ \operatorname{int} T_x(y^0)$ *for* $y^0 \in R(x)$.

Proof. Let an arbitrary $y^0 \in \mathbb{R}^n$ be given.

1.) Because of Lemma 3.2, $\Psi_D(y^0)$ has finite cardinality and there exists some $\varepsilon > 0$ with $\Psi_D(y^0) \supseteq \Psi_D(y)$ for all $y \in U_\varepsilon(y^0)$. With $\varphi(y) \in \{g(x,y)\}_{x\in\Psi_D(y)}$ it follows $\varphi(y) \in \{g(x,y)\}_{x\in\Psi_D(y^0)} \quad \forall y \in U_\varepsilon(y^0)$.

2.) Let $\bar{y} \in \operatorname{int} Y_x(y^0)$. Then there exists some $\delta > 0$ with $U_\delta(\bar{y}) \subseteq Y_x(y^0)$. Thus, $\bar{y} \in U_\varepsilon(y^0)$ and $U_\delta(\bar{y}) \subseteq R(x)$, i.e. $\bar{y} \in U_\varepsilon(y^0) \cap \operatorname{int} R(x)$.

Let $\bar{y} \in U_\varepsilon(y^0) \cap \operatorname{int} R(x)$. Then there exists some $\delta > 0$ with $U_\delta(\bar{y}) \subseteq U_\varepsilon(y^0)$ and $U_\delta(\bar{y}) \subseteq \operatorname{int} R(x)$. From Lemma 3.2 it follows $\Psi_D(y) = \{x\}\ \forall y \in U_\delta(\bar{y})$. Thus, $\varphi(y) = g(x,y)\ \forall y \in U_\delta(\bar{y})$, i.e. $y \in Y_x(y^0)\ \forall y \in U_\delta(\bar{y})$. Therefore, $\bar{y} \in \operatorname{int} Y_x(y^0)$. This implies the first equation of part 2.

Now let $\bar{y} \in Y_x(y^0)$. This means $\bar{y} \in R(x)$, $\bar{y} \in U_\varepsilon(y^0)$ and $\varphi(\bar{y}) = g(x,\bar{y})$. Since $R(x)$ is convex with nonempty interior (cf. Lemma 3.1) there exists some sequence $\{y^k\} \in \operatorname{int} R(x)$ with $y^k \to \bar{y}$, $k \to \infty$. W.l.o.g. we can further assume that $y^k \in U_\varepsilon(y^0)$ $\forall k$. Consequently, $y^k \in \operatorname{int} Y_x(y^0)$ $\forall k$ and thus $\bar{y} \in \operatorname{cl}\operatorname{int} Y_x(y^0)$.

3.) Let an arbitrary $r \in T_x(y^0)$ be given. Then there exist sequences $\{y^s\} \subseteq Y_x(y^0)$ and $\{t_s\} \subseteq \mathbb{R}_+$ with $y^s \to y^0$, $t_s \downarrow 0$ and $\lim\limits_{s\to\infty} \frac{y^s - y^0}{t_s} = r$. We can assume w.l.o.g. that $t_s \in (0,1)$ $\forall s$.

Take any $\tilde{y} \in \operatorname{int} Y_x(y^0)$ and let $\hat{y}^s := t_s\tilde{y} + (1-t_s)y^0 = y^0 + t_s(\tilde{y} - y^0)$. Then, $\lim_{s\to\infty} \hat{y}^s = y^0$ and $\frac{\hat{y}^s - y^0}{t_s} = \tilde{y} - y^0 =: \tilde{r}$ $\forall s$. Since $R(x)$ is convex it follows easily that $\hat{y}^s \in \operatorname{int} Y_x(y^0)$ $\forall s$ and $\tilde{r} \in \operatorname{int} T_x(y^0)$.

Now consider $z^s_\lambda := \lambda y^s + (1-\lambda)\hat{y}^s$ with $\lambda \in (0,1)$. Since $R(x)$ is convex and $\hat{y}^s \in \operatorname{int} Y_x(y^0)$ it follows $z^s_\lambda \in \operatorname{int} Y_x(y^0)$ $\forall \lambda$ $\forall s$. Then it holds $z^s_\lambda \to y^0$ for $s \to \infty$ and $\lim\limits_{s\to\infty} \frac{z^s_\lambda - y^0}{t_s} = \lambda r + (1-\lambda)\tilde{r} =: r_\lambda \in T_x(y^0)$ for all $\lambda \in (0,1)$. Moreover, $r_\lambda \to r$ for $\lambda \to 1$.

Now, from $z^s_\lambda \in \operatorname{int} Y_x(y^0)$ it follows easily that $z^s_\lambda - y^0 \in \operatorname{int} T_x(y^0)$ and thus $\frac{z^s_\lambda - y^0}{t_s} \in \operatorname{int} T_x(y^0)$ $\forall s$ $\forall \lambda \in (0,1)$.
Hence, $r_\lambda \in \operatorname{cl}\operatorname{int} T_x(y^0)$ $\forall \lambda \in (0,1)$. This together with $r_\lambda \to r$ for $\lambda \to 1$ implies $r \in \operatorname{cl}\operatorname{cl}\operatorname{int} T_x(y^0) = \operatorname{cl}\operatorname{int} T_x(y^0)$. □

Theorem 4.2 *$\operatorname{int} T_{x^1}(y^0) \cap \operatorname{int} T_{x^2}(y^0) = \emptyset$ for all $x^1, x^2 \in \Psi_D(y^0)$, $x^1 \neq x^2$.*

Proof. Let $r \in T_{x^1}(y^0) \cap T_{x^2}(y^0)$ be arbitrary. Due to $r \in T_{x^1}(y^0)$ there exists sequences $\{y^s\} \subseteq Y_{x^1}(y^0)$, $y^s \to y^0$ and $\{t_s\}$, $t_s \downarrow 0$ with $r^s := \frac{y^s - y^0}{t_s} \to r$.

From $y^s \in Y_{x^1}(y^0)$ $\forall s$ it follows $y^s \in R(x^1)$ $\forall s$, i.e. $F(x^1) - {y^s}^\top x^1 \leq F(x^2) - {y^s}^\top x^2$. Since $x^1, x^2 \in \Psi_D(y^0)$ it holds $F(x^1) - {y^0}^\top x^1 = F(x^2) - {y^0}^\top x^2$. Hence,

$$\begin{aligned} {y^s}^\top (x^1 - x^2) \geq F(x^1) - F(x^2) = {y^0}^\top (x^1 - x^2) \\ (y^s - y^0)^\top (x^1 - x^2) \geq 0 \quad \forall s \\ {r^s}^\top (x^1 - x^2) \geq 0 \quad \forall s. \\ \text{With } r^s \to r \text{ this yields } r^\top (x^1 - x^2) \geq 0. \end{aligned}$$

From $r \in T_{x^2}(y^0)$ it follows analogously $(x^1 - x^2)^\top r \leq 0$. Therefore it holds

$$(x^2 - x^1)^\top r = 0 \text{ for all } r \in T_{x^1}(y^0) \cap T_{x^2}(y^0).$$

Assume that there exists some $r \in \operatorname{int} T_{x^1}(y^0) \cap \operatorname{int} T_{x^2}(y^0)$. Then for all $t \in \mathbb{R}^n$, $\|t\| = 1$ there exists a real number $\delta > 0$ with $r + \delta t \in T_{x^1}(y^0) \cap T_{x^2}(y^0)$, i.e.

$$\begin{aligned} (x^2 - x^1)^\top (r + \delta t) &= 0 \\ \delta (x^2 - x^1)^\top t &= 0 \\ (x^2 - x^1)^\top t &= 0 \quad \forall t \end{aligned}$$

and therefore $x^1 = x^2$. □

Next we show that the function φ is not a GPC^1-function (cf. [2],[6]). For GPC^1-functions one requires additionally to the results in the Theorems 4.1 and 4.2 that there exists a number $\delta > 0$ such that for all $r \in T_{x^1}(y^0) \cap T_{x^2}(y^0)$, $\|r\| = 1$, $x^1 \neq x^2$ some $t_0 = t(r) \geq \delta$ can be found with $y^0 + tr \in Y_{x^1}(y^0)$ or $y^0 + tr \in Y_{x^2}(y^0)$ $\forall t \in (0, t_0)$. We will show that the functions φ usually do not have this property.

Example 4.1 Consider the lower level problem in (1) with the feasible set $S_D = \{x^1 = (0,0,0)^\top, x^2 = (1,0,0)^\top, x^3 = (0,1,0)^\top\}$ and $f(x,y) = \frac{1}{2}x^\top x - x^\top y$. Then we obtain the following regions of stability:

$$\begin{aligned} R(x^1) &= \{y \in \mathbb{R}^3 : y_1 \leq 1/2, y_2 \leq 1/2\} \\ R(x^2) &= \{y \in \mathbb{R}^3 : y_1 \geq 1/2, y_2 \leq y_1\} \\ R(x^3) &= \{y \in \mathbb{R}^3 : y_2 \geq 1/2, y_2 \geq y_1\}. \end{aligned}$$

Let $g(x,y) = (1/2, -1, 0)^\top x$ be the objective function of the upper level problem. Then,

$$\varphi_o(y) = \begin{cases} -1 & y \in R(x^3) \\ 0 & y \in R(x^1) \setminus R(x^3) \\ 1/2 & else \end{cases} .$$

Set $r = (0,0,1)^\top$ and $y^0 = (1/2, 1/2, 0)^\top$.
Further, $y^1(\varepsilon) := (1/2 - \varepsilon^2, 1/2 - \varepsilon^2, \varepsilon + \varepsilon^2)^\top \in Y_{x^1}(y^0)$ $\forall \varepsilon > 0$. Then

$$\begin{aligned} &\lim_{\varepsilon \to 0} y^1(\varepsilon) = (1/2, 1/2, 0)^\top = y^0, \\ \lim_{\varepsilon \to 0} &\frac{y^1(\varepsilon) - y^0}{\varepsilon} = \lim_{\varepsilon \to 0} (-\varepsilon, -\varepsilon, 1 + \varepsilon)^\top = r, \text{ i.e. } r \in T_{x^1}(y^0). \end{aligned}$$

Analogously $y^2(\varepsilon) := (1/2 + \varepsilon^2, 1/2, \varepsilon + \varepsilon^2)^\top \in Y_{x^2}(y^0)$ $\forall \varepsilon > 0$. Then

$$\begin{aligned} &\lim_{\varepsilon \to 0} y^2(\varepsilon) = (1/2, 1/2, 0)^\top = y^0, \\ \lim_{\varepsilon \to 0} &\frac{y^2(\varepsilon) - y^0}{\varepsilon} = \lim_{\varepsilon \to 0} (\varepsilon, 0, 1 + \varepsilon)^\top = r, \text{ i.e. } r \in T_{x^2}(y^0). \end{aligned}$$

Therefore, $r \in T_{x^1}(y^0) \cap T_{x^2}(y^0)$, $\|r\| = 1$, $x^1 \neq x^2$ but $\varphi_o(y^0 + tr) = -1 < g(x^i, y^0 + tr)$, $i = 1, 2$, $\forall t > 0$, i.e. $y^0 + tr \notin Y_{x^1}(y^0)$ and $y^0 + tr \notin Y_{x^2}(y^0)$ for all $t > 0$. △

Until now the description of the contingent cones has been more theoretical. Thus, for calculation we will need some better formula. In [10] many statements are given concerning contingent cones to closed convex sets. But, in general the sets $Y_x(y^0)$ are neither convex nor closed. Using

$$T_x(y^0) \subseteq \text{cl}\{r \in \mathbb{R}^n : \exists t_0 > 0 \text{ with } y^0 + tr \in Y_x(y^0)\ \forall t \in [0, t_0]\}$$

we obtain the following Lemma:

Lemma 4.1 *Let $\bar{x} \in \Psi_D(y^0)$. Then it holds*

$$T_{\bar{x}}(y^0) = \{r \in \mathbb{R}^n : 0 \le (\bar{x} - x)^\top r \quad \forall x \in \Psi_D(y^0)\}.$$

Proof. Let $r \in T_{\bar{x}}(y^0)$. Then there exists some sequence $\{r^k\}$ with

$$\lim_{k\to\infty} r^k = r \text{ and } y^0 + tr^k \in R(\bar{x}) \text{ for all } k$$

if $t > 0$ is sufficiently small. Hence,

$$F(\bar{x}) - (y^0 + tr^k)^\top \bar{x} \le F(x) - (y^0 + tr^k)^\top x \quad \forall x \in S_D$$
$$F(\bar{x}) - {y^0}^\top \bar{x} - t{r^k}^\top \bar{x} \le F(x) - {y^0}^\top x - t{r^k}^\top x \quad \forall x \in S_D.$$

On the other hand it holds $F(\bar{x}) - {y^0}^\top \bar{x} = F(x) - {y^0}^\top x \; \forall x \in \Psi_D(y^0)$. Thus, ${r^k}^\top (\bar{x} - x) \ge 0 \; \forall k \, \forall x \in \Psi_D(y^0)$. Consequently it holds $r^\top (\bar{x} - x) \ge 0 \; \forall x \in \Psi_D(y^0)$.

Let $0 \le (\bar{x} - x)^\top r \; \forall x \in \Psi_D(y^0)$. Then it holds

$$F(\bar{x}) - (y^0 + tr)^\top \bar{x} \le F(x) - (y^0 + tr)^\top x \quad \forall x \in \Psi_D(y^0) \; \forall t \ge 0.$$

Further there exists some $\varepsilon > 0$ with $\Psi_D(y) \subseteq \Psi_D(y^0) \; \forall y \in U_\varepsilon(y^0)$. Thus, for all $t \in (0, \varepsilon/\|r\|)$ it holds $F(\bar{x}) - (y^0 + tr)^\top \bar{x} \le F(x) - (y^0 + tr)^\top x \quad \forall x \in \Psi_D(y^0 + tr)$, i.e. $y^0 + tr \in R(\bar{x}) \; \forall t \in (0, \varepsilon/\|r\|)$. Now we will show that $r \in T_{\bar{x}}(y^0)$. Let $\tilde{y} = y^0 + t_0 r$ for some fixed $t_0 \in (0, \varepsilon/\|r\|)$. Since $R(\bar{x})$ is convex with nonempty interior there exists some sequence $\{y^s\} \in \operatorname{int} R(\bar{x})$ with $y^s \to \tilde{y}$, $s \to \infty$ and $y^s \in U_\varepsilon(y^0)$. Then it holds $(y^s - y^0)\lambda + y^0 \in \operatorname{int} Y_{\bar{x}}(y^0) \; \forall \lambda \in (0,1) \; \forall s$. Consequently, $y^s - y^0 \in T_{\bar{x}}(y^0) \; \forall s$. Since $T_{\bar{x}}(y^0)$ is a closed cone and $\tilde{y} - y^0 = \lim_{s\to\infty} y^s - y^0$ it follows $\tilde{y} - y^0 = t_0 r \in T_{\bar{x}}(y^0)$, i.e. it holds $r \in T_{\bar{x}}(y^0)$. □

Consequently, the cones $T_x(y^0)$ are polyhedral cones with nonempty interior for all $x \in \Psi_D(y^0)$.

5 The radial-directional derivative

In the following we formulate criteria for local optimality. For this we want to use the radial-directional derivative which was introduced by Recht [13]. Such kind of considerations have even been done for GPC^1-functions [2, 6]. But as shown, although our functions φ have some properties in common with GPC^1-functions they are in general not GPC^1-functions.

Definition 5.1 *Let $U \subseteq \mathbb{R}^n$ be an open set, $y^0 \in U$ and $\varphi : U \to \mathbb{R}$. We say that φ is radial-continuous at y^0 in direction $r \in \mathbb{R}^n$, $\|r\| = 1$, if there exists a real number $\varphi(y^0; r)$ such that*

$$\lim_{t \downarrow 0} \varphi(y^0 + tr) = \varphi(y^0; r).$$

If the radial limit $\varphi(y^0; r)$ exists for all $r \in \mathbb{R}^n$, $\|r\| = 1$, φ is called radial-continuous at y^0.

φ is radial-directionally differentiable at y^0, if there exists a positively homogeneous function $d\varphi_{y^0} : \mathbb{R}^n \to \mathbb{R}$ such that for all $r \in \mathbb{R}^n$, $\|r\| = 1$ and all $t > 0$ it holds

$$\varphi(y^0 + tr) - \varphi(y^0; r) = td\varphi_{y^0}(r) + o(y^0, tr)$$

with $\lim_{t \downarrow 0} o(y^0, tr)/t = 0$. Obviously, $d\varphi_{y^0}$ is uniquely defined and is called the radial-directional derivative of φ at y^0.

Theorem 5.1 *Both the optimistic solution function φ_o and the pessimistic solution function φ_p are radial-continuous and radial-directionally differentiable.*

Proof. Consider y^0 and some direction $r \in \mathbb{R}^n$, $\|r\| = 1$. Further let

$$I_r(y^0) := \{x \in \Psi_D(y^0) : \forall \varepsilon > 0 \, \exists t \in (0, \varepsilon) \text{ with } y^0 + tr \in Y_x(y^0)\}$$

and $G(y^0 + tr) := \min_{x \in I_r(y^0)} g(x, y^0 + tr)$.

Since $\Psi_D(y^0)$ has finite cardinality and the sets $R(x)$ are convex it holds $\varphi_o(y^0 + tr) = G(y^0 + tr)$ for all sufficiently small real numbers $t > 0$. Since the function $G(\cdot)$ is the minimum function of finitely many continuously differentiable functions it is continuous and quasidifferentiable (cf. [7]) and thus directionally differentiable in $t = 0$. Therefore the limits

$$\lim_{t \downarrow 0} G(y^0 + tr) = G(y^0) \text{ and } \lim_{t \downarrow 0} \frac{G(y^0 + tr) - G(y^0)}{t} = G'(y^0; r)$$

exist. Moreover, since for all $x \in I_r(y^0)$ it exists some sequence $\{t_k\} \downarrow 0$: $y^0 + t_k r \in Y_x(y^0)$ and

$$\lim_{t \downarrow 0} G(y^0 + tr) = \lim_{k \to \infty} G(y^0 + t_k r) = \lim_{k \to \infty} g(x, y^0 + t_k r) = g(x, y^0)$$

we derive

$$\varphi_o(y^0; r) = \lim_{t \downarrow 0} G(y^0 + tr) = G(y^0) = g(x, y^0) \; \forall x \in I_r(y^0). \tag{4}$$

Concerning the radial-directional derivative we obtain

$$\begin{aligned} d\varphi_{o y^0}(r) &= \lim_{t \downarrow 0} \frac{\varphi_o(y^0 + tr) - \varphi_o(y^0; r)}{t} = \lim_{t \downarrow 0} \frac{G(y^0 + tr) - G(y^0)}{t} \\ &= \nabla_y g(x, y^0) r \; \forall x \in I_r(y^0) \end{aligned} \tag{5}$$

since g is continuously differentiable with respect to y.

For $\varphi_p(y)$ we can prove the assertions analogously. □

Example 5.1 Let some feasible set $S_D = \{x^1 = (0,0)^\top, x^2 = (0,1)^\top, x^3 = (-1,0)^\top\}$ be given with functions $f(x,y) = \frac{1}{2}x^\top x - x^\top y$ and

$$g(x,y) = x_1 + x_2 \cdot \begin{cases} y_1^3 \sin \frac{1}{y_1} & y_1 > 0 \\ 0 & y_1 \le 0\,. \end{cases}$$

Then the function $g(x,y)$ is continuously differentiable with respect to y. The regions of stability are

$$\begin{aligned} R(x^1) &= \{y \in \mathbb{R}^2 :\; y_1 \ge -0.5,\; y_2 \le 0.5\} \\ R(x^2) &= \{y \in \mathbb{R}^2 :\; y_1 + y_2 \ge 0,\; y_2 \ge 0.5\} \\ R(x^3) &= \{y \in \mathbb{R}^2 :\; y_1 \le -0.5,\; y_1 + y_2 \le 0\}. \end{aligned}$$

Let $y^0 = (0, \frac{1}{2})^\top$ and $r = (1,0)^\top$. Then $I_r(y^0) = \{x^1, x^2\}$ for both the optimistic and the pessimistic solution function. Thus it holds

$$\varphi_o(y^0; r) = \varphi_p(y^0; r) = g(x^1, y^0) = g(x^2, y^0) = 0$$

and

$$\varphi_{o_{y^0}}(r) = \varphi_{p_{y^0}}(r) = \nabla_y g(x^i, y^0) r,\; i = 1,2.$$

Further it holds $\varphi_o(y^0) = \varphi_p(y^0) = 0$. Remarkable in this example is the fact that for all $\varepsilon > 0$ there exists some $t \in (0, \varepsilon)$ with either $\varphi(y^0 + tr) \ne g(x^1, y^0 + tr)$ or $\varphi(y^0 + tr) \ne g(x^2, y^0 + tr)$.

Now let $\bar{y} = (-\frac{1}{2}, \frac{1}{2})^\top$ and $r = (-1,1)^\top$. Then, for the optimistic solution function it holds

$$I_r(\bar{y}) = \{x^3\} \text{ and } \varphi_o(\bar{y}) = \varphi_o(\bar{y}; r) = -1$$

and for the pessimistic solution function it holds

$$I_r(\bar{y}) = \{x^2\} \text{ and } \varphi_p(\bar{y}) = \varphi_p(\bar{y}; r) = 0.$$

Considering the direction $r = (0,1)$ we obtain $I_r(\bar{y}) = \{x^2\}$ and $\varphi(\bar{y}; r) = 0$ for both the optimistic and the pessimistic case, but $\varphi_o(\bar{y}) = -1 \ne 0 = \varphi_p(\bar{y})$.

△

Lemma 5.1 *For all $y^0 \in \mathbb{R}^n$ and for all $r \in \mathbb{R}^n$ it holds:*

1. $\varphi_o(y^0) \le \varphi_o(y^0; r)$
2. $\varphi_p(y^0) \ge \varphi_p(y^0; r)$

Proof. Assume there exists some y^0 and some r with $\varphi_o(y^0) > \varphi_o(y^0; r)$. Then from $I_r(y^0) \subseteq \Psi_D(y^0)$ and the proof of Theorem 5.1 it follows that there exists some $x \in \Psi_D(y^0)$ with $\varphi_o(y^0; r) = g(x, y^0)$. Hence, $\varphi_o(y^0) > g(x, y^0)$ for some $x \in \Psi_D(y^0)$. This is a contradiction to the definition of φ_o.

The proof for φ_p is similar. □

6 Optimality criteria based on the radial-directional derivative

Let $\operatorname{locmin}\{\varphi(y) : y \in Y\}$ denote the set of all local minima of the function $\varphi(\cdot)$ over the region $Y \subseteq \mathbb{R}^n$.

Theorem 6.1 *It holds*

$$\operatorname{locmin}\{\varphi_p(y) : y \in \mathbb{R}^n\} \subseteq \operatorname{locmin}\{\varphi_o(y) : y \in \mathbb{R}^n\}.$$

Proof. Arguing by contradiction we assume that there is some y^0 with $y^0 \in \operatorname{locmin}\{\varphi_p(y) : y \in \mathbb{R}^n\}$ but $y^0 \notin \operatorname{locmin}\{\varphi_o(y) : y \in \mathbb{R}^n\}$. Then there exists some sequence $\{y^k\} \subseteq \mathbb{R}^n$ with $y^k \to y^0, k \to \infty$ and $\varphi_o(y^k) < \varphi_o(y^0)$. Since $\Psi_D(y^0)$ has finite cardinality and $\Psi_D(y^0) \supseteq \Psi_D(y)$ for all y in a neighborhood of y^0 we can assume w.l.o.g. that there exists some $x \in \Psi_D(y^0)$ with $x \in \Psi_D(y^k)$ and $\varphi_o(y^k) = g(x, y^k)\ \forall k$, i.e. $y^k \in Y_x(y^0)\ \forall k$. Since $g(x, \cdot)$ is differentiable with respect to y and $Y_x(y^0) \subseteq \operatorname{cl}\operatorname{int} Y_x(y^0)$ we can further assume that $y^k \in \operatorname{int} Y_x(y^0)\ \forall k$. Thus it holds $\Psi_D(y^k) = \{x\}\ \forall k$, i.e. $\varphi_o(y^k) = \varphi_p(y^k) = g(x, y^k)\ \forall k$. Consequently,

$$\varphi_p(y^k) = \varphi_o(y^k) < \varphi_o(y^0) \leq \varphi_p(y^0) \quad \forall k.$$

This is a contradiction to $y^0 \in \operatorname{locmin}\{\varphi_p(y) : y \in \mathbb{R}^n\}$. □

Thus, if $Y = \mathbb{R}^n$ and we know that some point y^0 is a local pessimistic solution then clearly y^0 is a local optimistic solution, too.

Further, for $y \in \operatorname{int} Y$ and $y \in \operatorname{locmin}\{\varphi_p(y) : y \in Y\}$ it follows analogously that $y \in \operatorname{locmin}\{\varphi_o(y) : y \in Y\}$. As the next example will show we indeed need the condition $y \in \operatorname{int} Y$.

Example 6.1 Let

$$\begin{aligned} S_D &= \{(0,1)^\top, (0,-1)^\top\}, \\ Y &= \{y \in \mathbb{R}^2 : y_2 \geq 0\}, \\ f(x,y) &= \frac{1}{2} x^\top x - x^\top y \text{ and} \\ g(x,y) &= x_2 y_1^2 + y_2. \end{aligned}$$

Then it holds for $y \in Y$

$$\varphi_p(y) = y_1^2 + y_2 \qquad \text{and} \qquad \varphi_o(y) = \begin{cases} y_1^2 + y_2 \text{ if } y_2 > 0 \\ -y_1^2 + y_2 \text{ if } y_2 = 0 \,. \end{cases}$$

Thus, $y^0 = (0,0)^\top$ is a local pessimistic but not a local optimistic solution. Moreover, y^0 is a global pessimistic solution. Some local optimistic solution does not exist. △

For further considerations we will need the contingent cone of the set Y. For each given point $y^0 \in Y$ this cone is defined as follows:

$$T_Y(y^0) := \{r \in \mathbb{R}^n : \exists\{y^s\} \subseteq Y \ \exists\{t_s\} \downarrow 0 : \ y^s \to y^0, s \to \infty, \\ \text{with } \lim_{s\to\infty} \frac{y^s - y^0}{t_s} = r\}.$$

The set $T_Y(y^0)$ is a convex, closed, nonempty cone [10]. Since Y is convex it holds

$$T_Y(y^0) = \text{cl } \mathbb{T}_Y(y^0)$$

with

$$\mathbb{T}_Y(y^0) = \{r \in \mathbb{R}^n : \ \exists t_0 > 0 \text{ with } y^0 + tr \in Y \ \forall t \in [0, t_0]\}.$$

Theorem 6.2 *Let $y^0 \in \mathbb{R}^n$ and let $\varphi : \mathbb{R}^n \to \mathbb{R}$ denote the optimistic or the pessimistic solution function. Then $y^0 \notin \text{locmin}\{\varphi(y) : \ y \in Y\}$ if there exists some $r \in \mathbb{T}_Y(y^0)$, $\|r\| = 1$, such that one of the following conditions 1,2 is satisfied:*

1. *$d\varphi_{y^0}(r) < 0$ and $\varphi(y^0; r) = \varphi(y^0)$*
2. *$\varphi(y^0; r) < \varphi(y^0)$*

Proof. Let the vector $r^0 \in \mathbb{T}_Y(y^0)$ with $\|r^0\| = 1$ satisfy condition 1. That means $d\varphi_{y^0}(r) = \lim_{t\downarrow 0} t^{-1}(\varphi(y^0 + tr^0) - \varphi(y^0; r^0)) < 0$. Then there exists some $t' \in (0, t_0)$ such that $\varphi(y^0 + tr^0) < \varphi(y^0; r^0)$ and $y^0 + tr^0 \in Y \ \forall t \in (0, t')$. Because of $\varphi(y^0; r^0) = \varphi(y^0)$ we have $\varphi(y^0 + tr^0) < \varphi(y^0)$ and $y^0 + tr^0 \in Y$ for all $t \in (0, t')$. Thus, y^0 cannot be a local minimum of φ.

Now let the vector $r^0 \in \mathbb{T}_Y(y^0)$ with $\|r^0\| = 1$ satisfy condition 2. Then it holds

$$\varphi(y^0) - \varphi(y^0; r^0) = \varphi(y^0) - \lim_{t\downarrow 0} \varphi(y^0 + tr^0) > 0.$$

Hence there exists some $t' \in (0, t_0)$ such that $y^0 + tr^0 \in Y$ and $\varphi(y^0) > \varphi(y^0 + tr^0)$ for all $t \in (0, t')$. Thus, y^0 cannot be a local minimum of φ. □

Since $d\varphi_{y^0}(\cdot)$ is not continuous it is indeed necessary to consider only the set $\mathbb{T}_Y(y^0)$. The consideration of $T_Y(y^0)$ would not lead to correct results as we will see in the next example.

Example 6.2 Let

$$\begin{aligned} S_D &= \{(-1,0)^\top, (1,0)^\top\}, \\ Y &= \{y \in \mathbb{R}^2 : \ (y_1 - 1)^2 + y_2^2 \leq 1\}, \\ f(x,y) &= \frac{1}{2} x^\top x - x^\top y \text{ and} \\ g(x,y) &= (x_1 + 1)(y_1^2 + y_2^2) + (x_1 - 1)y_2. \end{aligned}$$

Then $T_Y(y^0) = \{y \in \mathbb{R}^2 : y_1 \geq 0\}$ and for $r^0 = (0,1)^\top$ it holds $\varphi_o(y^0) = \varphi_o(y^0; r^0) = 0$ and $d\varphi_{o y^0}(r^0) = -2 < 0$. Thus, condition 1 is satisfied for φ_o and r^0 but $y^0 = (0,0)^\top$ is a global optimistic and pessimistic optimal solution.
△

Specifying the conditions of Theorem 6.2 by using Lemma 5.1 we obtain the following **necessary optimality conditions:**

Let $y^0 \in \text{locmin}\{\varphi_p(y) : y \in Y\}$. Then it holds

$$\varphi_p(y^0) = \varphi_p(y^0; r) \text{ and } d\varphi_{p y^0}(r) \geq 0 \quad \forall r \in \mathbb{T}_Y(y^0).$$

Let $y^0 \in \text{locmin}\{\varphi_o(y) : y \in Y\}$. Then for all $r \in \mathbb{T}_Y(y^0)$ it holds

$$\varphi_o(y^0) < \varphi_o(y^0; r) \text{ or } d\varphi_{o y^0}(r) \geq 0.$$

To prove the next theorem we will need the following lemma.

Lemma 6.1 *Assume it holds $\varphi_o(y^0) = g(x^0, y^0)$ for $y^0 \in \mathbb{R}^n$ and $x^0 \in \Psi_D(y^0)$. Then $r \in T_{x^0}(y^0)$ implies*

$$\varphi_o(y^0) = \varphi_o(y^0; r).$$

Proof. Since φ_o is radial-continuous there exists some $\tilde{x} \in \Psi_D(y^0)$ and some sequence $\{t_k\} \downarrow 0$ with $\tilde{x} \in \Psi_D(y^0 + t_k r)$, $\varphi_o(y^0 + t_k r) = g(\tilde{x}, y^0 + t_k r)$ and $\varphi_o(y^0; r) = \lim_{k \to \infty} \varphi_o(y^0 + t_k r) = \lim_{k \to \infty} g(\tilde{x}, y^0 + t_k r) = g(\tilde{x}, y^0)$. Clearly it holds $r \in T_{\tilde{x}}(y^0) \cap T_{x^0}(y^0)$. Then from the proof of Theorem 4.2 it follows that $r^\top(x^0 - \tilde{x}) = 0$. Further we know that $\tilde{x}, x^0 \in \Psi_D(y^0)$ and thus $F(x^0) - {x^0}^\top y^0 = F(\tilde{x}) - \tilde{x}^\top y^0$. Consequently, $F(x^0) - {x^0}^\top (y^0 + t_k r) = F(\tilde{x}) - \tilde{x}^\top (y^0 + t_k r)\ \forall k$, i.e. $x^0 \in \Psi_D(y^0 + t_k r)\ \forall k$. Thus we obtain $\varphi_o(y^0 + t_k r) \leq g(x^0, y^0 + t_k r)\ \forall k$, i.e.

$$\varphi_o(y^0; r) = \lim_{k \to \infty} \varphi_o(y^0 + t_k r) \leq \lim_{k \to \infty} g(x^0, y^0 + t_k r) = g(x^0, y^0) = \varphi_o(y^0).$$

Now from Lemma 5.1 it follows the equality. □

Theorem 6.3 *Assume that $y^0 \in Y$ is a point which satisfies one of the following two conditions:*

1. *$\varphi(y^0) < \varphi(y^0; r)\ \forall r \in T_Y(y^0)$*
2. *$\varphi(y^0) \leq \varphi(y^0; r)\ \forall r \in T_Y(y^0)$ and $d\varphi_{y^0}(r) > \gamma\ \forall r \in T_Y(y^0) : \varphi(y^0) = \varphi(y^0; r)$, $\|r\| = 1$ with $\gamma = 0$ in the optimistic case and $\gamma > 0$ in the pessimistic case.*

Then, φ achieves a local minimum at y^0.

Proof. Suppose $y^0 \in Y$ satisfies one of the two conditions of the theorem. Arguing by contradiction we assume that there is a sequence $\{y^k\}_{k\geq 1}$ with $y^k \to y^0, k \to \infty$ and $\varphi(y^k) < \varphi(y^0)\ \forall k$. Since $\Psi_D(y^0) \supseteq \Psi_D(y)$ for all y in a neighborhood of y^0 and $\Psi_D(y^0)$ has finite cardinality there exists some $x^0 \in \Psi_D(y^0)$ such that $Y_{x^0}(y^0)$ contains infinitely many of the points y^k, i.e. $\varphi(y^k) = g(x^0, y^k)$. In the following we consider the sequence $\{y^k\} \cap Y_{x^0}(y^0)$ and denote it by $\{y^k\}$ again. Because of the continuity of $g(x^0, \cdot)$ it follows

$$g(x^0, y^0) = \lim_{k\to\infty} g(x^0, y^k) = \lim_{k\to\infty} \varphi(y^k) \leq \varphi(y^0). \tag{6}$$

Let $r^k := \frac{y^k - y^0}{\|y^k - y^0\|}$, $k = 1, \ldots, \infty$. Then it holds $r^k \in \mathbb{T}_Y(y^0) \cap T_{x^0}(y^0)$. Further, let $\hat{r}$ an accumulation point of the sequence $\{r^k\}$. Clearly, $\hat{r} \in T_Y(y^0) \cap T_{x^0}(y^0)$.

i) Let φ denote the optimistic solution function. Then inequality (6) yields $g(x^0, y^0) = \varphi_o(y^0)$ since $x^0 \in \Psi_D(y^0)$. Now from Lemma 6.1 it follows $\varphi_o(y^0) = \varphi_o(y^0; r^k)\ \forall k$ and $\varphi_o(y^0) = \varphi_o(y^0; \hat{r})$. Thus, the first condition does not hold. Then the second condition must be satisfied. Since $\hat{r} \in T_{x^0}(y^0)$ it holds $y^0 + t\hat{r} \in R(x^0)$ for all $t \in [0, \varepsilon)$. Therefore, $\varphi_o(y^0 + t\hat{r}) \leq g(x^0, y^0 + t\hat{r})$ for all $t \in [0, \varepsilon)$. Hence,

$$\begin{aligned} 0 < d\varphi_{y^0}(\hat{r}) &= \lim_{t\downarrow 0} \frac{\varphi_o(y^0 + t\hat{r}) - \varphi_o(y^0; \hat{r})}{t} \\ &= \lim_{t\downarrow 0} \frac{\varphi_o(y^0 + t\hat{r}) - \varphi_o(y^0)}{t} \\ &\leq \lim_{t\downarrow 0} \frac{g(x^0, y^0 + t\hat{r}) - g(x^0, y^0)}{t} \\ &= \nabla_y g(x^0, y^0)\hat{r}. \end{aligned}$$

On the other hand it holds

$$\varphi_o(y^0) > \varphi_o(y^k) = g(x^0, y^0) + \|y^k - y^0\| \nabla_y g(x^0, y^0) r^k + o(\|y^k - y^0\|)$$

which together with $g(x^0, y^0) = \varphi_o(y^0)$ and $\lim\limits_{k\to\infty} \frac{o(\|y^k - y^0\|)}{\|y^k - y^0\|} = 0$ leads to

$$\nabla_y g(x^0, y^0)\hat{r} \leq 0.$$

But this is a contradiction, i.e. if y^0 is no local optimistic solution none of the two conditions holds.

ii) Let φ denote the pessimistic solution function. Then from Lemma 5.1 it follows $\varphi_p(y^0) \geq \varphi_p(y^0; r)$ for all $r \in T_Y(y^0)$, i.e. the first condition is not satisfied. Then the second condition must be satisfied, i.e. it holds $\varphi_p(y^0) = \varphi_p(y^0; r)$ and $d\varphi_{p_{y^0}}(r) > \gamma > 0$ for all $r \in T_Y(y^0)$.

Since φ_p is radial-continuous and radial differentiable for all k there exists some $x^k \in I_{r^k}(y^0)$. Because of $I_{r^k}(y^0) \subseteq \Psi_D(y^0)\ \forall k$ and the finite cardinality of $\Psi_D(y^0)$ we can assume w.l.o.g. that there exists some $\bar{x} \in \Psi_D(y^0)$ with $\bar{x} \in I_{r^k}(y^0)\ \forall k$. Thus, for all k it holds

$$\varphi_p(y^0; r^k) = \varphi_p(y^0) = g(\bar{x}, y^0) \quad \text{and} \quad 0 < \gamma < d\varphi_{p_{y^0}}(r^k) = \nabla_y g(\bar{x}, y^0) r^k.$$

Since $\hat{r}$ is an accumulation point of $\{r^k\}$ we obtain

$$0 < \gamma \leq \nabla_y g(\bar{x}, y^0)\hat{r}.$$

Further we have $r^k \in T_{x^0}(y^0) \cap T_{\bar{x}}(y^0)$. Thus, $y^0 + tr^k \in R(x^0) \cap R(\bar{x})\ \forall t \in [0, \varepsilon)$ which yields $\varphi_p(y^k) \geq g(\bar{x}, y^0)\ \forall k$. Consequently, for all k it holds

$$\varphi_p(y^0) > \varphi_p(y^k) = g(\bar{x}, y^k) = g(\bar{x}, y^0) + \|y^k - y^0\| \nabla_y g(x^0, y^0) r^k + o(\|y^k - y^0\|)$$

which together with $g(\bar{x}, y^0) = \varphi_p(y^0)$ and $\lim\limits_{k\to\infty} \frac{o(\|y^k - y^0\|)}{\|y^k - y^0\|} = 0$ leads to

$$\nabla_y g(\bar{x}, y^0)\hat{r} \leq 0.$$

But this is a contradiction, i.e. if y^0 is no local pessimistic solution none of the two conditions holds. □

Specifying the conditions of Theorem 6.3 by using Lemma 5.1 we obtain the following **sufficient optimality conditions:**

Let

$$\varphi_p(y^0) = \varphi_p(y^0; r) \text{ and } d\varphi_{p_{y^0}}(r) > \gamma > 0 \quad \forall r \in T_Y(y^0).$$

Then $y^0 \in \text{locmin}\{\varphi_p(y) : y \in Y\}$.

Let

$$\varphi_o(y^0) < \varphi_o(y^0; r) \text{ or } d\varphi_{o_{y^0}}(r) > 0 \quad \forall r \in T_Y(y^0).$$

Then $y^0 \in \text{locmin}\{\varphi_o(y) : y \in Y\}$.

Example 6.3 Consider the bilevel programming problem

$$\begin{cases} \min\{g(x,y) : y \in \mathbb{R}^2,\ x \in \Psi_D(y)\} \\ \Psi_D(y) = \underset{x}{\text{argmin}}\,\{\frac{1}{2}\|x\|^2 - y^\top x : x_1 \leq 0, x_2 \geq 0, -x_1 + x_2 \leq 1,\ x \in \mathbb{Z}^2\} \end{cases}$$

with $g(x, y) = x_2(y_2 - (y_1 + 0.5)^2 - 0.5) + (1 - x_2)(y_1 - y_2 + 1) + x_1(3y_1 + 1.5)$. We obtain

$$S_D = \left\{ x^1 = \begin{pmatrix} 0 \\ 1 \end{pmatrix}, x^2 = \begin{pmatrix} 0 \\ 0 \end{pmatrix}, x^3 = \begin{pmatrix} -1 \\ 0 \end{pmatrix} \right\} \text{ with}$$

$$\begin{aligned} R(x^1) &= \{y \in \mathbb{R}^2 : y_2 \geq 0.5,\ y_1 + y_2 \geq 0\}, \\ R(x^2) &= \{y \in \mathbb{R}^2 : y_2 \leq 0.5,\ y_1 \geq -0.5\} \text{ and} \\ R(x^3) &= \{y \in \mathbb{R}^2 : y_1 \leq -0.5,\ y_1 + y_2 \leq 0\}. \end{aligned}$$

Then we have

$$\varphi_p(y) = \begin{cases} y_2 - (y_1 + 0.5)^2 - 0.5 & \text{if } y_2 > 0.5,\ y_1 + y_2 > 0 \\ y_1 - y_2 + 1 & \text{if } y_2 \leq 0.5,\ y_1 \geq -0.5 \\ -2y_1 - y_2 - 0.5 & \text{if } y_1 + y_2 \leq 0,\ y_1 < -0.5\,. \end{cases}$$

Let $y^0 = (-1/2, 1/2)^\top$. Then it holds $\varphi_p(y^0) = \varphi_p(y^0; r) = 0\ \forall r \in \mathbb{R}^2$ and

$$0 < d\varphi_{p_{y^0}}(r) = \begin{cases} r_2 & \text{if } r_2 > 0,\ r_1 + r_2 > 0 \\ r_1 - r_2 & \text{if } r_2 \leq 0,\ r_1 \geq 0 \\ -2r_1 - r_2 & \text{if } r_1 < 0,\ r_1 + r_2 \leq 0\,. \end{cases}$$

However y^0 is no local minimum of φ_p since $y(t) = (t - 0.5, 0.5(1+t^2))^\top \to y^0$ for $t \downarrow 0$ but $\varphi_p(y(t)) = -\frac{1}{2}t^2 < \varphi_p(y^0)\ \forall t > 0$. This is no contradiction to Theorem 6.3 since there does not exist any $\gamma > 0$ with $\gamma < d\varphi_{p_{y^0}}(r)\ \forall r$. $\triangle$

7 Optimality criteria using radial subdifferential

Definition 7.1 *Let $U \subseteq \mathbb{R}^n$, $y^0 \in U$ and $\varphi : U \to \mathbb{R}$ be radial-directionally differentiable at y^0. We say that $d \in \mathbb{R}^n$ is a radial subgradient of φ at y^0 if*

$$\varphi(y^0) + \langle r, d \rangle \leq \varphi(y^0; r) + d\varphi_{y^0}(r)$$

is satisfied for all $r : \varphi(y^0) \geq \varphi(y^0; r)$.

The set of all subgradients is called subdifferential and is denoted by $\partial_{rad}\varphi(y^0)$.

The following necessary criterion for the existence of a radial subgradient is valid:

Theorem 7.1 ([6]) *If there exists some $r \in \mathbb{R}^n$ with $\varphi(y^0; r) < \varphi(y^0)$ then it holds $\partial_{rad}\varphi(y^0) = \emptyset$.*

With this theorem we get the following equivalent definition of the radial subgradient:

$$\partial_{rad}\varphi(y^0) = \{d \in \mathbb{R}^n : \langle r, d \rangle \leq d\varphi_{y^0}(r)\,\forall r \text{ satisfying } \varphi(y^0) = \varphi(y^0; r)\},$$

if there is no direction such that the radial limit in this direction is less than the function value.

Using Lemma 5.1 we obtain that for the pessimistic solution function either $\partial_{rad}\varphi_p(y^0) = \emptyset$ if there exists some r with $\varphi_p(y^0) > \varphi_p(y^0; r)$ or $\partial_{rad}\varphi_p(y^0) = \{d \in \mathbb{R}^n : \langle d, r \rangle \leq d\varphi_{p_{y^0}}(r)\ \forall r\}$.
For the optimistic solution function the condition of Theorem 7.1 is never valid.

Thus,

$$\partial_{rad}\varphi_o(y^0) = \{d \in \mathbb{R}^n : \langle r, d \rangle \leq d\varphi_{o_{y^0}}(r)\,\forall r \text{ satisfying } \varphi_o(y^0) = \varphi_o(y^0; r)\}$$

and

$$\partial_{rad}\varphi_P(y^0) = \{d \in \mathbb{R}^n : \langle r, d\rangle \le d\varphi_{P_{y^0}}(r)\, \forall r\}$$

if there is no r such that $\varphi_P(y^0; r) < \varphi_P(y^0)$.

Next we want to give further descriptions for the set $\partial_{rad}\varphi(y^0)$ by using equation (5). To do this we will need the following notations:

$$\mathcal{T}(y^0) := \{r \in \mathbb{R}^n : \varphi(y^0) = \varphi(y^0; r)\}$$
$$I(y^0) := \bigcup_{r\in\mathcal{T}(y^0)} I_r(y^0)$$

Then the following Lemma holds:

Lemma 7.1 *1.* $I(y^0) = \{x \in \Psi_D(y^0) : g(x, y^0) = \varphi(y^0)\}$
2. $\text{cl}\ \mathcal{T}(y^0) = \bigcup_{x\in I(y^0)} T_x(y^0)$
3. $\partial_{rad}\varphi(y^0) = \bigcap_{x\in I(y^0)}\{d \in \mathbb{R}^n : \langle d, r\rangle \le \nabla_y g(x, y^0)r\ \forall r \in T_x(y^0)\}$

Proof. 1. Let $x \in I(y^0)$. Then there exists some $r \in \mathcal{T}(y^0)$ with $x \in I_r(y^0)$. Because of the definitions of $I_r(y^0)$ and $\mathcal{T}(y^0)$ it holds $\varphi(y^0) = \varphi(y^0; r) = g(x, y^0)$ and $x \in \Psi_D(y^0)$, i.e. $x \in \{x \in \Psi_D(y^0) : g(x, y^0) = \varphi(y^0)\}$.
Let $x^0 \in \Psi_D(y^0)$ with $g(x^0, y^0) = \varphi(y^0)$. Let $r = \nabla F(x^0)^\top - y^0$. Since $\nabla F(x^0)^\top \in \text{int}\ R(x^0)$ and $y^0 \in R(x^0)$ it holds $\lambda\nabla F(x^0)^\top + (1-\lambda)y^0 \in \text{int}\ R(x^0)\ \forall\lambda \in (0, 1)$, i.e. $y^0 + \lambda r \in \text{int}\ R(x^0)\ \forall\lambda \in (0, 1)$. Consequently, $\varphi(y^0 + \lambda r) = g(x^0, y^0 + \lambda r)\ \forall\lambda \in (0, 1)$. But this means $x^0 \in I_r(y^0)$ and $\varphi(y^0; r) = g(x^0, y^0) = \varphi(y^0)$, i.e. $r \in \mathcal{T}(y^0)$ and thus $x^0 \in I(y^0)$.

2. Let $\hat{r} \in \text{cl}\ \mathcal{T}(y^0)$. Then there exists some sequence $\{r^k\}_{k=1}^\infty \subseteq \mathcal{T}(y^0)$ with $\lim_{k\to\infty} r^k = \hat{r}$. Since $I(y^0) \subseteq \Psi_D(y^0)$ and card $\Psi_D(y^0) < \infty$ there exists w.l.o.g. some $x \in I(y^0)$ with $x \in I_{r^k}(y^0)\ \forall k$, i.e. $r^k \in T_x(y^0)\ \forall k$. Then from $T_x(y^0)$ being closed it follows that $\hat{r} \in T_x(y^0)$, i.e. $\hat{r} \in \bigcup_{x\in I(y^0)} T_x(y^0)$.
Let $\hat{r} \in \bigcup_{x\in I(y^0)} T_x(y^0)$. Then there exists some $x \in I(y^0)$ with $\hat{r} \in T_x(y^0)$. Since $T_x(y^0) \subseteq \text{cl int}\ T_x(y^0)$ there exists some sequence $\{r^k\}_{k=1}^\infty \subseteq \text{int}\ T_x(y^0)$ with $\lim_{k\to\infty} r^k = \hat{r}$. Thus, for all k it holds $y^0 + tr^k \in \text{int}\ R(x)$ for all $t > 0$ being sufficiently small. This means $x \in I_{r^k}(y^0)$ and $r^k \in \mathcal{T}(y^0)$ for all k. Consequently, $\hat{r} \in \text{cl}\ \mathcal{T}(y^0)$.

3. Let $d \in \bigcap_{x\in I(y^0)}\{d \in \mathbb{R}^n : \langle d, r\rangle \le \nabla_y g(x, y^0)r\ \forall r \in T_x(y^0)\}$. Then for all $x \in I(y^0)$ it holds $\langle d, r\rangle \le \nabla_y g(x, y^0)r = d\varphi_{y^0}(r)\ \forall r \in T_x(y^0)$. Thus, $\langle d, r\rangle \le d\varphi_{y^0}(r)\ \forall r \in \bigcup_{x\in I(y^0)} T_x(y^0) \supseteq \mathcal{T}(y^0)$, i.e. $d \in \partial_{rad}\varphi(y^0)$.
Let $d \in \partial_{rad}\varphi(y^0)$. Now consider some arbitrary $x \in I(y^0)$ and some $r \in T_x(y^0)$. Then there exist some sequence $\{r^k\}_{k=1}^\infty \subseteq \text{int}\ T_x(y^0)$ and $\lim_{k\to\infty} r^k = r$. Since $\text{int}\, T_x(y^0) \subseteq \mathcal{T}(y^0)$ and $d \in \partial_{rad}\varphi(y^0)$ it follows $\langle d, r^k\rangle \le d\varphi_{y^0}(r^k) = \nabla_y g(x, y^0)r^k\ \forall k$ and thus $\langle d, r\rangle \le \nabla_y g(x, y^0)r$. Consequently, $d \in \bigcap_{x\in I(y^0)}\{d \in \mathbb{R}^n : \langle d, r\rangle \le \nabla_y g(x, y^0)r\ \forall r \in T_x(y^0)\}$.

□

Lemma 7.2 *For all points $y^0 \in \mathbb{R}^n$ and $\bar{x} \in \Psi_D(y^0)$ the set*

$$N_{\bar{x}}(y^0) := \text{cone}\,\{(x - \bar{x}) : \; x \in \Psi_D(y^0)\}$$

is the normal cone of the contingent cone $T_{\bar{x}}(y^0)$. Further it holds

$$\partial_{rad}\varphi(y^0) = \bigcap_{x \in I(y^0)} (N_x(y^0) + \nabla_y g(x, y^0)^\top).$$

Proof. We know from Lemma 4.1 that the contingent cone $T_{\bar{x}}(y^0)$ is equal to

$$T_{\bar{x}}(y^0) = \{r \in \mathbb{R}^n : \; (x - \bar{x})^\top r \le 0 \quad \forall x \in \Psi_D(y^0)\}.$$

Obviously it is the normal cone of the polyhedral cone $N_{\bar{x}}(y^0)$. Since $N_{\bar{x}}(y^0)$ is convex and closed the normal cone of $T_{\bar{x}}(y^0)$ is $N_{\bar{x}}(y^0)$ again.

Let $d \in \partial_{rad}\varphi(y^0)$. Because of Lemma 7.1 for all $x \in I(y^0)$ it holds

$$\langle d, r \rangle \le d\varphi_{y^0}(r) = \nabla_y g(x, y^0) r \qquad \forall r \in T_x(y^0).$$

Consequently, $d - \nabla_y g(x, y^0)^\top$ lies in the normal cone of $T_x(y^0)$ for all $x \in I(y^0)$, i.e. $d - \nabla_y g(x, y^0)^\top \in N_x(y^0)$ for all $x \in I(y^0)$. But this means

$$d \in N_x(y^0) + \nabla_y g(x, y^0)^\top \text{ for all } x \in I(y^0).$$

Thus, $\partial_{rad}\varphi(y^0) \subseteq \bigcap_{x \in I(y^0)}(N_x(y^0) + \nabla_y g(x, y^0)^\top)$. The reverse inclusion follows analogously. □

Example 7.1 Let $y^0 = (0, 0)^\top$ and

$$S_D = \left\{x^1 = \begin{pmatrix}1\\1\end{pmatrix}, x^2 = \begin{pmatrix}1\\-1\end{pmatrix}, x^3 = \begin{pmatrix}-1\\1\end{pmatrix}, x^4 = \begin{pmatrix}-1\\-1\end{pmatrix}\right\}$$

$$f(x, y) = \frac{1}{2}\|x\|^2 - x^\top y$$

$$g(x, y) = \left(\frac{3 + x_1}{2}\right) y_1 - 2y_2 + (x_1 - x_2)^2.$$

Then it holds

$$\begin{aligned}
R(x^1) &= \{y : \; y_1 \ge 0, y_2 \ge 0\},\; g(x^1, y) = 2y_1 - 2y_2 \\
R(x^2) &= \{y : \; y_1 \ge 0, y_2 \le 0\},\; g(x^2, y) = 2y_1 - 2y_2 + 4 \\
R(x^3) &= \{y : \; y_1 \le 0, y_2 \ge 0\},\; g(x^3, y) = y_1 - 2y_2 + 4 \\
R(x^4) &= \{y : \; y_1 \le 0, y_2 \le 0\},\; g(x^4, y) = y_1 - 2y_2.
\end{aligned}$$

Consequently, for the optimistic solution function it holds $\varphi_o(y^0) = 0$ and $I(y^0) = \{x^1, x^4\}$. Further, since $N_{x^1}(y^0) = R(x^4)$, $N_{x^4}(y^0) = R(x^1)$ and $\nabla_y g(x^1, y^0) = (2, -2)$, $\nabla_y g(x^4, y^0) = (1, -2)$ it holds

$$\begin{aligned}
\partial_{rad}\varphi(y^0) &= (N_{x^1}(y^0) + \nabla_y g(x^1, y^0)^\top) \cap (N_{x^4}(y^0) + \nabla_y g(x^4, y^0)^\top) \\
&= \{d \in \mathbb{R}^2 : \; d_1 \le 2, \; d_2 \le -2\} \cap \{d \in \mathbb{R}^2 : \; d_1 \ge 1, \; d_2 \ge -2\} \\
&= [1, 2] \times \{-2\}.
\end{aligned}$$

△

Now we derive optimality criteria in connection with the radial subdifferential.

Assume some point $y^0 \in \operatorname{locmin}\{\varphi(y) : y \in Y\}$ is given. Then we know from Theorem 6.2 that for all $r \in \mathbb{T}_Y(y^0)$, $\|r\| = 1$ it holds $\varphi(y^0; r) \geq \varphi(y^0)$ and $d\varphi_{y^0}(r) \geq 0$ if $\varphi(y^0; r) = \varphi(y^0)$. Consequently,

$$
\begin{aligned}
& 0 \leq \nabla_y g(x, y^0) r \quad \forall x \in I(y^0) \quad \forall r \in T_x(y^0) \cap \mathbb{T}_Y(y^0) \\
\text{and thus} \quad & 0 \leq \nabla_y g(x, y^0) r \quad \forall x \in I(y^0) \quad \forall r \in T_x(y^0) \cap T_Y(y^0).
\end{aligned}
$$

This means that $-\nabla_y g(x, y^0)^\top$ lies in the normal cone of $T_x(y^0) \cap T_Y(y^0)$ for all $x \in I(y^0)$. Let $\bar{I}(y^0) := \{x \in I(y^0) : \text{ relint } T_Y(y^0) \cap \text{relint } T_x(y^0) \neq \emptyset\}$. Since both cones are convex and closed the normal cone of $T_x(y^0) \cap T_Y(y^0)$ is equal to $N_Y(y^0) + N_x(y^0)$ for $x \in \bar{I}(y^0)$ where $N_Y(y^0)$ denotes the normal cone of $T_Y(y^0)$. Consequently,

$$
\begin{aligned}
-\nabla_y g(x, y^0)^\top & \in N_Y(y^0) + N_x(y^0) \quad \forall x \in \bar{I}(y^0) \\
0 & \in N_Y(y^0) + (N_x(y^0) + \nabla_y g(x, y^0)^\top) \quad \forall x \in \bar{I}(y^0) \\
0 & \in \bigcap_{x \in \bar{I}(y^0)} \left[N_Y(y^0) + (N_x(y^0) + \nabla_y g(x, y^0)^\top)\right].
\end{aligned}
$$

If it holds $y^0 \in \operatorname{int} Y$ we have $N_Y(y^0) = \{0\}$, $T_Y(y^0) = \mathbb{R}^n$ and $I(y^0) = \bar{I}(y^0)$. Thus, it holds the following theorem:

Theorem 7.2 *Let φ denote the optimistic or pessimistic solution function for the bilevel programming problem (1). If $y^0 \in \operatorname{locmin}\{\varphi(y) : y \in Y\}$ then*

$$
0 \in \bigcap_{x \in \bar{I}(y^0)} \left[N_Y(y^0) + (N_x(y^0) + \nabla_y g(x, y^0)^\top)\right].
$$

If additionally $y^0 \in \operatorname{int} Y$ then $0 \in \partial_{rad} \varphi(y^0)$.

Example 7.2 Let S_D denote the vertex set of a regular hexagon with radius 2, i.e. let S_D be equal to

$$
\left\{x^1 = \begin{pmatrix} 2 \\ 0 \end{pmatrix}, x^2 = \begin{pmatrix} 1 \\ \sqrt{3} \end{pmatrix}, x^3 = \begin{pmatrix} -1 \\ \sqrt{3} \end{pmatrix}, x^4 = \begin{pmatrix} -2 \\ 0 \end{pmatrix}, x^5 = \begin{pmatrix} -1 \\ -\sqrt{3} \end{pmatrix}, x^6 = \begin{pmatrix} 1 \\ \sqrt{3} \end{pmatrix}\right\}.
$$

Further let

$$
g(x, y) = \begin{cases} y_1 + 4y_2 & \text{if } x = x^1 \\ 2y_1 - y_2 & \text{if } x = x^2 \\ 1 & \text{else}. \end{cases}
$$

Consider the optimistic solution function $\varphi_o(y)$ and the set $Y = \{y \in \mathbb{R}^2 : y_2 \geq 0, y_2 \leq y_1, y_1 \leq 1\}$. Then $y^0 = (0, 0)^\top$ is an optimistic optimal solution. It holds $I(y^0) = \{x^1, x^2\} = \bar{I}(y^0)$. Further,

$$N_{x^1}(y^0) = \{y : y_2 \le -\sqrt{3}y_1, y_2 \ge \sqrt{3}y_1\},$$
$$N_{x^2}(y^0) = \{y : y_2 \le -\sqrt{3}y_1, y_2 \le 0\}$$
$$\text{and} \quad N_Y(y^0) = \{y : y_1 + y_2 \le 0, y_1 \le 0\}.$$

Then it holds

$$\begin{aligned} 0 &\in N_Y(y^0) + N_{x^1}(y^0) + \nabla_y g(x^1, y^0)^\top \\ &= \{y \in \mathbb{R}^2 : \ y_1 \le 1, \ y_2 \le 4 - (y_1 - 1)\sqrt{3}\}, \\ 0 &\in N_Y(y^0) + N_{x^2}(y^0) + \nabla_y g(x^2, y^0)^\top \\ &= \{y \in \mathbb{R}^2 : \ y_1 + y_2 \le 1, \ y_2 \le -1 - (y_1 - 2)\sqrt{3}\}. \end{aligned}$$

Thus, the conditions of Theorem 7.2 are satisfied. Further,

$$\partial_{rad}\varphi_o(y^0) = \{d \in \mathbb{R}^2 : \ d_2 \le -1, \ d_2 \ge 4 + (d_1 - 1)\sqrt{3}\}.$$

In optimization one has very often necessary optimality criteria of the form $0 \in \partial\varphi(y^0) + N_Y(y^0)$. Such kind of necessary optimality criterium is usually not fulfilled for our problem. For instance in this example it holds $0 \notin \partial_{rad}\varphi_o(y^0) + N_Y(y^0)$. △

Theorem 7.3 *Let φ denote the optimistic or pessimistic solution function for the bilevel programming problem (1). If $0 \in int\,(\partial_{rad}\varphi(y^0) + N_Y(y^0))$ then φ achieves at y^0 a local minimum.*

Proof. Clearly $\partial_{rad}\varphi(y^0) \ne \emptyset$. Thus it holds $\varphi(y^0) \le \varphi(y^0; r)\ \forall r \in \mathbb{R}^n, \|r\| = 1$ because of Theorem 7.1.

Let $0 \in int\,(\partial_{rad}\varphi(y^0) + N_Y(y^0))$. Then there exists some $\gamma > 0$ such that for all $r \in \mathbb{R}^n$, $\|r\| = 1$ it holds $\gamma r \in (\partial_{rad}\varphi(y^0) + N_Y(y^0))$. Now fix some $\hat{r} \in \mathcal{T}(y^0) \cap T_Y(y^0)$, $\|\hat{r}\| = 1$. Then there exists some $s \in N_Y(y^0)$ with $(\gamma\hat{r} - s) \in \partial_{rad}\varphi(y^0)$. Using the definition of $\partial_{rad}\varphi(y^0)$ we obtain

$$\gamma\langle \hat{r}, r\rangle - \langle s, r\rangle = \langle \gamma\hat{r} - s, r\rangle \le d\varphi_{y^0}(r) \quad \forall r \in \mathcal{T}(y^0).$$

Because of $\hat{r} \in T_Y(y^0)$ and $s \in N_Y(y^0)$ it holds $\langle \hat{r}, s\rangle \le 0$ and thus

$$0 < \gamma \le \gamma\|\hat{r}\|^2 - \langle s, \hat{r}\rangle \le d\varphi_{y^0}(\hat{r}).$$

Thus, since $\hat{r}$ was arbitrary the sufficient optimality criterium is satisfied (Theorem 6.3), i.e.

$$0 < \gamma \le d\varphi_{y^0}(r) \qquad \forall r \in \mathcal{T}(y^0) \cap T_Y(y^0).$$

Hence, $y^0 \in \text{locmin}\{\varphi(y) : \ y \in Y\}$. □

References

1. J. F. Bard, Practical Bilevel Optimization: Algorithms and Applications, Kluwer Academic Publishers, Dordrecht, 1998
2. S. Dempe, Foundations of Bilevel Programming, Kluver Academic Publishers, Dordrecht, 2002
3. S. Dempe, Annotated Bibliography on Bilevel Programming and Mathematical Programs with Equilibrium Constraints, Optimization, 2003, Vol. 52, pp. 333-359
4. S. Dempe, V. Kalashnikov and Roger Z. Ríos-Mercado, Discrete Bilevel Programming: Application to a Natural Gas Cash-Out Problem, European Journal of Operational Research, 2005, Vol. 166, pp. 469-488
5. S. Dempe and K. Richter, Bilevel programming with knapsack constraints, Central European Journal of Operations Research, 2000, Vol. 8, pp. 93-107
6. S. Dempe, T. Unger, Generalized PC^1-Functions, Optimization, 1999, Vol. 46, pp. 311-326
7. V. F. Dem'yanov, A. N. Rubinov, Quasidifferential Calculus, Optimization Software, Publications Division, New York, 1986
8. T. Edmunds and J.F. Bard, An algorithm for the mixed-integer nonlinear bilevel programming problem, Annals of Operations Research, 1992, Vol. 34, pp. 149–162
9. R.-H. Jan and M.-S. Chern, Nonlinear integer bilevel programming, European Journal of Operational Research, 1994, Vol. 72, pp. 574-587
10. J.-P. Hiriart-Urruty and C. Lemarechal, Convex Analysis and Minimization Algorithms, Vol. 1, Springer-Verlag, Berlin et. al., 1993
11. P. Loridan and J. Morgan, On strict ε-solutions for a two-level optimization problem, In: W. Buhler et at. (eds.), Proceedings of the International Conference on Operations Research 90, Springer Verlag, Berlin, 1992, pp. 165-172
12. P. Loridan and J. Morgan, Weak via strong Stackelberg problem: New results, Journal of Global Optimization, 1996, Vol. 8, pp. 263-287
13. P. Recht, Generalized Derivatives: An Approach to a New Gradient in Nonsmooth Optimization, volume 136 of Mathematical Systems in Economics, Anton Hain, Frankfurt am Main, 1993
14. G. Savard and J. Gauvin, The steepest descent direction for the nonlinear bilevel programming problem, Operations Research Letters, 1994, Vol. 15, pp. 265-272
15. L. N. Vicente and G. Savard and J. J. Judice, The discrete linear bilevel programming problem, Journal of Optimization Theory and Applications, 1996, Vol. 89, pp. 597-614
16. U. Wen and Y. Yang, Algorithms for solving the mixed integer two-level linear programming problem, Computers and Operations Research, 1990, Vol. 17, pp. 133–142

On approximate mixed Nash equilibria and average marginal functions for two-stage three-players games

Lina Mallozzi[1] and Jacqueline Morgan[2]

[1] Dipartimento di Matematica e Applicazioni, Università di Napoli "Federico II", Via Claudio, 21 - 80125 Napoli (Italy) mallozzi@unina.it
[2] Dipartimento di Matematica e Statistica, Università di Napoli "Federico II", Compl. M. S. Angelo, Via Cintia - 80126 Napoli (Italy) morgan@unina.it

Summary. In this paper we consider a two-stage three-players game: in the first stage one of the players chooses an optimal strategy knowing that, at the second stage, the other two players react by playing a noncooperative game which may admit more than one Nash equilibrium. We investigate continuity properties of the set-valued function defined by the Nash equilibria of the (second stage) two players game and of the marginal functions associated to the first stage optimization problem. By using suitable approximations of the mixed extension of the Nash equilibrium problem, we obtain without convexity assumption the lower semicontinuity of the set-valued function defined by the considered approximate Nash equilibria and the continuity of the associate approximate average marginal functions when the second stage corresponds to a particular class of noncooperative games called antipotential games.

Key Words: mixed strategy, Radon probability measure, ε-approximate Nash equilibrium, marginal functions, noncooperative games, two-stage three-players game, antipotential game.

1 Introduction

Let X, Y_1, Y_2 be compact subsets of metric spaces and f_1, f_2 be two real valued functions defined on $X \times Y_1 \times Y_2$. Consider the parametric noncooperative two players game $\Gamma(x) = \{Y_1, Y_2, f_1(x,\cdot,\cdot), f_2(x,\cdot,\cdot)\}$ where $x \in X$ and f_1, f_2 are the payoff functions of players P_1 and P_2. Any player is assumed to minimize his own payoff function called cost function. For all $x \in X$, we denote by $N(x)$ the set of the Nash equilibria ([19]) of the game $\Gamma(x)$, i.e. the set of the solutions to the following problem $\mathcal{N}(x)$

S. Dempe and V. Kalashnikov (eds.), *Optimization with Multivalued Mappings*, pp. 97-107
©2006 Springer Science + Business Media, LLC

$$\begin{cases} \text{find } (\bar{y}_1, \bar{y}_2) \in Y_1 \times Y_2 \text{ such that} \\ f_1(x, \bar{y}_1, \bar{y}_2) = \inf_{y_1 \in Y_1} f_1(x, y_1, \bar{y}_2) \\ f_2(x, \bar{y}_1, \bar{y}_2) = \inf_{y_2 \in Y_2} f_2(x, \bar{y}_1, y_2). \end{cases}$$

When the set $N(x)$ has more than one element for at least one $x \in X$, one can investigate some continuity properties of the set-valued function defined by the set $N(x)$ for all $x \in X$. These properties could be useful, from theoretical and numerical point of view, in problems involving the so-called *marginal functions.* More precisely, let l be a real valued function defined on $X \times Y_1 \times Y_2$. For any $x \in X$, one can consider the following functions associated to optimization problems in which the constraints describe the set of Nash equilibria of the game $\Gamma(x)$:

$$w(x) = \sup_{(y_1, y_2) \in N(x)} l(x, y_1, y_2)$$

$$u(x) = \inf_{(y_1, y_2) \in N(x)} l(x, y_1, y_2)$$

These marginal functions, called respectively *sup-marginal function* and *inf-marginal function*, are concerned in many applicative situations as illustrated in the following.

• The multi-stage problem involving the marginal function $w(x)$

$$\begin{cases} \text{find } \bar{x} \in X \text{ such that} \\ \\ \inf_{x \in X} \sup_{(y_1, y_2) \in N(x)} l(x, y_1, y_2) = \sup_{(y_1, y_2) \in N(\bar{x})} l(\bar{x}, y_1, y_2) = w(\bar{x}) \end{cases} \tag{1}$$

corresponds to a two-stage game with the three players P_0, P_1, P_2 and l cost function of P_0. In the first stage, player P_0 (called the leader) chooses an optimal strategy knowing that, at the second stage, two players P_1 and P_2 (called the followers) react by playing a non cooperative game. When there exists more than one Nash equilibrium at the second stage for at least one strategy of P_0, if it is assumed that the leader cannot influence the choice of the followers, then the followers can react to a leader's strategy by choosing a Nash equilibrium which can hurt him as much as possible. Therefore, P_0 will choose a security strategy which minimizes the worst, assuming that he has no motivation to restrict his worst case design to a particular subset of the Nash equilibria. The hierarchical problem (1) is called "Weak Hierarchical Nash Equilibrium Problem", in line with the terminology used in previous papers on hierarchical problems (see, for example, [5], [11]). Economic examples of such games can be found in [21], [22], [14] where the supply side of an oligopolistic market supplying a homogeneous product non cooperatively is modelled and in [20], where in a two country imperfect competition model the firms face three different types of decisions. In the setting of transportation and telecommunications see, for example, [15] and [1].

- The multi-stage problem involving the marginal function $u(x)$

$$\begin{cases} \text{find } \bar{x} \in X \text{ such that} \\ \\ \inf_{x \in X} \inf_{(y_1,y_2) \in N(x)} l(x, y_1, y_2) = \inf_{(y_1,y_2) \in N(\bar{x})} l(\bar{x}, y_1, y_2) = u(\bar{x}) \end{cases} \tag{2}$$

corresponds to a two-stage three-players game when there exists again more than one Nash equilibrium at the second stage for at least one strategy of P_0, and it is assumed now that the leader can force the choice of the followers to choose the Nash equilibrium that is the best for him. The hierarchical problem (2) is called "Strong Hierarchical Nash Equilibrium Problem" in line with the terminology used in previous papers on hierarchical problems ([5], [11]). It is also known as a mathematical programming problem with equilibrium constraints (MPEC) in line with the terminology used in [12], [6], [7], where one can find applications and references.

Remember that, if l is a continuous real valued function defined on $X \times Y_1 \times Y_2$ and N is a sequentially lower semicontinuous and sequentially closed graph set-valued function on X, then the marginal functions w and u are continuous on X ([9]). A set-valued function T is said to be sequentially lower semicontinuous at $x \in X$ if for any sequence $(x_n)_n$ converging to x in X and for any $y \in T(x)$, there exists a sequence (y_n) converging to y in Y such that $y_n \in T(x_n)$ for n sufficiently large (see, for example, [2], [9]). The set-valued function T is said to be sequentially closed graph at $x \in X$ if for any sequence $(x_n)_n$ converging to x in X and for any sequence $(y_n)_n$ converging to y in Y such that $y_{n_k} \in T(x_{n_k})$ for a selection of integers $(n_k)_k$, we have $y \in T(x)$ (see, for example, [2], [9]). For simplicity in the following the word "sequentially" will be omitted.

Unfortunately, the set-valued function N can be non lower semicontinuous even when smooth data are present (see, for example, [19], [17]). So, in [17] a suitable approximate Nash equilibrium concept has been introduced which guarantees lower semicontinuity results under some convexity assumption on the cost functions. When these convexity assumptions are not satisfied, as in the case of zero-sum games previously investigated by the authors ([13]), one can consider mixed strategies for P_1 and P_2 and the mixed extension of the parametric Nash equilibrium problem $\mathcal{N}(x)$. More precisely, let $\overline{M}(Y_1)$, $\overline{M}(Y_2)$ be the sets of Radon probability measures on Y_1 and Y_2 ([4], [23]) and assume that the cost functions of P_1 and P_2, respectively $f_1(x, \cdot, \cdot)$, $f_2(x, \cdot, \cdot)$, are continuous functions on $Y_1 \times Y_2$ for all $x \in X$. The *average cost functions* of players P_1 and P_2 are defined by (see, for example, [3]):

$$\hat{f}_i(x, \mu_1, \mu_2) = \int_{Y_1} \int_{Y_2} f_i(x, y_1, y_2) \, d\mu_1(y_1) \, d\mu_2(y_2)$$

for $i = 1, 2$. We denote by $\hat{N}(x)$ the set of Nash equilibria of the extended game defined by $\hat{\Gamma}(x) = \{\overline{M}(Y_1), \overline{M}(Y_2), \hat{f}_1(x, \cdot, \cdot) \hat{f}_2(x, \cdot, \cdot)\}$, i.e. the set of mixed Nash equilibria of the game $\Gamma(x)$. Assuming that $l(x,\cdot,\cdot)$ is a continuous function on $Y_1 \times Y_2$, for all $x \in X$, one can consider the average cost function for P_0 defined by

$$\hat{l}(x,\mu_1,\mu_2)= \int_{Y_1}\int_{Y_2} l(x,y_1,y_2)\; d\mu_1(y_1)\; d\mu_2(y_2).$$

The following real functions defined on X by

$$\hat{w}(x) = \sup_{(\mu_1,\mu_2)\in \hat{N}(x)} \hat{l}(x,\mu_1,\mu_2)$$

$$\hat{u}(x) = \inf_{(\mu_1,\mu_2)\in \hat{N}(x)} \hat{l}(x,\mu_1,\mu_2)$$

will be called respectively *sup-average marginal function* and *inf-average marginal function.* Having in mind to obtain the continuity of the average marginal functions, now we look for the lower semicontinuity of the set-valued function defined, for all $x \in X$, by the set $\hat{N}(x)$ of Nash equilibria of the game $\hat{\Gamma}(x)$ (i.e. mixed Nash equilibria of the game $\Gamma(x)$).

Unfortunately, the following example deals with a game where the set-valued function defined by $\hat{N}(x)$ is not lower semicontinuous on X.

Example 1.1 Let $X = [0,1]$ be the set of parameters and $Y_1 = \{\alpha_1,\beta_1\}$, $Y_2 = \{\alpha_2,\beta_2\}$ be the strategy sets of P_1, P_2 respectively. For any $x \in X$, we have the following bimatrix game:

	α_2	β_2
α_1	$-1,x$	$0,2x$
β_1	$0,-1$	$0,x$

Then:

$$N(x) = \begin{cases} \{(\alpha_1,\alpha_2),(\alpha_1,\beta_2)\} & \text{if } x = 0 \\ \{(\alpha_1,\alpha_2)\} & \text{if } x \neq 0. \end{cases}$$

Here $\overline{M}(Y_i)$ is the set of the discrete probability measures on Y_i $(i = 1,2)$. The extended cost functions are:

$$\hat{f}_1(x,\mu_1,\mu_2)= -\, pq$$

$$\hat{f}_2(x,\mu_1,\mu_2)=xp - xq + x - q + pq$$

where $\mu_1 = p\delta(\alpha_1)+(1-p)\delta(\beta_1) \in \overline{M}(Y_1)$, $\mu_2 = q\delta(\alpha_2)+(1-q)\delta(\beta_2) \in \overline{M}(Y_2)$ and $p,q \in [0,1]$. δ is the Dirac measure and μ_1 means that the strategy α_1 is chosen with probability p and the strategy β_1 is chosen with probability $1-p$, for $p \in [0,1]$.

In this case we have:

$$\hat{N}(x) = \begin{cases} \{(1,q),\ q \in [0,1]\} & \text{if } x = 0 \\ \{(1,1)\} & \text{if } x \neq 0 \end{cases}$$

which is not a lower semicontinuous set-valued function at $x = 0$.

However, by considering suitable approximations of the mixed extension of the Nash equilibrium problem, we will prove, without any convexity assumption, that the set-valued function defined by the considered approximate Nash equilibria is lower semicontinuous and that the corresponding approximate average marginal functions are continuous functions. More precisely, in Section 2 we introduce two concepts of approximate Nash equilibria for the extended game $\hat{\Gamma}(x)$ and we investigate the properties of lower semicontinuity and closedness of the set-valued functions defined by these approximate Nash equilibria. In Section 3 continuity of the associate approximate average marginal functions is obtained.

2 ε-approximate Nash equilibria

In line with the approximate solution concept introduced in [10] and in [17], we introduce a concept of approximate mixed Nash equilibrium:

Definition 2.1 *Let $x \in X$ and $\varepsilon > 0$; a strict ε-approximate mixed Nash equilibrium is a solution to the problem $\breve{\mathcal{N}}(x, \varepsilon)$:*

$$\begin{cases} \text{find } (\bar{\mu}_1, \bar{\mu}_2) \in \overline{M}(Y_1) \times \overline{M}(Y_2) \text{ s.t.} \\ \hat{f}_1(x, \bar{\mu}_1, \bar{\mu}_2) + \hat{f}_2(x, \bar{\mu}_1, \bar{\mu}_2) \\ \qquad < \inf\limits_{\mu_1 \in \overline{M}(Y_1)} \hat{f}_1(x, \mu_1, \bar{\mu}_2) + \inf\limits_{\mu_2 \in \overline{M}(Y_2)} \hat{f}_2(x, \bar{\mu}_1, \mu_2) + \varepsilon \end{cases}$$

The set of solutions to the problem $\breve{\mathcal{N}}(x, \varepsilon)$ will be denoted by $\breve{N}(x, \varepsilon)$.

Remark 2.1 *For all $x \in X$, the set of the strict ε-approximate mixed Nash equilibria $\breve{N}(x, \varepsilon)$ is not empty, differently from the set of the strict ε-approximate Nash equilibria $\tilde{N}(x, \varepsilon)$ ([18]) defined by*

$$\tilde{N}(x, \varepsilon) = \{(\bar{y}_1, \bar{y}_2) \in Y_1 \times Y_2 : \ f_1(x, \bar{y}_1, \bar{y}_2) + f_2(x, \bar{y}_1, \bar{y}_2) <$$

$$\inf_{y_1 \in Y_1} f_1(x, y_1, \bar{y}_2) + \inf_{y_2 \in Y_2} f_2(x, \bar{y}_1, y_2) + \varepsilon\}$$

which can be empty. In fact, in the matching pennies example $\tilde{N}(x, \varepsilon) = \emptyset$ but $\breve{N}(x, \varepsilon)$ is an open nonempty square. More precisely, let X be the set of parameters and $Y_1 = \{\alpha_1, \beta_1\}$, $Y_2 = \{\alpha_2, \beta_2\}$ be the strategy sets of P_1, P_2 respectively. For any $x \in X$, we have the following bimatrix game:

	α_2	β_2
α_1	1,−1	−1,1
β_1	−1,1	1,−1

Then for any $x \in X$ *we have that* $N(x) = \emptyset$ *and* $\tilde{N}(x, \varepsilon) = \emptyset$ *for any* $\varepsilon > 0$. *If mixed strategies are considered* $\overline{M}(Y_i)$ *(i=1,2),* $\hat{f}_1(x, \mu_1, \mu_2) = 4pq - 2p - 2q + 1$, $\hat{f}_2 = -\hat{f}_1$ *for* $p, q \in [0,1]$ *and* $\hat{N}(x) = \{(1/2, 1/2)\}$, $\breve{N}(x, \varepsilon) = \{(p,q) \in [0,1]^2 \ : \ p \in]1/2 - \varepsilon/2, 1/2 + \varepsilon/2[, \ q \in]|p - 1/2| + (1-\varepsilon)/2, -|p - 1/2| + (1+\varepsilon)/2[\}$.

Obviously, the set-valued function defined by the set of the strict ε-approximate mixed Nash equilibrium of a game is not always closed graph on X. The following theorem gives sufficient conditions for its lower semicontinuity on X and will be used later on.

Theorem 2.1 *Assume that* f_1, f_2 *are continuous functions on* $X \times Y_1 \times Y_2$. *Then, for all* $\varepsilon > 0$, *the set-valued function* $\breve{N}(\cdot, \varepsilon)$ *is lower semicontinuous on* X.

Proof. We have to prove that for all $x \in X$, for all (x_n) converging to x and for all $(\mu_1, \mu_2) \in \breve{N}(x, \varepsilon)$, there exists a sequence $(\mu_{1,n}, \mu_{2,n})$ converging to (μ_1, μ_2) s.t. $(\mu_{1,n}, \mu_{2,n}) \in \breve{N}(x_n, \varepsilon)$ for n sufficiently large.

Let (x_n) be a sequence converging to x and $(\bar{\mu}_1, \bar{\mu}_2) \in \breve{N}(x, \varepsilon)$. Then

$$\hat{f}_1(x, \bar{\mu}_1, \bar{\mu}_2) + \hat{f}_2(x, \bar{\mu}_1, \bar{\mu}_2) < \inf_{\mu_1 \in \overline{M}(Y_1)} \hat{f}_1(x, \mu_1, \bar{\mu}_2) + \inf_{\mu_2 \in \overline{M}(Y_2)} \hat{f}_2(x, \bar{\mu}_1, \mu_2) + \varepsilon. \tag{3}$$

Since $\hat{f}_1, \hat{f}_2$ are continuous, for all sequences $(\bar{\mu}_{1,n})$ converging to $\bar{\mu}_1$ and $(\bar{\mu}_{2,n})$ converging to $\bar{\mu}_2$ we have that

$$\lim_{n \to +\infty} (\hat{f}_1(x_n, \bar{\mu}_{1,n}, \bar{\mu}_{2,n}) + \hat{f}_2(x_n, \bar{\mu}_{1,n}, \bar{\mu}_{2,n})) = \hat{f}_1(x, \bar{\mu}_1, \bar{\mu}_2) + \hat{f}_2(x, \bar{\mu}_1, \bar{\mu}_2). \tag{4}$$

Since $\overline{M}(Y_1)$, $\overline{M}(Y_2)$ are compact, $\inf_{\mu_1 \in \overline{M}(Y_1)} \hat{f}_1(\cdot, \mu_1, \cdot)$ and $\inf_{\mu_2 \in \overline{M}(Y_2)} \hat{f}_2(\cdot, \cdot, \mu_2)$ are lower semicontinuous functions (Proposition 4.1.1 in [9]). Therefore, in light of (3) and (4)

$$\lim_{n \to +\infty} (\hat{f}_1(x_n, \bar{\mu}_{1,n}, \bar{\mu}_{2,n}) + \hat{f}_2(x_n, \bar{\mu}_{1,n}, \bar{\mu}_{2,n})) = \hat{f}_1(x, \bar{\mu}_1, \bar{\mu}_2) + \hat{f}_2(x, \bar{\mu}_1, \bar{\mu}_2) <$$

$$\inf_{\mu_1 \in \overline{M}(Y_1)} \hat{f}_1(x, \mu_1, \bar{\mu}_2) + \inf_{\mu_2 \in \overline{M}(Y_2)} \hat{f}_2(x, \bar{\mu}_1, \mu_2) + \varepsilon \leq$$

$$\lim_{n\to+\infty}\Big(\inf_{\mu_1\in\overline{M}(Y_1)}\hat{f}_1(x_n,\mu_1,\bar{\mu}_{2,n})+\inf_{\mu_2\in\overline{M}(Y_2)}\hat{f}_2(x_n,\bar{\mu}_{1,n},\mu_2)\Big)+\varepsilon.$$

For n sufficiently large, we can deduce:

$$\begin{aligned}&\hat{f}_1(x_n,\bar{\mu}_{1,n},\bar{\mu}_{2,n})+\hat{f}_2(x_n,\bar{\mu}_{1,n},\bar{\mu}_{2,n})\\ <&\inf_{\mu_1\in\overline{M}(Y_1)}\hat{f}_1(x_n,\mu_1,\bar{\mu}_{2,n})+\inf_{\mu_2\in\overline{M}(Y_2)}\hat{f}_2(x_n,\bar{\mu}_{1,n},\mu_2)+\varepsilon\end{aligned}$$

that is $(\bar{\mu}_{1,n},\bar{\mu}_{2,n})\in\breve{N}(x_n,\varepsilon)$.

Remark 2.2 *Let us note that Theorem 2.1 can be applied also in the case where $N(x)=\emptyset$ for some $x\in X$. In fact $\emptyset\neq\hat{N}(x)\subseteq\breve{N}(x,\varepsilon)$ for all $x\in X$ and $\varepsilon>0$.*

Having in mind to obtain closedness and lower semicontinuity simultaneously, we introduce now a suitable concept of approximate mixed Nash equilibrium.

Definition 2.2 *Let $x\in X$ and $\varepsilon>0$; an ε-approximate mixed Nash equilibrium is a solution to the problem $\hat{\mathcal{N}}(x,\varepsilon)$:*

$$\begin{cases}\text{find } (\bar{\mu}_1,\bar{\mu}_2)\in\overline{M}(Y_1)\times\overline{M}(Y_2)\\ \text{s.t.}\hat{f}_1(x,\bar{\mu}_1,\bar{\mu}_2)+\hat{f}_2(x,\bar{\mu}_1,\bar{\mu}_2)\\ \qquad\qquad\le\inf_{\mu_1\in\overline{M}(Y_1)}\hat{f}_1(x,\mu_1,\bar{\mu}_2)+\inf_{\mu_2\in\overline{M}(Y_2)}\hat{f}_2(x,\bar{\mu}_1,\mu_2)+\varepsilon\end{cases}$$

The set of solutions to the problem $\hat{\mathcal{N}}(x,\varepsilon)$ will be denoted by $\hat{N}(x,\varepsilon)$.

Remark 2.3 *It is easy to see that if f_1,f_2 are continuous functions on $X\times Y_1\times Y_2$, then the set-valued function $\hat{N}(\cdot,\varepsilon)$ is closed graph at x, for all $x\in X$.*

Example 2.1 In Example 1.1 we have that $\inf_{\mu_1\in\overline{M}(Y_1)}\hat{f}_1(x,\mu_1,\mu_2)=-q$ and that $\inf_{\mu_2\in\overline{M}(Y_2)}\hat{f}_2(x,\mu_1,\mu_2)=xp-1+p$. The set of the ε-approximate mixed Nash equilibria is:
for $x\le\varepsilon$

$$\hat{N}(x,\varepsilon)=\{(p,q)\in[0,1]^2 \ s.\ t.\ p\in[1-\varepsilon+x-xq,1],q\in[0,1]\},$$

for $x>\varepsilon$

$$\hat{N}(x,\varepsilon)=\{(p,q)\in[0,1]^2 \ s.\ t.\ p\in[1-\varepsilon+x-xq,1],q\in[1-(\varepsilon/x),1]\}.$$

Note that the set-valued function $x\in X\mapsto\hat{N}(x,\varepsilon)$ is closed graph and lower semicontinuous on X.

The bimatrix game in Example 1.1 has a special structure connected with the definition of exact potential games ([16]). Recall that the two players game $\{A, B, K, L\}$, where K, L are real valued functions defined on $A \times B$, is called an *exact potential game* if there is a potential function $P: A \times B \mapsto \mathcal{R}$ such that
$K(a_2, b) - K(a_1, b) = P(a_2, b) - P(a_1, b)$, for all $a_1, a_2 \in A$ and for each $b \in B$
$L(a, b_1) - L(a, b_2) = P(a, b_1) - P(a, b_2)$, for each $a \in A$ and for all $b_1, b_2 \in B$.
In exact potential games, information concerning Nash equilibria are incorporated into a real-valued function that is the potential function.

The following theorem gives a lower semicontinuity result for the set-valued function defined by the set of the ε-approximate mixed Nash equilibria.

Theorem 2.2 *Assume that f_1, f_2 are continuous functions on $X \times Y_1 \times Y_2$ and that the game $\Omega(x) = \{Y_1, Y_2, f_1(x, \cdot, \cdot), -f_2(x, \cdot, \cdot)\}$ is an exact potential game for all $x \in X$ ($\Gamma(x)$ will be said to be an antipotential game for all $x \in X$). Then, for all $\varepsilon > 0$, the set-valued function $\hat{N}(\cdot, \varepsilon)$ is lower semicontinuous on X.*

Proof. Since $\Omega(x)$ is an exact potential game, according to [8], there exists a potential function P defined on $X \times Y_1 \times Y_2$ such that

$$f_1(x, y_1, y_2) = P(x, y_1, y_2) + h(x, y_2)$$

$$-f_2(x, y_1, y_2) = P(x, y_1, y_2) + k(x, y_1)$$

where h, k are real valued functions defined and continuous on $X \times Y_2, X \times Y_1$ respectively. By considering the mixed extensions of Y_1, Y_2, the function $\hat{f}_1 + \hat{f}_2 = \hat{h} - \hat{k}$ is convex on $\overline{M}(Y_1) \times \overline{M}(Y_2)$ and one can apply Corollary 3.1 in [18] to get the lower semicontinuity of $\hat{N}(\cdot, \varepsilon)$ on X. For the sake of completeness we give the proof.

Let $(\overline{\mu}_1, \overline{\mu}_2) \in \hat{N}(x, \varepsilon)$ such that $(\overline{\mu}_1, \overline{\mu}_2) \notin \breve{N}(x, \varepsilon)$. Since $\breve{N}(x, \varepsilon) \neq \emptyset$, there exists $(\breve{\mu}_1, \breve{\mu}_2) \in \breve{N}(x, \varepsilon)$ and consider the sequence $\overline{\mu}_{i,n} = (1/n)\breve{\mu}_i + (1 - 1/n)\overline{\mu}_i$ $(i = 1, 2)$ for $n \in \mathcal{N}$. We have that $\overline{\mu}_{i,n} \mapsto \overline{\mu}_i$, $i = 1, 2$ and

$$\hat{f}_1(x, \overline{\mu}_{1,n}, \overline{\mu}_{2,n}) + \hat{f}_2(x, \overline{\mu}_{1,n}, \overline{\mu}_{2,n}) < (1/n)\big[\hat{v}_1(x, \breve{\mu}_2) + \hat{v}_2(x, \breve{\mu}_1) + \varepsilon\big] +$$

$$(1 - 1/n)\big[\hat{v}_1(x, \overline{\mu}_2) + \hat{v}_2(x, \overline{\mu}_1) + \varepsilon\big] \leq \hat{v}_1(x, \overline{\mu}_{2,n}) + \hat{v}_2(x, \overline{\mu}_{1,n}) + \varepsilon$$

being $\hat{v}_1(x, \mu_2) = \inf_{\mu_1 \in \overline{M}(Y_1)} \hat{f}_1(x, \mu_1, \mu_2)$ and $\hat{v}_2(x, \mu_1) = \inf_{\mu_2 \in \overline{M}(Y_2)} \hat{f}_2(x, \mu_1, \mu_2)$.
This means that $(\overline{\mu}_{1,n}, \overline{\mu}_{2,n}) \in \breve{N}(x, \varepsilon)$ and then $\hat{N}(x, \varepsilon) \subseteq cl\breve{N}(x, \varepsilon)$, where $cl\breve{N}(x, \varepsilon)$ is the sequential closure of $\breve{N}(x, \varepsilon)$. By Theorem 2.1 for all sequences $(x_n)_n$ converging to x we have $\breve{N}(x, \varepsilon) \subseteq \operatorname{Liminf}_n \breve{N}(x_n, \varepsilon)$. Therefore

$$\hat{N}(x, \varepsilon) \subseteq cl\breve{N}(x, \varepsilon) \subseteq cl \operatorname{Liminf}_n \breve{N}(x_n, \varepsilon) = \operatorname{Liminf}_n \breve{N}(x_n, \varepsilon) \subseteq \operatorname{Liminf}_n \hat{N}(x_n, \varepsilon)$$

Remark that, since $\overline{M}(Y_1)$ and $\overline{M}(Y_2)$ are first countable topological spaces, $\operatorname{Liminf}_n \breve{N}(x_n, \varepsilon)$ is a closed subset in $\overline{M}(Y_1) \times \overline{M}(Y_2)$.

Example 2.2 Note that in Example 1.1,

$$\Omega(x) = \{Y_1, Y_2, f_1(x, \cdot, \cdot), -f_2(x, \cdot, \cdot)\}$$

is an exact potential game with potential

	α_2	β_2
α_1	$-x$	$-2x$
β_1	$1-x$	$-2x$

Remark 2.4 *Theorem 2.2 extends Theorem 3.1 in [13] where existence of approximate mixed strategies for zero-sum games is obtained without convexity assumptions.*

3 Continuity properties of the approximate average marginal functions

By using the concepts of approximate mixed Nash equilibria given in Section 2, we give the continuity results for the following approximate average marginal functions.

Definition 3.1 *Let $x \in X$ and $\varepsilon > 0$; the following real functions defined on X:*

$$\hat{w}(x, \varepsilon) = \sup_{(\mu_1, \mu_2) \in \hat{N}(x, \varepsilon)} \hat{l}(x, \mu_1, \mu_2)$$

$$\hat{u}(x, \varepsilon) = \inf_{(\mu_1, \mu_2) \in \hat{N}(x, \varepsilon)} \hat{l}(x, \mu_1, \mu_2)$$

will be called ε-approximate sup-average marginal function and ε-approximate inf-average marginal function respectively.

So, we have the following theorem.

Theorem 3.1 *Assume that l, f_1, f_2 are continuous functions on $X \times Y_1 \times Y_2$ and that $\Gamma(x)$ is an antipotential game for all $x \in X$. Then, for all $\varepsilon > 0$, the ε-approximate average marginal functions $\hat{w}(\cdot, \varepsilon)$ and $\hat{u}(\cdot, \varepsilon)$ are continuous on X.*

Proof. In light of the assumptions $\hat{l}$ is continuous on $X \times \overline{M}(Y_1) \times \overline{M}(Y_2)$, the set-valued function $\hat{N}(\cdot, \varepsilon)$ is lower semicontinuous and closed graph on X. We obtain the proof by using the results given in [9] on the inf-marginal function in a sequential setting.

Example 3.1 In Example 1.1, let l be defined as follows:

	α_2	β_2
α_1	$x-1$	x
β_1	0	x

In this case the inf-marginal function $u(x) = x - 1$ is continuous on $[0, 1]$, while the sup-marginal function

$$w(x) = \begin{cases} 0 & \text{if } x = 0 \\ x-1 & \text{if } x \neq 0 \end{cases}$$

is not lower semicontinuous at $x = 0$. Even if we use mixed Nash equilibria of the game $\Gamma(x)$, the sup-average marginal function may be not continuous. In fact $\hat{l}(x, \mu_1, \mu_2) = (x - 1)pq + x(1 - q)$ and

$$\hat{w}(x) = \sup_{(\mu_1,\mu_2)\in\hat{N}(x)} \hat{l}(x, \mu_1, \mu_2) = w(x) = \begin{cases} 0 & \text{if } x = 0 \\ x-1 & \text{if } x \neq 0 \end{cases}$$

so $\hat{w}$ is not continuous at $x = 0$.

However, by considering for $\varepsilon > 0$ the set of the ε-approximate mixed Nash equilibria, the ε-approximate inf-average marginal function

$$\hat{w}(x, \varepsilon) = \begin{cases} x & \text{if } 0 \leq x \leq \varepsilon \\ x - 1 + \varepsilon/x & \text{if } x > \varepsilon \end{cases}$$

is continuous on $[0, 1]$.

References

1. Altman, E., Boulogne, T., El-Azouzi, R. and Jimenez, T. (2004), A survey on networking games in telecommunications, Computers and Operation Research.
2. Aubin, J. P. and Frankowska, H. (1990) *Set-valued Analysis*, Birkhauser, Boston.
3. Basar, T. and Olsder, G.J. (1995) *Dynamic Noncooperative Game Theory*, Academic Press. New York.
4. Borel, E. (1953) The theory of play and integral equations with skew symmetric kernels, *Econometrica* vol. 21, pp. 97-100.
5. Breton, M., Alj, A. and Haurie, A. (1988), Sequential Stackelberg equilibria in two-person games, *Journal of Optimization Theory and Applications* vol. 59, pp. 71-97.
6. Dempe, S. (2002) *Foundations of Bilevel Programming*, Kluwer Academic Publishers, Dordrecht.
7. Dempe, S. (2003) Annotated bibliography on bilevel programming and mathematical programs with equilibrium constraints, *Optimization* vol. 52, pp. 333-359.

8. Facchini, G., Van Megen, F., Borm, P. and Tijs, S. (1997), Congestion models and weighted Bayesian potential games, *Theory and Decision* vol. 42, pp. 193-206.
9. Lignola, M.B. and Morgan, J. (1992) Semi-continuities of marginal functions in a sequential setting, *Optimization* vol. 24, pp. 241-252.
10. Loridan, P. and Morgan, J. (1989) On strict ε-solutions for a two-level optimization problem, *Proceedings of the international Conference on Operation Research 90 in Vienna*, Ed. By W. Buhler, G. Feichtinger, F. Harti, F.J. Radermacher, P. Stanley, Springer Verlag, Berlin, pp. 165-172.
11. Loridan, P. and Morgan, J. (1996) Weak via strong Stackelberg problems: new results, *Journal of Global Optimization* vol. 8, pp. 263-287.
12. Luo, Z.-Q., Pang, J.-S., Ralph, D. (1996) *Mathematical programs with equilibrium constraints*, Cambridge University Press, Cambridge.
13. Mallozzi, L. and Morgan, J. (2001) Mixed strategies for hierarchical zero-sum games. In: Advances in dynamic games and applications (Maastricht, 1998), *Annals of the International Society on Dynamic Games*, Birkhauser Boston MA vol. 6, pp. 65-77.
14. Mallozzi, L. and Morgan, J. (2005) On equilibria for oligopolistic markets with leadership and demand curve having possible gaps and discontinuities, *Journal of Optimization Theory and Applications* vol. 125, n.2, pp. 393-407.
15. Marcotte, P. and Blain, M. (1991) A Stackelberg-Nash model for the design of deregulated transit system, *Dynamic Games in Economic Analysis*, Ed. by R.H. Hamalainen and H.K. Ethamo, Lecture Notes in Control and Information Sciences, Springer Verlag, Berlin, vol. 157.
16. Monderer, D. and Shapley, L.S. (1996) Potential games, *Games and Economic Behavior* vol. 14, pp. 124-143.
17. Morgan, J. and Raucci, R. (1999) New convergence results for Nash equilibria, *Journal of Convex Analysis* vol. 6, n. 2, pp. 377-385.
18. Morgan, J. and Raucci, R. (2002) Lower semicontinuity for approximate social Nash equilibria, *International Journal of Game Theory* vol. 31, pp. 499-509.
19. Nash, J. (1951) Non-cooperative games, *Annals of Mathematics* vol. 54, pp. 286-295.
20. Petit, M.L. and Sanna-Randaccio, F. (2000) Endogenous R&D and foreign direct investment in international oligopolies, *International Journal of Industrial Organization* vol. 18, pp. 339-367.
21. Sheraly, H.D., Soyster, A.L. and Murphy, F.H. (1983) Stackelberg-Nash-Cournot Equilibria: characterizations and computations, *Operation Research* vol. 31, pp. 253-276.
22. Tobin, R.L. (1992) Uniqueness results and algorithm for Stackelberg-Cournot-Nash equilibria, *Annals of Operation Research* vol. 34, pp. 21-36.
23. von Neumann, J. and Morgenstern, O. (1944) *Theory of Games and Economic Behavior*, New York Wiley.

Part II

Mathematical Programs with Equilibrium Constraints

A direct proof for M-stationarity under MPEC-GCQ for mathematical programs with equilibrium constraints

Michael L. Flegel and Christian Kanzow

University of Würzburg, Institute of Applied Mathematics and Statistics, Am Hubland, 97074 Würzburg, Germany
{flegel,kanzow}@mathematik.uni-wuerzburg.de

Summary. Mathematical programs with equilibrium constraints are optimization problems which violate most of the standard constraint qualifications. Hence the usual Karush-Kuhn-Tucker conditions cannot be viewed as first order optimality conditions unless relatively strong assumptions are satisfied. This observation has lead to a number of weaker first order conditions, with M-stationarity being the strongest among these weaker conditions. Here we show that M-stationarity is a first order optimality condition under a very weak Guignard-type constraint qualification. We present a short and direct approach.

Key Words. Mathematical programs with equilibrium constraints, M-stationarity, Guignard constraint qualification.

1 Introduction

We consider the following program, known across the literature as a *mathematical program with complementarity*—or often also *equilibrium—constraints*, *MPEC* for short:

$$
\begin{aligned}
&\min f(z) \\
&\text{s.t. } g(z) \le 0, \;\; h(z) = 0, \\
&\qquad G(z) \ge 0, \, H(z) \ge 0, \, G(z)^T H(z) = 0,
\end{aligned}
\tag{1}
$$

where $f : \mathbb{R}^n \to \mathbb{R}$, $g : \mathbb{R}^n \to \mathbb{R}^m$, $h : \mathbb{R}^n \to \mathbb{R}^p$, $G : \mathbb{R}^n \to \mathbb{R}^l$, and $H : \mathbb{R}^n \to \mathbb{R}^l$ are continuously differentiable.

It is easily verified that the standard Mangasarian-Fromovitz constraint qualification is violated at every feasible point of the program (1), see, e.g., [2]. The weaker Abadie constraint qualification can be shown to only hold in restrictive circumstances, see [16, 3]. A still weaker CQ, the Guignard CQ, has

S. Dempe and V. Kalashnikov (eds.), *Optimization with Multivalued Mappings*, pp. 111-122
©2006 Springer Science + Business Media, LLC

a chance of holding, see [3]. Any of the classic CQs imply that a Karush-Kuhn-Tucker point (called a strongly stationary point by the MPEC community, see the discussion in [3]) is a necessary first order condition.

However, because only the weakest constraint qualifications have a chance of holding, new constraint qualifications tailored to MPECs, and with it new stationarity concepts, have arisen, see, e.g., [11, 19, 16, 14, 15, 6, 22].

One of the stronger stationarity concepts introduced is M-stationarity [14] (see (8)). It is second only to strong stationarity. Weaker stationarity concepts like A- and C-stationarity have also been introduced [4, 19], but it is commonly held that these are too weak since such points allow for trivial descent directions to exist.

M-stationary points also play an important role for some classes of algorithms for the solution of MPECs. For example, Scholtes [20] has introduced an algorithm which, under certain assumptions to the MPEC (1), converges to an M-stationary point, but not in general to a strongly stationary point. Later, Hu and Ralph [9] proved a generalization of this result by showing that a limit point of a whole class of algorithms is an M-stationary point of the MPEC (1).

Hence it is of some importance to know when an M-stationary point is in fact a first order condition. This paper is dedicated to answering that question. We will show M-stationarity to be a necessary first order condition under MPEC-GCQ, an MPEC variant of the classic Guignard CQ. This result has recently been established in [7], using a very general approach involving disjunctive optimization problems. The aim of this paper is to present a very direct and short proof, focussing on the MPEC (1).

The organization of this paper is as follows: In Section 2 we introduce some concepts and results necessary for proving our main result. This is done in Section 3, referring to Section 2 and introducing additional concepts as needed.

A word on notation. Given two vectors x and y, we use $(x, y) := (x^T, y^T)^T$ for ease of notation. Comparisons such as $\leq$ and $\geq$ are understood componentwise. Given a vector $a \in \mathbb{R}^n$, a_i denotes the i-th component of that vector. Given a set $\nu \subseteq \{1, \dots, n\}$ we denote by $x_\nu \in \mathbb{R}^{|\nu|}$ that vector which consists of those components of $x \in \mathbb{R}^n$ which correspond to the indices in ν. Furthermore, we denote the set of all partitions of ν by $\mathcal{P}(\nu) := \{(\nu_1, \nu_2) \mid \nu_1 \cup \nu_2 = \nu,\ \nu_1 \cap \nu_2 = \emptyset\}$. By $\mathbb{R}^l_+ := \{x \in \mathbb{R}^l \mid x \geq 0\}$ we mean the nonnegative orthant of $\mathbb{R}^l$. Finally, the graph of a multifunction (set-valued function) $\Phi : \mathbb{R}^m \rightrightarrows \mathbb{R}^n$ is defined as $\operatorname{gph} \Phi := \{(v, w) \in \mathbb{R}^{m+n} \mid w \in \Phi(v)\}$.

2 Preliminaries

We will now introduce some notation and concepts in the context of MPECs which we will need for the remainder of this paper.

From the complementarity term in (1) it is clear that for a feasible point z^*, either $G_i(z^*)$, or $H_i(z^*)$, or both must be zero. To differentiate between these cases, we divide the indices of G and H into three sets:

$$\alpha := \alpha(z^*) := \{i \mid G_i(z^*) = 0,\ H_i(z^*) > 0\}, \tag{2a}$$

$$\beta := \beta(z^*) := \{i \mid G_i(z^*) = 0,\ H_i(z^*) = 0\}, \tag{2b}$$

$$\gamma := \gamma(z^*) := \{i \mid G_i(z^*) > 0,\ H_i(z^*) = 0\}. \tag{2c}$$

The set β is called the *degenerate set.*

The standard Abadie and Guignard CQs are defined using the tangent cone of the feasible set of a mathematical program. The MPEC variants of these CQs (see Definition 2.1) also make use of this tangent cone. If we denote the feasible set of (1) by $\mathcal{Z}$, the tangent cone of (1) in a feasible point z^* is defined by

$$\mathcal{T}(z^*) := \left\{d \in \mathbb{R}^n \,\middle|\, \exists \{z^k\} \subset \mathcal{Z}, \exists t_k \searrow 0 \,:\, z^k \to z^* \text{ and } \frac{z^k - z^*}{t_k} \to d\right\}. \tag{3}$$

Note that the tangent cone is closed, but in general not convex.

For the standard Abadie and Guignard CQs, the constraints of the mathematical program are linearized. This makes less sense in the context of MPECs because information we keep for G and H, we throw away for the complementarity term (see also [3]). Instead, the authors proposed the *MPEC-linearized* tangent cone in [6],

$$\begin{aligned} \mathcal{T}^{lin}_{\text{MPEC}}(z^*) := \{d \in \mathbb{R}^n \mid \nabla g_i(z^*)^T d \le 0, \quad & \forall i \in \mathcal{I}_g, \\ \nabla h_i(z^*)^T d = 0, \quad & \forall i = 1, \dots, p, \\ \nabla G_i(z^*)^T d = 0, \quad & \forall i \in \alpha, \\ \nabla H_i(z^*)^T d = 0, \quad & \forall i \in \gamma, \\ \nabla G_i(z^*)^T d \ge 0, \quad & \forall i \in \beta, \\ \nabla H_i(z^*)^T d \ge 0, \quad & \forall i \in \beta, \\ (\nabla G_i(z^*)^T d) \cdot (\nabla H_i(z^*)^T d) = 0, \quad & \forall i \in \beta\}, \end{aligned} \tag{4}$$

where $\mathcal{I}_g := \{i \mid g_i(z^*) = 0\}$ is the set of active inequality constraints at z^*. Note that here, the component functions of the complementarity term have been linearized separately, so that we end up with a *quadratic term* in (4).

Similar to the classic case, it holds that

$$\mathcal{T}(z^*) \subseteq \mathcal{T}^{lin}_{\text{MPEC}}(z^*)$$

(see [6]).

Guignard CQ is often stated using the so-called duals of the tangent and linearized tangent cones (see, e.g., [1]). We therefore introduce the concept of the dual cone. Given an arbitrary cone $\mathcal{C} \subseteq \mathbb{R}^n$, its *dual cone* $\mathcal{C}^*$ is defined as follows:

$$\mathcal{C}^* := \{v \in \mathbb{R}^n \mid v^T d \geq 0 \quad \forall d \in \mathcal{C}\}. \tag{5}$$

Together with the introduction of the MPEC-linearized tangent cone (4), this inspires the following variants of the Abadie and Guignard CQs for MPECs.

Definition 2.1 *The MPEC (1) is said to satisfy* MPEC-Abadie CQ, *or* MPEC-ACQ, *at a feasible vector* z^* *if*

$$\mathcal{T}(z^*) = \mathcal{T}^{lin}_{\text{MPEC}}(z^*) \tag{6}$$

holds. It is said to satisfy MPEC-Guignard CQ, *or* MPEC-GCQ, *at a feasible vector* z^* *if*

$$\mathcal{T}(z^*)^* = \mathcal{T}^{lin}_{\text{MPEC}}(z^*)^* \tag{7}$$

holds.

We refer the reader to [6] for a rigorous discussion of MPEC-ACQ.

Note that obviously MPEC-ACQ in z^* implies MPEC-GCQ in z^*. The converse is not true, in general, as can be seen from the following example.

Example 2.1 Consider the MPEC

$$\min z_1^2 + z_2^2 \quad \text{s.t.} \quad z_1^2 \geq 0, z_2^2 \geq 0, z_1^2 z_2^2 = 0.$$

Then $z^* := 0$ is the unique minimizer of this program, and a simple calculation shows that $\mathcal{T}(z^*) = \{z \in \mathbb{R}^2 \,|\, z_1 z_2 = 0\}, \mathcal{T}^{lin}_{\text{MPEC}}(z^*) = \mathbb{R}^2$, hence MPEC-ACQ does not hold. On the other hand, this implies $\mathcal{T}(z^*)^* = \{0\}$ and $\mathcal{T}^{lin}_{\text{MPEC}}(z^*)^* = \{0\}$, i.e., MPEC-GCQ is satisfied. △

We will therefore present our main result using MPEC-GCQ (see Theorem 3.1). Naturally, it holds under MPEC-ACQ as well.

As mentioned in the introduction, various stationarity concepts have arisen for MPECs. Though we only need M-stationarity, we also state A-, C- and strong stationarity for completeness' sake, see [19, 16, 4] for more detail.

Let $z^* \in \mathcal{Z}$ be feasible for the MPEC (1). We call z^* *M-stationary* if there exists λ^g, λ^h, λ^G, and λ^H such that

$$\begin{aligned} 0 = \nabla f(z^*) + \sum_{i=1}^{m} \lambda_i^g \nabla g_i(z^*) + \sum_{i=1}^{p} \lambda_i^h \nabla h_i(z^*) - \\ - \sum_{i=1}^{l} \left[\lambda_i^G \nabla G_i(z^*) + \lambda_i^H \nabla H_i(z^*)\right], \end{aligned} \tag{8}$$

$$\begin{array}{lll} \lambda_\alpha^G \ \text{free}, & & \lambda_\gamma^G = 0, \\ & (\lambda_i^G > 0 \wedge \lambda_i^H > 0) \ \vee \ \lambda_i^G \lambda_i^H = 0 \quad \forall i \in \beta & \\ \lambda_\gamma^H \ \text{free}, & & \lambda_\alpha^H = 0, \end{array}$$

$$g(z^*) \leq 0, \qquad \lambda^g \geq 0, \qquad g(z^*)^T \lambda^g = 0.$$

The other stationarity concepts differ from M-stationarity only in the restriction imposed upon λ_i^G and λ_i^H for $i \in \beta$, as detailed in the following list:

- strong stationarity [19, 16]: $\lambda_i^G \geq 0 \wedge \lambda_i^H \geq 0 \; \forall i \in \beta$;
- M-stationarity [14]: $(\lambda_i^G > 0 \wedge \lambda_i^H > 0) \vee \lambda_i^G \lambda_i^H = 0 \; \forall i \in \beta$;
- C-stationarity [19]: $\lambda_i^G \lambda_i^H \geq 0 \; \forall i \in \beta$;
- A-stationarity [4]: $\lambda_i^G \geq 0 \vee \lambda_i^H \geq 0 \; \forall i \in \beta$.

Note that the intersection of A- and C-stationarity yields M-stationarity and that strong stationarity implies M- and hence A- and C-stationarity. Also note that Pang and Fukushima [16] call a strongly stationary point a *primal-dual stationary point.* The "C" and "M" stand for Clarke and Mordukhovich, respectively, since they occur when applying the Clarke or Mordukhovich calculus to the MPEC (1). The "A" might stand for "alternative" because that describes the properties of the Lagrange multipliers, or "Abadie" because it first occured when MPEC-ACQ was applied to the MPEC (1), see [6].

We will now introduce some normal cones, which will become important in our subsequent analysis. For more detail on the normal cones we use here, see [12, 10, 18].

Definition 2.2 *Let $\Omega \subseteq \mathbb{R}^l$ be nonempty and closed, and $v \in \Omega$ be given. We call*

$$\hat{N}(v, \Omega) := \{w \in \mathbb{R}^l \mid \limsup_{\substack{v^k \to v \\ \{v^k\} \subset \Omega \setminus \{v\}}} w^T(v^k - v)/\|v^k - v\| \leq 0\} \tag{9}$$

the Fréchet normal cone *or* regular normal cone *[18] to Ω at v, and*

$$N(v, \Omega) := \{\lim_{k \to \infty} w^k \mid \exists \{v^k\} \subset \Omega : \lim_{k \to \infty} v^k = v, \; w^k \in \hat{N}(v^k, \Omega)\} \tag{10}$$

the limiting normal cone *to Ω at v.*

By convention, we set $\hat{N}(v, \Omega) = N(v, \Omega) := \emptyset$ if $v \notin \Omega$. By $N_\Omega^\times : \mathbb{R}^l \rightrightarrows \mathbb{R}^l$ we denote the multifunction that maps $v \mapsto N^\times(v, \Omega)$, where $\times$ is a placeholder for one of the normal cones defined above.

Note that if v is in the interior of Ω, both normal cones reduce to $\{0\}$, as is well known.

Since the limiting normal cone is the most important one in our subsequent analysis, we did not furnish it with an index to simplify notation.

To cope with the complementarity term in the constraints of the MPEC (1), we recall the following result which investigates the limiting normal cone to a complementarity set. This result was originally stated in a slightly different format by Outrata in [14, Lemma 2.2], see also [21, Proposition 3.7]. A proof for this particular formulation may be found in [5, Proposition 2.5].

Proposition 2.1 *Let the set*

$$\mathcal{C} := \{(a, b) \in \mathbb{R}^{2l} \mid a \geq 0,\ b \geq 0\ a^T b = 0\} \tag{11}$$

be given. Then, for an arbitrary but fixed $(a, b) \in \mathcal{C}$, define

$$\mathcal{I}_a = \{i \mid a_i = 0, b_i > 0\},\ \mathcal{I}_b = \{i \mid a_i > 0, b_i = 0\},\ \mathcal{I}_{ab} = \{i \mid a_i = 0, b_i = 0\}.$$

Then the limiting normal cone to $\mathcal{C}$ in (a, b) is given by

$$\begin{aligned} N((a, b), \mathcal{C}) = \{(x, y) \in \mathbb{R}^{2l} \mid\ & x_{\mathcal{I}_b} = 0, y_{\mathcal{I}_a} = 0, \\ & (x_i < 0 \wedge y_i < 0) \vee x_i y_i = 0\ \forall i \in \mathcal{I}_{ab}\}. \end{aligned} \tag{12}$$

Another important set is a polyhedral convex set, whose limiting normal cone at the origin will be needed in our subsequent analysis and is therefore given in the following lemma.

Lemma 2.1 *Let vectors $a_i \in \mathbb{R}^n$, $i = 1, \ldots, k$ and $b_i \in \mathbb{R}^n$, $i = 1, \ldots, l$ be given and define the convex set*

$$\begin{aligned} \mathcal{D} := \{d \in \mathbb{R}^n \mid\ & a_i^T d \leq 0, \quad \forall i = 1, \ldots, k, \\ & b_j^T d = 0, \quad \forall j = 1, \ldots, l\}. \end{aligned} \tag{13}$$

Then the limiting normal cone of $\mathcal{D}$ at 0 is given by

$$\begin{aligned} N(0, \mathcal{D}) = \{v \in \mathbb{R}^n \mid\ & v = \sum_{i=1}^{k} \alpha_i a_i + \sum_{j=1}^{l} \beta_j b_j \\ & \alpha_i \geq 0, \quad \forall i = 1, \ldots, k\}, \end{aligned} \tag{14}$$

Proof. Since $\mathcal{D}$ is convex, [18, Theorem 6.9] may be invoked and the statement of this lemma is given by Theorem 3.2.2 and its proof in [1]. □

3 M-Stationarity

We start off this section by stating our main result. The remainder of the paper is dedicated to proving this result. Note that a similar result has been stated and proved in [22], though under the stronger MPEC-ACQ.

Theorem 3.1 *Let z^* be a local minimizer of the MPEC (1) at which MPEC-GCQ holds. Then there exists a Lagrange multiplier λ^* such that (z^*, λ^*) satisfies the conditions for M-stationarity (8).*

The fundamental idea of the proof is due to Ye, see [22, Theorem 3.1]. It is based on the fact that, under MPEC-GCQ, the tangent cone is described by some linear equations and inequalities, and a linear complementarity problem.

Using the tangent cone as the feasible set of a mathematical program yields an MPEC. We are then able to glean the conditions for M-stationarity for the original MPEC (1) from this "affine" MPEC. In the following, we make this idea more precise.

Since z^* is a local minimum of (1), we get

$$\nabla f(z^*)^T d \geq 0 \qquad \forall d \in \mathcal{T}(z^*)$$

from standard optimization theory, cf. [13]. Using the definition of the dual cone, this may be expressed as

$$\nabla f(z^*) \in \mathcal{T}(z^*)^*,$$

and since MPEC-GCQ holds, i.e. $\mathcal{T}(z^*)^* = \mathcal{T}^{lin}_{\text{MPEC}}(z^*)^*$, this is equivalent to

$$\nabla f(z^*) \in \mathcal{T}^{lin}_{\text{MPEC}}(z^*)^*.$$

Resolving the definition of the dual cone, we obtain

$$\nabla f(z^*)^T d \geq 0 \qquad \forall d \in \mathcal{T}^{lin}_{\text{MPEC}}(z^*). \tag{15}$$

This, in turn, is equivalent to $d^* = 0$ being a minimizer of

$$\begin{aligned} &\min_{d} \ \nabla f(z^*)^T d \\ &\text{s.t. } d \in \mathcal{T}^{lin}_{\text{MPEC}}(z^*). \end{aligned} \tag{16}$$

This is a *mathematical program with affine equilibrium constraints*, or MPAEC.

It is easily verified that $d^* = 0$ being a minimizer of (16) is equivalent to $(d^*, \xi^*, \eta^*) = (0, 0, 0)$ being a minimizer of

$$\begin{aligned} &\min_{(d,\xi,\eta)} \ \nabla f(z^*)^T d \\ &\text{s.t. } (d, \xi, \eta) \in \mathcal{D} := \mathcal{D}_1 \cap \mathcal{D}_2 \end{aligned} \tag{17}$$

with

$$\begin{aligned} \mathcal{D}_1 := \{(d, \xi, \eta) \mid \ & \nabla g_i(z^*)^T d \leq 0, && \forall i \in \mathcal{I}_g, \\ & \nabla h_i(z^*)^T d = 0, && \forall i = 1, \ldots, p, \\ & \nabla G_i(z^*)^T d = 0, && \forall i \in \alpha, \\ & \nabla H_i(z^*)^T d = 0, && \forall i \in \gamma, \\ & \nabla G_i(z^*)^T d - \xi_i = 0, && \forall i \in \beta, \\ & \nabla H_i(z^*)^T d - \eta_i = 0, && \forall i \in \beta\} \end{aligned} \tag{18}$$

and

$$\mathcal{D}_2 := \{(d, \xi, \eta) \mid \xi \geq 0, \ \eta \geq 0, \ \xi^T \eta = 0\}. \tag{19}$$

Once more, since $(0, 0, 0)$ is a minimizer of (17), B-stationarity holds, which in this case means that

$$(\nabla f(z^*), 0, 0)^T w \geq 0 \qquad \forall w \in \mathcal{T}((0,0,0), \mathcal{D}),$$

where $\mathcal{T}((0,0,0), \mathcal{D})$ denotes the tangent cone to the set $\mathcal{D}$ in the point $(0,0,0)$. By virtue of [18, Proposition 6.5], this is equivalent to

$$(-\nabla f(z^*), 0, 0) \in \hat{N}((0,0,0), \mathcal{D}) \subseteq N((0,0,0), \mathcal{D}). \tag{20}$$

Note, once again, that the limiting normal cone $N(\cdot, \cdot)$ is equal to the limit of the Fréchet normal cone $\hat{N}(\cdot, \cdot)$.

In order to calculate $N((0,0,0), \mathcal{D})$ in a fashion conducive to our goal, we need to consider the normal cones $\mathcal{D}_1$ and $\mathcal{D}_2$ separately. To be able to do this, we need some auxiliary results. We start off with the definition of a polyhedral multifunction (see [17]).

Definition 3.1 *We say that a multifunction* $\Phi : \mathbb{R}^n \rightrightarrows \mathbb{R}^m$ *is a* polyhedral multifunction *if its graph is the union of finitely many polyhedral convex sets.*

We now show that a certain multifunction, which is defined using $\mathcal{D}_1$ and $\mathcal{D}_2$, is a polyhedral multifunction. We will need this to apply a result by Henrion, Jourani and Outrata [8].

Lemma 3.1 *Let the multifunction* $\Phi : \mathbb{R}^{n+2|\beta|} \rightrightarrows \mathbb{R}^{n+2|\beta|}$ *be given by*

$$\Phi(v) := \{w \in \mathcal{D}_1 \mid v + w \in \mathcal{D}_2\}. \tag{21}$$

Then Φ is a polyhedral multifunction.

Proof. Since the graph of Φ may be expressed as

$$\begin{aligned}
\operatorname{gph} \Phi = \{(d^v, \xi^v, \eta^v, d^w, \xi^w, \eta^w) \in \mathbb{R}^{2(n+2|\beta|)} \mid \\
&\nabla g_i(z^*)^T d^w \leq 0, && \forall i \in \mathcal{I}_g, \\
&\nabla h_i(z^*)^T d^w = 0, && \forall i = 1, \ldots, p, \\
&\nabla G_i(z^*)^T d^w = 0, && \forall i \in \alpha, \\
&\nabla H_i(z^*)^T d^w = 0, && \forall i \in \gamma, \\
&\nabla G_i(z^*)^T d^w - \xi_i^w = 0, && \forall i \in \beta, \\
&\nabla H_i(z^*)^T d^w - \eta_i^w = 0, && \forall i \in \beta, \\
&\xi^v + \xi^w \geq 0,\ \eta^v + \eta^w \geq 0,\ (\xi^v + \xi^w)^T(\eta^v + \eta^w) = 0\}
\end{aligned}$$

$$
\begin{aligned}
= \bigcup_{(\nu_1,\nu_2)\in\mathcal{P}(\{1,\ldots,|\beta|\})} \{(d^v,\xi^v,\eta^v,d^w,\xi^w,\eta^w) \in \mathbb{R}^{2(n+2|\beta|)} \mid \\
&\nabla g_i(z^*)^T d^w \le 0, && \forall i \in \mathcal{I}_g, \\
&\nabla h_i(z^*)^T d^w = 0, && \forall i = 1,\ldots,p, \\
&\nabla G_i(z^*)^T d^w = 0, && \forall i \in \alpha, \\
&\nabla H_i(z^*)^T d^w = 0, && \forall i \in \gamma, \\
&\nabla G_i(z^*)^T d^w - \xi_i^w = 0, && \forall i \in \beta, \\
&\nabla H_i(z^*)^T d^w - \eta_i^w = 0, && \forall i \in \beta, \\
&\xi^v_{\nu_1} + \xi^w_{\nu_1} = 0, \quad \xi^v_{\nu_2} + \xi^w_{\nu_2} \ge 0, \\
&\eta^v_{\nu_1} + \eta^w_{\nu_1} \ge 0, \quad \eta^v_{\nu_2} + \eta^w_{\nu_2} = 0 \},
\end{aligned}
$$

it is obviously the union of finitely many polyhedral convex sets. By Definition 3.1, Φ is therefore a polyhedral multifunction. □

Since Φ defined in (21) is a polyhedral multifunction, [17, Proposition 1] may be invoked to show that Φ is locally upper Lipschitz continuous at every point $v \in \mathbb{R}^{n+2|\beta|}$. It is therefore in particular calm at every $(v, w) \in \operatorname{gph}\Phi$ in the sense of [8]. By invoking [8, Corollary 4.2] we see that (20) implies

$$
(-\nabla f(z^*), 0, 0) \in N((0,0,0), \mathcal{D}_1) + N((0,0,0), \mathcal{D}_2). \tag{22}
$$

Now, the limiting normal cone of $\mathcal{D}_1$ is given by Lemma 2.1. This yields that there exist λ^g, λ^h, λ^G and λ^H with $\lambda^g_{\mathcal{I}_g} \ge 0$ such that

$$
\begin{aligned}
\begin{pmatrix} -\nabla f(z^*) \\ 0 \\ 0 \end{pmatrix} \in & \sum_{i\in\mathcal{I}_g} \lambda_i^g \begin{pmatrix} \nabla g_i(z^*) \\ 0 \\ 0 \end{pmatrix} + \sum_{i=1}^{p} \lambda_i^h \begin{pmatrix} \nabla h_i(z^*) \\ 0 \\ 0 \end{pmatrix} \\
& - \sum_{i\in\alpha} \lambda_i^G \begin{pmatrix} \nabla G_i(z^*) \\ 0 \\ 0 \end{pmatrix} - \sum_{i\in\gamma} \lambda_i^H \begin{pmatrix} \nabla H_i(z^*) \\ 0 \\ 0 \end{pmatrix} \\
& - \sum_{i\in\beta} \left[\lambda_i^G \begin{pmatrix} \nabla G_i(z^*) \\ -e^i \\ 0 \end{pmatrix} + \lambda_i^H \begin{pmatrix} \nabla H_i(z^*) \\ 0 \\ -e^i \end{pmatrix} \right] \\
& + N((0,0,0), \mathcal{D}_2),
\end{aligned} \tag{23}
$$

where e^i denotes that unit vector in $\mathbb{R}^{|\beta|}$ which corresponds to the position of i in β. Note that since the signs in the second and third lines of (23) are arbitrary, they were chosen to facilitate the notation of the proof.

First, we take a look at the second and third components in (23). To this end, we rewrite the normal cone to $\mathcal{D}_2$ in the following fashion:

$$
\begin{aligned}
N((0,0,0), \mathcal{D}_2) &= N(0, \mathbb{R}^n) \times N((0,0), \{(\xi,\eta) \mid \xi \ge 0, \eta \ge 0, \xi^T\eta = 0\}) \\
&= \{0\} \times N((0,0), \{(\xi,\eta) \mid \xi \ge 0, \eta \ge 0, \xi^T\eta = 0\}).
\end{aligned} \tag{24}
$$

Here the first equality is due to the Cartesian product rule (see, e.g., [12] or [18, Proposition 6.41]). The second equality uses that 0 is in the interior of $\mathbb{R}^n$, and hence any normal cone reduces to $\{0\}$.

Substituting (24) into (23) yields

$$(-\lambda_\beta^G, -\lambda_\beta^H) \in N((0,0), \{(\xi,\eta) \mid \xi \geq 0, \eta \geq 0, \xi^T\eta = 0\}).$$

Applying Proposition 2.1, we obtain that

$$(\lambda_i^G > 0 \land \lambda_i^H > 0) \quad \lor \quad \lambda_i^G \lambda_i^H = 0$$

for all $i \in \beta$. Note that since we need to determine the limiting normal cone in the point $(0,0)$, it holds that $\mathcal{I}_a = \mathcal{I}_b = \emptyset$ in Proposition 2.1.

Finally, we set $\lambda_\gamma^G := 0$, $\lambda_\alpha^H := 0$, and $\lambda_i^g := 0$ for all $i \notin \mathcal{I}_g$ and have thus acquired the conditions for M-stationarity (8) with $\lambda^* := (\lambda^g, \lambda^h, \lambda^G, \lambda^H)$, completing the proof of Theorem 3.1. Note that even though we derived our conditions using the MPAEC (16), we have in fact acquired the conditions for M-stationarity of our original MPEC (1).

Remark 3.1 *We wish to draw attention to two fundamental ideas used in the proof of Theorem 3.1. The first is due to Ye and entails introducing an MPAEC (see (16) and the discussion preceeding it).*

The second idea is that we can separate the benign constraints from the complementarity constraints (divided here into $\mathcal{D}_1$ and $\mathcal{D}_2$) and consider the two types of constraints separately. We are able to do this because

$$N((0,0,0), \mathcal{D}) \subseteq N((0,0,0), \mathcal{D}_1) + N((0,0,0), \mathcal{D}_2)$$

(see (20) and (22)) holds due to a result by Henrion, Jourani and Outrata (see [8, Corollary 4.2]). Note that this does not hold in general, but is a direct consequence of our MPAEC (16) having constraints characterized by affine functions.

Note that the MPEC-Guignard constraint qualification is satisfied not only under MPEC-ACQ, but also under many other conditions like the MPEC-MFCQ assumption or an MPEC-variant of a Slater-condition, see [6], as well as a number of other constraint qualifications, see [22]. Hence all these stronger constraint qualifications imply that M-stationarity is a necessary first order optimality condition. In particular, a local minimizer is an M-stationary point under the MPEC-MFCQ assumption used in [19]. However, the authors of [19] were only able to prove C-stationarity to be a necessary first order condition under MPEC-MFCQ.

We also note that the MPEC-Guignard constraint qualification does not guarantee that a local minimizer is a strongly stationary point. This follows from the observation that even the stronger MPEC-MFCQ condition does not imply strong stationarity, see [19] for a counterexample.

4 Conclusion

We proved that a very weak assumption, the MPEC-Guignard constraint qualification, implies that a local minimum satisfies the relatively strong first order optimality condition, M-stationarity, in the framework of mathematical programs with equilibrium constraints. The proof was obtained using a relatively direct approach, whereas the same result was obtained in [7] as a special case of a much more general approach. We also note that Theorem 3.1 improves on several existing results such as those found in [22, 5].

5 Acknowledgements

The authors wish to thank Jiři Outrata for pointing out a mistake and possible improvements in the initial draft of our approach. Also, our gratitude goes out to Jane Ye, who was always helpful when questions pertaining to her approach arose.

References

1. M. S. Bazaraa and C. M. Shetty, *Foundations of Optimization*, vol. 122 of Lecture Notes in Economics and Mathematical Systems, Springer-Verlag, Berlin, Heidelberg, New York, 1976.
2. Y. Chen and M. Florian, *The nonlinear bilevel programming problem: Formulations, regularity and optimality conditions*, Optimization, 32 (1995), pp. 193–209.
3. M. L. Flegel and C. Kanzow, *On the Guignard constraint qualification for mathematical programs with equilibrium constraints*, Optimization, to appear.
4. ——, *A Fritz John approach to first order optimality conditions for mathematical programs with equilibrium constraints*, Optimization, 52 (2003), pp. 277–286.
5. ——, *On M-stationarity for mathematical programs with equilibrium constraints*, Journal of Mathematical Analysis and Applications, 310 (2005), pp. 286-302.
6. ——, *Abadie-type constraint qualification for mathematical programs with equilibrium constraints*, Journal of Optimization Theory and Applications, 124 (2005), pp. 595–614.
7. M. L. Flegel, C. Kanzow, and J. Outrata, *Optimality conditions for disjunctive programs with application to mathematical programs with equilibrium constraints.* Institute of Applied Mathematics and Statistics, University of Würzburg, Preprint, October 2004.
8. R. Henrion, A. Jourani, and J. Outrata, *On the calmness of a class of multifunctions*, SIAM Journal on Optimization, 13 (2002), pp. 603–618.
9. X. Hu and D. Ralph, *Convergence of a penalty method for mathematical programming with complementarity constraints*, Journal of Optimization Theory and Applications, 123 (2004), pp. 365-390.

10. P. D. Loewen, *Optimal Control via Nonsmooth Analysis*, vol. 2 of CRM Proceedings & Lecture Notes, American Mathematical Society, Providence, RI, 1993.
11. Z.-Q. Luo, J.-S. Pang, and D. Ralph, *Mathematical Programs with Equilibrium Constraints*, Cambridge University Press, Cambridge, UK, 1996.
12. B. S. Mordukhovich, *Generalized differential calculus for nonsmooth and set-valued mappings*, Journal of Mathematical Analysis and Applications, 183 (1994), pp. 250–288.
13. J. Nocedal and S.J. Wright, *Numerical Optimization*, Springer, New York, NY, 1999.
14. J. V. Outrata, *Optimality conditions for a class of mathematical programs with equilibrium constraints*, Mathematics of Operations Research, 24 (1999), pp. 627–644.
15. ——, *A generalized mathematical program with equilibrium constraints*, SIAM Jounral of Control and Optimization, 38 (2000), pp. 1623–1638.
16. J.-S. Pang and M. Fukushima, *Complementarity constraint qualifications and simplified B-stationarity conditions for mathematical programs with equilibrium constraints*, Computational Optimization and Applications, 13 (1999), pp. 111–136.
17. S. M. Robinson, *Some continuity properties of polyhedral multifunctions*, Mathematical Programming Study, 14 (1981), pp. 206–214.
18. R. T. Rockafellar and R. J.-B. Wets, *Variational Analysis*, vol. 317 of A Series of Comprehensive Studies in Mathematics, Springer, Berlin, Heidelberg, 1998.
19. H. Scheel and S. Scholtes, *Mathematical programs with complementarity constraints: Stationarity, optimality, and sensitivity*, Mathematics of Operations Research, 25 (2000), pp. 1–22.
20. S. Scholtes, *Convergence properties of a regularization scheme for mathematical programs with complementarity constraints*, SIAM Journal on Optimization, 11 (2001), pp. 918–936.
21. J. J. Ye, *Constraint qualifications and necessary optimality conditions for optimization problems with variational inequality constraints*, SIAM Journal on Optimization, 10 (2000), pp. 943–962.
22. ——, *Necessary and sufficient optimality conditions for mathematical programs with equilibrium constraints*, Journal on Mathematical Analysis and Applications, 307 (2005), pp. 350-369.

On the use of bilevel programming for solving a structural optimization problem with discrete variables

Joaquim J. Júdice[1], Ana M. Faustino[2], Isabel M. Ribeiro[2] and A. Serra Neves[3]

[1] Departamento de Matemática da Universidade de Coimbra and Instituto de Telecomunicações, Coimbra, Portugal joaquim.judice@co.it.pt
[2] Secção de Matemática do Departamento de Engenharia Civil, Faculdade de Engenharia da Universidade do Porto, Porto, Portugal {afausti, iribeiro}@fe.up.pt
[3] Secçáo de Materiais de Construçáo do Departamento de Engenharia Civil, Faculdade de Engenharia da Universidade do Porto, Porto, Portugal asneves@fe.up.pt

Summary. In this paper, a bilevel formulation of a structural optimization problem with discrete variables is investigated. The bilevel programming problem is transformed into a Mathematical Program with Equilibrium (or Complementarity) Constraints (MPEC) by exploiting the Karush-Kuhn-Tucker conditions of the follower's problem.

A complementarity active-set algorithm for finding a stationary point of the corresponding MPEC and a sequential complementarity algorithm for computing a global minimum for the MPEC are analyzed. Numerical results with a number of structural problems indicate that the active-set method provides in general a structure that is quite close to the optimal one in a small amount of effort. Furthermore the sequential complementarity method is able to find optimal structures in all the instances and compares favorably with a commercial integer program code for the same purpose.

Key Words: Structural optimization, mixed integer programming, global optimization, complementarity.

1 Introduction

In the last few decades, Structural Optimization has become an area of increasing interest and intense research [1, 3, 5, 10, 9, 12, 20, 22, 23, 25]. These models are formulated as challenging optimization problems representing the elastoplastic laws of mechanics and searching for a structure with the least

S. Dempe and V. Kalashnikov (eds.), *Optimization with Multivalued Mappings*, pp. 123-142
©2006 Springer Science + Business Media, LLC

volume. A quite general structural optimization model has been introduced in [8] whose formulation leads into a bilinear program with linear and bilinear constraints. The variables of this optimization problem are associated to the coordinates at each node of the structure and the cross sectional areas. The latter should belong to a fixed set of admissible values. Furthermore each feasible solution is characterized by a vector x, whose components are 1 or 0, depending on the corresponding bar to be or not to be included in the optimal structure.

As discussed in [8], this bilinear program with discrete variables can be reduced into a mixed integer zero-one linear program. Computational experience reported in [8] shows that the model is quite appropriate for finding a structure that requires small amount of material. A commercial code, such as OSL [18], can in general find an optimal solution for the optimization problem when the number of nodes and pre-fixed values for the cross-sectional areas are small. However, the algorithm faces difficulties in finding such a solution when the dimension of the problem increases.

A mixed integer zero-one linear program can be shown to be equivalent to a Linear Bilevel Programming Problem [2]. By exploiting the Karush-Kuhn-Tucker conditions of the follower's problem it is possible to reduce this bilevel program into a Mathematical Programming Problem with Equilibrium (or Complementarity) Constraints of the following form

$$\begin{array}{ll} \text{MPEC: } \textit{Minimize} & c^T z + d^T y \\ \qquad \textit{subject to} & Ew = q + Mz + Ny \\ & z \geq 0, \quad w \geq 0 \\ & y \in K_y \\ & z^T w = 0 \end{array} \tag{1}$$

where $q \in \mathbb{R}^p$, c, $z \in \mathbb{R}^n$, d, $y \in \mathbb{R}^m$, M, $E \in \mathbb{R}^{p \times n}$, $N \in \mathbb{R}^{p \times m}$ and

$$K_y = \{y \in \mathbb{R}^m : \ Cy = b, \ y \geq 0\}$$

with $C \in \mathbb{R}^{l \times m}$ and $b \in \mathbb{R}^l$.

Due to its structure, an active-set methodology seems to be quite appropriate to process this MPEC. A complementarity active-set (CASET) algorithm has been introduced in [16] to find a stationary point for the MPEC. The procedure maintains complementarity during the entire process and has been shown to converge to a stationary point under reasonable hypotheses. Computational experience reported in [16] has shown that the proposed algorithm is in general quite efficient to process moderate and even large MPECs.

A Sequential Linear Complementarity (SLCP) algorithm has been introduced in [14] to find a global minimum for a linear MPEC. The algorithm finds a sequence of stationary points of the MPEC with strictly decreasing value. The last stationary point of this sequence is shown to be a global minimum of the MPEC. Computational experience reported in [13, 14, 15] indicates that the algorithm is quite efficient to find a stationary point that is a global minimum of the MPEC, but faces difficulties in establishing that such a global minimum has been achieved.

In practice, engineers search for a structure that serves their purposes, that is, a feasible solution of the mixed integer program with a small objective function value is requested. As each stationary point of the MPEC corresponds to a feasible solution of its equivalent zero-one integer program, then both the CASET and SLCP algorithms seem to be valid approaches to find a good structure for the structural model. In this paper we investigate how these two algorithms perform for a number of structures presented in [8]. The experiments indicate that the CASET algorithm is able to find in general a good structure in a small amount of effort. On the other hand, the SLCP algorithm has always found a global optimal structure for the model. Furthermore the computational effort required by the SLCP algorithm tends to become much smaller than the one needed by an integer program code as the dimension of this problem increases.

The organization of the paper is as follows. In Section 2 the structural model and its formulation are introduced. Section 3 is devoted to the equivalence between a zero-one mixed integer program and an MPEC. The algorithms CASET and SLCP are briefly described in sections 4 and 5. Finally computational experience with these algorithms on a set of structural problems and some conclusions are included in the last two sections.

2 A topological optimization model

The admissible structural domain is referenced by a bidimensional cartesian system Oxy, in which the various alternative solutions for the problem under consideration can be developed. A discretisation [26] of this domain is then considered in which the mesh is composed by bar elements joined at the nodal points.

The structural domain is submitted to the various actions defined in the safety code [6] such as the structural self-weight, wind, earthquake and so on. These actions lead to different l loading conditions, each of them is represented by nodal point loads

$$f^l = \begin{bmatrix} f_x^l \\ f_y^l \end{bmatrix}.$$

Some of these loads are reactions r^l, when the associated nodes are connected to the exterior. The nodal displacements

$$u^l = \begin{bmatrix} u_x^l \\ u_y^l \end{bmatrix}$$

are associated to these nodal forces. The stress field within each bar element i for loading condition l can be determined from its axial load e_i^l, while the strain field is given by the axial deformation d_i^l.

The fundamental conditions to be satisfied in the serviceability limit states are equilibrium, compatibility, boundary conditions and elastic constitutive relations of the structural material.

Equilibrium has to be verified at a nodal level and relates the elastic axial bar forces e_e^l with support reactions r_e^l and applied nodal loads f^l by

$$C^T e_e^l - Br_e^l - f^l = 0, \tag{2}$$

where C and B are matrices depending on the structural topology.

The compatibility conditions imply equal displacement for all the bar ends joining at the same node and can be expressed as

$$d_e^l = Cu^l, \tag{3}$$

where d_e^l is the bar deformation vector, u^l is the nodal displacement vector and C is the connectivity matrix already used in (2).

The forces e_e^l in the structural bars are related to the bar deformations d_e^l by linear elastic constitutive relations given by the so-called Hooke's Law

$$e_e^l = KD_A d_e^l, \tag{4}$$

where $D_A = diag\{A_i\}$, with A_i a discrete variable associated to the cross-sectional area of bar i and $K = diag\{E_i h_i^{-1}\}$, with $E_i > 0$ the Young's modulus of bar i and h_i its length . It follows from (2), (3) and (4) that

$$C^T KD_A Cu^l - Br_e^l - f^l = 0. \tag{5}$$

The structural boundary conditions are given by

$$u_m^l = 0 \tag{6}$$

for the nodes m connected to supports with zero displacement.

The nodal displacements should comply with the upper and lower bounds defined in the safety codes

$$u_{min} \leq u^l \leq u_{max}. \tag{7}$$

The ultimate limit states can be considered on the basis of the Plasticity Theory. According to the Static Theorem, the fundamental conditions to be fulfilled are equilibrium, plasticity conditions and boundary conditions.

The equilibrium conditions are given in a similar form to (2) by

$$C^T e_p^l - Br_p^l - \lambda f^l = 0, \tag{8}$$

where e_p^l is the plastic force vector, r_p^l the plastic reaction vector and λ is a partial safety majoration factor for the nodal forces corresponding to the applied actions, prescribed in structural safety codes [6, 7].

The plasticity conditions can be expressed as

$$e_{min} \leq e_p^l \leq e_{max}, \tag{9}$$

where e_{min} and e_{max} are the minimum and maximum admissible values for the element forces defined in the code [7].

The conditions (5), (6), (7), (8) and (9) considered so far are satisfied by many solutions in which some bars have zero force. A vector x is further introduced in the model such that each variable x_i is associated with bar i and takes value 1 or 0, depending on the bar i to be or not to be included in the solution.

The force in a generic bar i can then be replaced by the product $x_i e_{p_i}^l$ yielding a null force in non-existing bars. So the axial bar force must verify the following conditions

$$D_x e_{min} \leq e_p^l \leq D_x e_{max}, \tag{10}$$

where

$$D_x = diag(x_i). \tag{11}$$

Furthermore the diagonal matrix D_A takes the form $D_A D_x$. The model seeks an optimal solution corresponding to the minimum use of structural material V. If A_i is the cross-sectional area of bar i and h_i is its length, then the objective function takes the form

$$V = \sum_i x_i A_i h_i. \tag{12}$$

The optimization problem described by the equations (2-11) consists of minimizing a bilinear function in variables x_i and A_j on a set of linear and bilinear constraints. Furthermore x_i are zero-one variables and the variables A_i can only assume values in a discrete set of positive fixed numbers A_{ik}, $k = 1, \ldots, N_i$. These variables can be transformed into a set of zero-one variables y_{ik} by using traditional manipulations, as described in [8]. On the other hand, bilinear terms such as $x_i y_{ik}$ can be transformed into variables by exploiting the so-called Reformulation-Linearization Technique RLT [8, 24]. These transformations lead into a zero-one mixed-integer program, as shown below.

Unfortunately optimal structures associated to the optimization problem may be not kinematically stable. In order to avoid such type of structures the so-called Grubler's Criterion [11] is exploited. As discussed in [8], this criterion can be analytically presented by some further linear constraints.

All these considerations lead into the following formulation of the structural model [8] under study in this paper.

OPT: $$Minimize\ V = \sum_{i=1}^{nb} \left(\sum_{k=1}^{N_i} A_{ik} y_{ik} \right) h_i$$

subject to

$$\sum_{i=1}^{nb} M_{ji} \left(\sum_{k=1}^{N_i} A_{ik} q_{ik}^l \right) - \sum_{m=1}^{na} B_{jm} r_{e_m}^l - f_j^l = 0 \quad (13)$$

$$d^l = Cu^l \quad (14)$$

$$u_{min} \leq u^l \leq u_{max} \quad (15)$$

$$d_{min_i} y_{ik} \leq q_{ik}^l \leq d_{max_i} y_{ik} \quad (16)$$

$$d_{min_i} \left(1 - \sum_{k=1}^{N_i} y_{ik} \right) \leq d_i^l - \sum_{k=1}^{N_i} q_{ik}^l \quad (17)$$

$$d_i^l - \sum_{k=1}^{N_i} q_{ik}^l \leq d_{max_i} \left(1 - \sum_{k=1}^{N_i} y_{ik} \right) \quad (18)$$

$$u_{j_m}^l = 0 \quad (19)$$

$$-C^T e_p^l + B r_p^l + \lambda f^l = 0 \quad (20)$$

$$t_{min_i} \sum_{k=1}^{N_i} A_{ik} y_{ik} \leq e_{p_i}^l \leq t_{max_i} \sum_{k=1}^{N_i} A_{ik} y_{ik} \quad (21)$$

$$z_n \leq \sum_{i \in I(n)} \sum_{k=1}^{N_i} y_{ik} \leq |I(n)| \, z_n \quad (22)$$

$$2 * \sum_{n=1}^{nn} z_n - \sum_{i=1}^{nb} \sum_{k=1}^{N_i} y_{ik} - \sum_{n=1}^{nn} s_n z_n \leq 0 \quad (23)$$

$$-C^T e_a + B r_a + f_a Z = 0 \quad (24)$$

$$t_{min_i} \sum_{k=1}^{N_i} A_{ik} y_{ik} \leq e_{a_i} \leq t_{max_i} \sum_{k=1}^{N_i} A_{ik} y_{ik} \quad (25)$$

$$y_{ik} \in \{0, 1\} \quad (26)$$

$$\sum_{k=1}^{N_i} y_{ik} \leq 1, \quad (27)$$

where $l = 1, \ldots, nc$, $j = 1, \ldots, 2nn$, $j_m = 1, \ldots, na$, $k = 1, \ldots, N_i$, $n = 1, \ldots, nn$ and $i = 1, \ldots, nb$.

The meanings of the parameters in this program are presented below:

nb	number of bars;
na	number of simple supports;
nn	number of nodes;
nc	number of loading conditions;
N_i	number of discrete sizes available for cross-sectional area of bar i;
A_{ik}	k-th discrete size for bar i;
C	$nb \times 2nn$ matrix of direction cosines relating bar forces with nodal directions;
B	$2nn \times na$ matrix of direction cosines relating nodal directions with nodal supports directions;
M	matrix $\left[C^T diag\left(\frac{E_i}{h_i}\right)\right]$;
E_i	Young's modulus of bar i;
h_i	length of bar i;
f_j^l	applied nodal loads in direction j for loading condition l;
$I(n)$	set of bars indices which occur in node n;
λ	safety factor;
$\lvert I(n)\rvert$	cardinal of set $I(n)$;
s_n	number of simple supports associated with node n;
Z	$2nn \times 2nn$ diagonal matrix, with z_{jj} equal to z_n of the node n associated to the direction j;
f_a	perturbed nodal load applied in all directions;
t_{min_i}, t_{max_i}	minimum and maximum stress in compression and tension, respectively, of bar i;
d_{min_i}, d_{max_i}	minimum and maximum elongation of bar i;
u_{min_j}, u_{max_j}	minimum and maximum nodal displacement in direction j.

The variables have the following meanings:

y_{ik_i}	$0-1$ variable stating whether the k-th discrete size for bar i is or not the cross-sectional area of bar i;
$e_{p_i}^l$	bar force of bar i for loading condition l;
$r_{p_m}^l$	plastic reaction in supports m for loading condition l;
$r_{e_m}^l$	elastic reaction in supports m for loading condition l;
d_i^l	deformation of bar i for loading condition l;
u_j^l	nodal displacement in the direction j for loading condition l;
$q_{ik_i}^l$	elongation of bar i corresponding to each discrete size k for bar i in loading condition l;
z_n	$0-1$ variable stating whether the node n exists or not;
e_{a_i}	bar force of bar i for the perturbed nodal load;
r_{a_m}	plastic reaction in supports m for the perturbed nodal load.

Thus the mixed–integer linear program (OPT) has

$$nc\times\left(4nn+5nb+2\sum_{i=1}^{nb}N_i\right)+3nb+4nn+1$$

constraints and

$$nc\times\left(2nb+2nn+2na+\sum_{i=1}^{nb}N_i\right)+\sum_{i=1}^{nb}N_i+nn+nb+na$$

variables.

3 Reduction to a Mathematical Program with Complementarity Constraints

In the previous section, the topological optimization model has been formulated as a mixed-integer linear program, which can be stated as

$$\begin{aligned} \text{PLI: } & \textit{Minimize } c^T x + d^T u \\ & \textit{subject to } Ax + Bu = g \\ & \qquad Fu = h \\ & \qquad u \geq 0, \quad x_i \in \{0,1\}, \quad i = 1,\dots,n. \end{aligned} \tag{28}$$

As discussed in [2], this mixed integer program can be shown to be equivalent to the following Bilevel Program

$$\begin{aligned} \text{BL: } & \textit{Minimize } c^T x + d^T u \\ & \textit{subject to } Ax + Bu = g \\ & \qquad 0 \leq x \leq e \\ & \qquad u \geq 0 \\ & \qquad Fu = h \\ & \qquad e^T v = 0, \; v \geq 0 \\ & \qquad \textit{Minimize} \quad - e^T v \\ & \qquad \textit{subject to} \quad v \leq x \\ & \qquad\qquad\qquad\qquad v \leq e - x, \end{aligned} \tag{29}$$

where $e \in \mathbb{R}^n$ is a vector of ones.

By exploiting the Karush-Kuhn-Tucker conditions of the follower's problem (29), it is possible to reduce the BL problem into the following Mathematical Programming Problem with Equilibrium (or Complementarity) Constraints

$$\text{MPEC:} \quad \begin{array}{ll} Minimize & c^T x + d^T u \\ subject\ to & \left.\begin{array}{l} Ax + Bu = g \\ 0 \leq x \leq e \\ u \geq 0 \\ Fu = h \\ \alpha + \beta = 1 \\ v + \tau - x = 0 \\ v + s + x = e \\ e^T v = 0 \\ \alpha,\ \beta,\ v,\ \tau,\ s \geq 0 \\ \alpha^T \tau = \beta^T s = 0 \end{array}\right\} \text{GLCP.} \end{array}$$

The constraints of this MPEC constitute a Generalized Complementarity Problem (GLCP), which can be written in the form

$$\begin{array}{l} Ew = q + Mz + Ny \\ w \geq 0,\ z \geq 0 \\ y \in K_y \\ z^T w = 0, \end{array} \tag{30}$$

where

$$K_y = \{y :\ y \geq 0,\ Cy = b\}. \tag{31}$$

GLCP can be processed by a direct or an iterative method provided the matrices E and M satisfy some nice properties. In particular [17], if E is the identity matrix and M is a Positive Semi-Definite (PSD) matrix, then the GLCP can be solved by finding a stationary point of the following quadratic program

$$\begin{array}{ll} \text{QP:}\ Minimize & z^T w \\ \quad subject\ to & Ew = q + Mz + Ny \\ & z \geq 0,\ w \geq 0 \\ & y \in K_y. \end{array} \tag{32}$$

Unfortunately, the GLCP under consideration does not satisfy this property, as the matrices E and M are not even square. An enumerative method [15, 21] is then required to process the GLCP. The algorithm searches for a solution of the GLCP by exploiting a binary tree that is constructed based on the dichotomy presented in the complementarity conditions $z_i w_i = 0$. At each node k, the algorithm computes a stationary point of a quadratic program QP(k) that is obtained from QP by adding the constraints

$$\begin{array}{l} z_i = 0,\ i \in L_k \\ w_j = 0,\ j \in W_k, \end{array}$$

where L_k and W_k are the set of the fixed variables at this node. The incorporation of such QP solver enables the enumerative algorithm to find a solution of the GLCP in a reasonable effort, even for large problems. In fact, a solution of the GLCP is exactly a stationary point of QP(k) such that $z^T w = 0$.

The enumerative algorithm faces difficulties when the GLCP is feasible (the linear constraints are consistent) but has no solution. In this case the last property does not hold and the method requires an exhaustive search in the tree to terminate.

As a final remark, it is important to add that the enumerative method can be implemented by using an active-set code such as MINOS [18]. A comparison of such an active-set implementation of the enumerative method with a reduced-gradient based version [15], shows that the former is in general more efficient to find a solution to the GLCP [21].

4 A Complementarity Active-Set Algorithm

The Complementarity Active-Set Algorithm [16] uses an active-set strategy [19] to find a stationary point of MPEC, that is, a solution satisfying the necessary first-order KKT conditions of the nonlinear program (NLP), that is obtained from MPEC (1) by considering the complementarity conditions $z_i w_i = 0$, $i = 1, \ldots, n$ as constraints. Thus this NLP has the following form

$$\text{NLP:} \quad \begin{array}{ll} \textit{Minimize} & c^T z + d^T y \\ \textit{subject to} & \left.\begin{array}{l} Ew = q + Mz + Ny \\ Cy = b \\ z \geq 0, \quad w \geq 0, \quad y \geq 0 \\ z_i w_i = 0, \; i = 1, \ldots, n. \end{array}\right] \text{GLCP} \end{array} \tag{33}$$

where $q \in \mathbb{R}^p$, $c, w, z \in \mathbb{R}^n$, $d, y \in \mathbb{R}^m$, $E, M \in \mathbb{R}^{p\times n}$, $N \in \mathbb{R}^{p\times m}$, $C \in \mathbb{R}^{l\times m}$ and $b \in \mathbb{R}^l$.

The algorithm consists essentially of using an active-set technique on the set of solutions of the GLCP given by the constraints of the MPEC. Thus at each iteration k, the iterates (w, z, y) satisfy the constraints of (1), and the set of the active constraints is given by

$$\left.\begin{array}{rl} Ew - Mz - Ny &= q \\ Cy &= b \\ w_i &= 0, \; i \in L_w \subseteq \{1, \ldots, n\} \\ z_i &= 0, \; i \in L_z \subseteq \{1, \ldots, n\} \\ y_i &= 0, \; i \in L_y \subseteq \{1, \ldots, m\} \end{array}\right\} \tag{34}$$

where L_z , L_y, and L_w are the sets of the currently active constraints corresponding to the nonnegative constraints on the variables z, y, and w, respectively and $L_z \cup L_w = \{1, \ldots, n\}$.

The active constraints (34) constitute a linear system of the form

$$D_k x = g^k,$$

where $x = (w^T, z^T, y^T)^T$ and $D_k \in \mathbb{R}^{t\times(2n+m)}$, with $t = l+p+ \mid L_w \mid + \mid L_z \mid + \mid L_y \mid$ and $\mid H \mid$ is the cardinality of the set H, where that p and l are the number of rows of the matrices A and $[E \quad -M - N]$ respectively.

The first-order optimality conditions for the problem

$$Minimize \{f(x) : \ D_k x = g^k\}$$

can be written in the form

$$\nabla f(x) = D_k^T \mu$$
$$D_k x = g^k.$$

In order to facilitate a unique set of Lagrange multipliers μ, the following condition is assumed to hold throughout the proposed procedure:

Nondegeneracy Assumption: $t \leq 2n + m$ and $\text{rank}(D_k) = t$.

This hypothesis is not restrictive under the usual full row rank of the matrices C and $[E, -M, -N]$. Consequently, the active-set is always linearly independent. Furthermore, let us partition the Lagrange multipliers vector μ into three subvectors denoted by

$\beta \rightarrow$ subvector associated the first set of equality constraints in (34)
$\vartheta \rightarrow$ subvector associated the second set of equality constraints in (34)
$\lambda_i^x \rightarrow$ subvector associated with $x_i = 0$ in the last three sets of equality constraints in (34).

The main steps of the complementary active-set algorithm are described below.

COMPLEMENTARITY ACTIVE-SET ALGORITHM - CASET

Step 0

Set $k = 1$ and find a solution x^k of the GLCP associated with MPEC. Let $D_k x = g^k$ be the set of active constraints at x^k and let L_y, L_z, and L_w be the index sets associated with the nonnegative active constraints $y_i = 0$, $z_i = 0$, and $w_i = 0$, respectively.

Step 1 *Optimality Conditions*

If x^k is not a stationary (KKT) point (see [4]) for the Equality Problem

$$\text{EP: } Minimize \ f(x) \quad subject \ to \ D_k x = g^k,$$

then go to Step 2. Otherwise, there exists a unique μ such that

$$D_k^T \mu = \nabla f(x^k),$$

and two cases can occur:

1. If
$$\begin{aligned} \lambda_i^y &\geq 0 \text{ for all } i \in L_y \\ \lambda_i^z &\geq 0 \text{ for all } i \in L_z \cap L_w \\ \lambda_i^w &\geq 0 \text{ for all } i \in L_z \cap L_w, \end{aligned}$$
stop: x^k is a stationary point for MPEC.
2. If there exists at least one i such that
$$\begin{aligned} \lambda_i^y &< 0 \text{ for } i \in L_y \\ \text{or } \lambda_i^z &< 0 \text{ for } i \in L_z \cap L_w \\ \text{or } \lambda_i^w &< 0 \text{ for } i \in L_z \cap L_w, \end{aligned}$$
remove an active constraint $y_i = 0$, or $z_i = 0$, or $w_i = 0$, associated with the most negative Lagrange multiplier. Let $D_{k_i}x = g_i^k$ be the row removed from $D_k x = g^k$, and rearrange the rows of $D_k x = g^k$ in the following way
$$D_k = \begin{bmatrix} \bar{D}_k \\ D_{k_i} \end{bmatrix}, \quad g^k = \begin{bmatrix} \bar{g}^k \\ g_i^k \end{bmatrix}.$$
Find a direction d such that $\nabla f(x^k)^T d < 0$, $\bar{D}_k d = 0$, and $D_{k_i} d > 0$. Replace D_k by $\bar{D}_k$ and go to Step 3.

Step 2 *Determination of Search Direction*

Find a descent direction for f in the set of active constraints, i.e, find d such that
$$\begin{aligned} \nabla f(x^k)^T d &< 0 \\ D_k d &= 0. \end{aligned}$$

Step 3 *Determination of Stepsize*

1. Find the largest value α_{max} of α such that
$$x^k + \alpha d \geq 0,$$
from
$$\alpha_{max} = min\left\{ \frac{x_i^k}{-d_i} : d_i < 0, \quad i \notin (L_z \cup L_w \cup L_y) \right\}.$$
2. Compute $0 < \alpha_k \leq \alpha_{max}$ such that
$$x^k + \alpha_k d$$
provides a sufficient decrease for f using any line search technique [4].

If $\alpha_k = +\infty$, stop; MPEC is unbounded.

Step 4 *Update of iterate*

Compute
$$x^{k+1} = x^k + \alpha_k d.$$

If $\alpha_k = \alpha_{max}$, add to the active set the constraints $x_i \geq 0$ for which α_{max} was attained such that the nondegeneracy assumption remains true. Return to Step 1 with $k := k + 1$.

As it is shown in [16], this algorithm possesses global convergence to a Stationary Point of the MPEC under a nondegenerate condition. The algorithm can also be extended to deal with degenerate cases and can be implemented by using an active-set code, such as MINOS. Computational experience reported in [16] on the solution of MPECs, taken from different sources, indicates that the algorithm CASET is quite efficient to find stationary points for MPECs of moderate and even large dimensions.

5 A Sequential Linear Complementarity Algorithm - SLCP

In this section, we briefly describe a Sequential Linear Complementarity (SLCP) algorithm [14] that finds a global minimum of the MPEC, by computing a set of stationary points with strictly reducing objective function values. To do this, in each iteration k of the algorithm the objective function is replaced by the cut

$$c^T z + d^T y \leq \lambda_k,$$

where λ_k is a constant to be defined later. So in each iteration a GLCP(λ_k) of the form below is solved first:

$$\begin{aligned} & Ew = q + Mz + Ny \\ & w \geq 0, \ z \geq 0 \\ & y \in K_y \\ & c^T z + d^T y \leq \lambda_k \\ & z^T w = 0. \end{aligned} \tag{35}$$

Let $(\bar{w}, \bar{z}, \bar{y})$ be such a solution. Then algorithm CASET with this initial point is applied to find a stationary point of the MPEC. To guarantee that the algorithm moves toward a global minimum of the MPEC, the sequence of step lengths $\{\lambda_k\}$ must be strictly decreasing. An obvious definition for λ_k is as below

$$\lambda_k = c^T z^{k-1} - d^T y^{k-1} - \gamma \mid c^T z^{k-1} + d^T y^{k-1} \mid$$

where γ is a small positive number and $(w^{k-1}, z^{k-1}, y^{k-1})$ is the stationary point found in the previous iteration.

Now consider the GLCP(λ_k) given by (35). Then there are two possible cases as stated below.

(i) GLCP(λ_k) has a solution that has been found by the enumerative method discussed before.
(ii) GLCP(λ_k) has no solution.

In the first case the algorithm uses this solution to find the stationary point of the MPEC associated to this iteration. In the last case, the stationary point $(\bar{z}^{k-1}, \bar{y}^{k-1})$ computed in iteration $(k-1)$ is an ε-global minimum for the MPEC, where

$$\varepsilon = \gamma \mid c^T \bar{z}^{k-1} + d^T \bar{y}^{k-1} \mid \tag{36}$$

and γ is a small positive tolerance.

The steps of the SLCP algorithm are presented below.

SEQUENTIAL LINEAR COMPLEMENTARY ALGORITHM - SLCP

Step 0 Set $k = 0$. Let $\gamma > 0$ a positive tolerance and $\lambda_0 = +\infty$.

Step 1 Solve GLCP(λ_k). If GLCP(λ_k) has no solution, go to Step 2. Otherwise, let (w^k, z^k, y^k) be a solution of GLCP(λ_k). Apply CASET algorithm with this starting point to find a stationary point of MPEC. Let $(\bar{w}^k, \bar{z}^k, \bar{y}^k)$ be such a point. Let

$$\lambda_{k+1} = c^T \bar{z}^k + d^T \bar{y}^k - \gamma \mid c^T \bar{z}^k + d^T \bar{y}^k \mid .$$

Set $k = k + 1$ and repeat the step.

Step 2 If $k = 0$, MPEC has no solution. Otherwise, $(\bar{z}^{k-1}, \bar{y}^{k-1})$ is an ε-global optimal solution for the MPEC, where ε is given by (36) (it is usually a global minimum of the MPEC).

As discussed in Section 3, the enumerative method faces great difficulties to show that the last GLCP has no solution. So the SLCP algorithm is able to find a global minimum, but it has difficulties to establish that such a solution has been found. Computational experience presented in [13, 14, 15] confirms this type of behavior in practice. It is also important to add that the SLCP algorithm can be implemented by using an active-set code such as MINOS. In fact the SLCP algorithm only uses the enumerative and the CASET methods, which are both implemented by using this type of methodology.

6 Computational experience

In this section some computational experience is reported on the solution of some structural models introduced in [8] by exploiting the MPEC formulation and using the algorithms CASET and SLCP. All the computations have been performed on a Pentium IV 2.4GHz machine having 256 MB of RAM.

(I) Test Problems

In each test problem the corresponding initial structure consists of nodal points and bars and takes a similar form to the type mesh displayed in Figure 1.

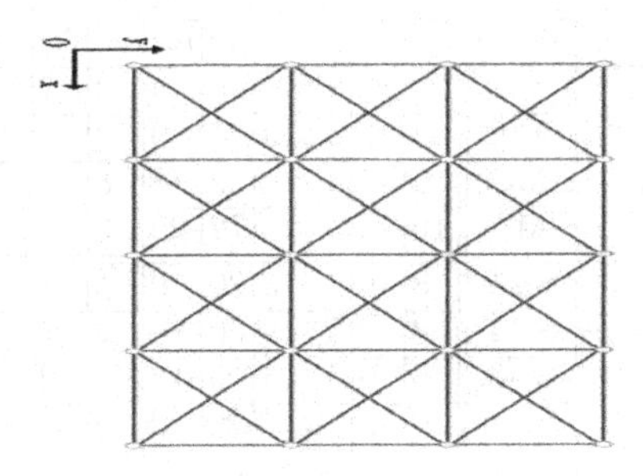

Fig. 1. Initial mesh

The main goal of this model is to find the set of included bars in the so–called optimal shape of the structure, which is given by the values of the $0-1$ variables x_i in the optimal solution of the problem.

Different types of sizes of initial meshes, as well as of applied nodal forces have been taken in consideration in the constitution of the test problems. Four sizes of initial meshes, M_i, $i = 0, \ldots, 3$, have been considered whose topologies are presented in Table 1 and that lead to five test problems PT0 to PT4, according to the following definitions:

- PT0 - mesh M0 and only one nodal load is applied ($f^1_{x_4} = 65$, $f^1_{y_4} = 0$).
- PT1, PT2 - mesh M1 and two types of applied nodal loads are applied. In PT1 is applied only one nodal load ($f^1_{x_8} = 0$, $f^1_{y_8} = -65$), while two nodal loads ($f^1_{x_8} = 0$, $f^1_{y_8} = -65$, $f^2_{x_9} - 40$, $f^2_{y_9} = -40$) are applied in PT2.
- PT3 - mesh M2 and two nodal loads are simultaneously applied ($f^1_{x_3}45.9619$, $f^1_{y_3} = -45.9619$, $f^1_{x_{12}} = 45.9619$, $f^1_{y_{12}} = 45.9619$).
- PT4 - mesh M3 and only one nodal load is applied ($f^1_{x_{23}} = 0$, $f^1_{y_{23}} = -65$).

In these definitions the following parameters are used:

$f^l_{x_n}$ nodal load in (kN) applied in node n in direction Ox for loads combination l;

$f^l_{y_n}$ nodal load in (kN) applied in node n in direction Oy for loads combination l.

In Table 1 are included the following notations:

nal dimension of the mesh in terms of number of nodal in Ox and Oy axes, respectively (in Figure 1, $nal = 5 \times 4$)

h_x total length (in m) to the Ox axis

h_y total length (in m) to the Oy axis

nb number of bars

nn number of nodes

na number of simple supports

S_i set of discrete sizes available for cross-sectional area of bar i (in cm^2)

N_i number of discrete sizes available for cross-sectional area of bar i

	Mesh	h_x	h_y	nal	nb	nn	na	N_i	S_i
Group I	M0	4	3	2×2	6	4	3	1	3
	M1	8	6	3×3	20	9	3	1	3
	M2	6	9	3×4	29	12	8	1	3
	M3	16	12	5×5	72	25	3	1	3
Group II	Sm1	8	6	3×3	20	9	3	2	0.5;3
	Sm2	8	6	3×3	20	9	3	3	0.5;1;2
	Sm3	6	9	3×4	29	12	8	2	0.5;3
	Sm4	6	9	3×4	29	12	8	3	0.5;2;3

Table 1. Test Problems Meshes

In the first group of test problems, structures have been considered for which a unique discrete value is available for cross-sectional area of each bar. In the second group it is allowed that each bar of the structure assumes one of the values in a finite set S_i of discrete sizes available for its cross-sectional area. This last group leads to four additional test problems, assigned for St1, St2, St3 and St4, and whose associated initial meshes are Sm1, Sm2, Sm3 and Sm4, respectively. The meshes Sm1 and Sm2 have the same dimensions of the M1 mesh, while Sm3 and Sm4 have the same dimensions of the ones in M2. The nodal loads applied in St1 and St2 are the same as in Pt1, while in St3 and St4 are the same as in Pt3. The number of constraints (nr) and the number of variables (nv) of formulation OPT associated to these test problems are presented in Table 2.

		OPT	
	Prob	nr	nv
Group I	Pt0	93	51
	Pt1	273	136
	Pt2	449	220
	Pt3	387	205
	Pt4	921	444
Group II	St1	313	176
	St2	353	216
	St3	445	263
	St4	503	321

Table 2. Dimensions of test problems

In all test problems the displacements and bars stress limits considered are $u_{max} = -u_{min} = 50cm$, $t_{max} = -t_{min} = 355MPa$, respectively and the partial safety factor λ is equal to 1.5.

(II) Solution of MPECs

This section reports the computational experience performed with the algorithms CASET and SLCP for the solution of MPECs associated with the integer linear program (OPT).

The dimensions of the resultant MPEC problems are included in Table 3, where *nr*, *nv* and *nvc* denote the number of constraints, number of variables and pairs of complementary variables, respectively.

		MPEC		
	PROB	*nr*	*nv*	*nvc*
GROUP I	PT0	124	111	30
	PT1	361	310	87
	PT2	537	394	87
	PT3	511	451	123
	PT4	1213	1026	291
GROUP II	ST1	461	470	147
	ST2	561	630	207
	ST3	656	683	210
	ST4	801	915	297

Table 3. Dimensions of MPEC test problems

Table 4 includes the performance of the integer program code OSL for finding a global minimum to the test problems [8]. In this table, as well as in the sequel, ND and NI are, respectively, the number of searched nodes and the number of iterations (pivot steps) performed by the process, T is the total CPU time in seconds for solving the optimization problem and OBJ. is the objective function value obtained by the algorithm. Note that for problem PT4, OSL code has not been able to terminate after 25000000 pivots steps.

The first computational experience has been performed with the CASET algorithm for finding a stationary point for the MPECs associated with the structural optimization problems and analyzing if this solution is near to the global optimal solution. Table 5 includes the performance of this algorithm on the solution of these test problems.

The numerical results clearly indicate that CASET algorithm has been able to find a structure with a volume close to the global optimal one. This is particularly evident for the problems with exactly one possible cross area for each bar. Furthermore this solution has been found in a quite small amount of effort as compared to that of the OSL code.

Table 6 includes the computational results achieved by the Sequential Complementary Algorithm on the solution of the test problems. In this table, besides the previously used parameters, IT represents the number of GLCP(λ_k) solved by algorithm SLCP, while NIS, TS and NDS are the number

	OSL			
PROB	NI	T	ND	OBJ (dm^3)
PT0	53	0.04	7	3.60
PT1	3033	0.69	311	10.80
PT2	5579	1.77	497	12.90
PT3	891143	325.64	82075	11.92
PT4	>25000000	15018.78	347541	27.30
ST1	64943	22.80	8132	7.05
ST2	57473	30.01	10052	4.90
ST3	4788682	3996.54	411084	6.29
ST4	20606789	61486.08	1496081	5.46

Table 4. Computation of Global Minimum of Integer Program OPT by using the OSL code

	CASET		
PROB	NI	T	OBJ
PT0	116	0.03	3.60
PT1	522	0.24	11.10
PT2	961	0.44	14.10
PT3	922	0.44	12.66
PT4	9859	9.00	27.90
ST1	726	0.36	7.10
ST2	820	0.56	8.75
ST3	1819	1.03	9.40
ST4	1731	1.44	9.80

Table 5. Computation of a stationary point of MPEC by using the CASET algorithm

	SLCP							
PROB	IT	NI	T	ND	NIS	TS	NDS	OBJ.
PT0	2	249	0.05	46	128	0.04	9	3.60
PT1	2	5572	2.30	907	494	0.17	27	10.80
PT2	3	28281	14.73	3828	1015	0.44	44	12.90
PT3	6	63210	30.78	8278	23843	12.16	3417	11.92
PT4(*)	4	140015	124.00	3884	39968	35.25	1073	27.30
ST1	2	11370	5.09	1177	691	0.28	27	7.05
ST2	16	35871	23.90	4723	12698	9.28	1779	4.90
ST3	15	498688	279.89	34578	160949	86.52	9719	6.29
ST4	16	1602271	1171.91	99962	421851	318.40	31144	5.46

Table 6. Application of SLCP algorithm to the structural problems

of pivot steps, the CPU time in seconds and the number of nodes searched until the optimal solution is obtained, respectively. Moreover, the notation (*)

is used in problem PT4 to indicate that the solution of the last GLCP was interrupted because the maximum limit of 100000 pivot steps was exceeded.

A comparison between the SLCP algorithm and the code OSL shows that the latter procedure performs better for problems of smaller dimensions. However, as the dimension increases the SLCP algorithm becomes more efficient to obtain a global minimum. It is important to add that the SLCP algorithm computes stationary points of the MPEC with strictly decreasing objective function value. Since each one of these stationary points corresponds to a feasible solution of the zero-one integer programming formulation of the structural model, then the engineer is able to receive a number of structures (equal to the number of iterations of the algorithm SLCP) in a reasonable amount of time. For instance for problem ST4 the algorithm SLCP requires only 421851 pivot steps to give the engineer 16 structures including the one given by the CASET method and the global optimal structure.

7 Conclusions

In this paper we have investigated the solution of a zero-one integer program associated with a structural model by using two MPEC techniques. A Complementarity Active-Set (CASET) algorithm for finding a stationary point of a MPEC and a Sequential Linear Complementarity (SLCP) algorithm for computing a global minimum have been considered in this study. Numerical results of some experiments with these techniques show that both procedures are in general efficient for their purposes. We believe that the results shown in this paper may influence the use of MPEC algorithms to process integer programming problems. This is a subject of future research.

Acknowledgement: Support for the first author was provided by *Instituto de Telecomunicações* and by *FCT* under grant POCTI/35059/MAT/2000.

References

1. I. Arora and M. Haung: Methods for optimization of nonlinear problems with discrete variables: a review. *Structural Optimization* 8(1994), 69-85.
2. C. Audet, P. Hansen, B. Jaumard and G. Savard: Links between the Linear Bilevel and Mixed 0-1 Programming Problems, *Journal of Optimization Theory and Applications* 93(1997), 273-300.
3. J. Bauer: A survey of methods for discrete optimum structural design, *Computer Assisted Mechanics and Engineering Sciences* 1(1994), 27-38.
4. M. Bazaraa, H. Sherali and C.Shetty: *Nonlinear Programming: Theory and Algorithms*, John Wiley & Sons, New York, 1993.
5. S. Bollapragada, O. Ghattas and J. Hoocker: Optimal Design of Truss Structures by Logic-Based Branch and Cut, *Operations Research* 49(2001), 42-51.

6. Eurocode 1 - EN 1991: *Basis of Design and Actions on Structures*, CEN, Brussels, 1998.
7. Eurocode 2 - EN 1992: *Design of Concrete Structures - Part1: General Rules and Rules for Buildings*, CEN, Brussels, 1999.
8. A. Faustino, J. Júdice, A. Neves and I. Ribeiro: An integer programming model for truss topology optimization, Working Paper, 2005.
9. A. Ghali, A. M. Nevilleand and T. G. Brown: *Structural analysis: A unified classical and matrix approach*, Spon Press, London, 2003.
10. O. Ghattas and I. Grossmann: MINLP and MILP strategies for discrete sizing structural optimization problems, In: Proceedings of ASCE 10th Conference on Electronic Computation, Indianapolis, 1991.
11. A. Ghosh and A. Mllik: *Theory of Mechanisms and Machines*, Affiliated East-West Press, New Delhi, 1988.
12. X. Guo, G. Cheng and K. Yamazaki: A new approach for the solution of singular optima in truss topology optimization with tress and local buckling constraints, *Structural and Multidisciplinary Optimization* 22(2001), 364-372.
13. J. Júdice and A. Faustino: A computational analysis of LCP methods for bilinear and concave quadratic programming, *Computers and Operations Research* 18(1991), 645-654.
14. J. Júdice and A. Faustino: A sequential LCP algorithm for bilevel linear programming, *Annals of Operations Research* 34(1992), 89-106.
15. J. Júdice, A. Faustino and I. Ribeiro: On the solution of NP-hard Linear Complementarity Problems, TOP-*Sociedad de Estatstica e Investigacion Operativa*, 10(2002)1, 125-145.
16. J. Júdice, H. Sherali, I. Ribeiro and A. Faustino: A Complementarity Active-Set Algorithm for Mathematical Programming Problems with Equilibrium Constraints, Working Paper, 2005.
17. J. Júdice and L. Vicente: On the solution and complexity of a generalized linear complementarity problem, *Journal of Global Optimization*, 4(1994), 415-424.
18. B. Murtagh and A. Saunders: MINOS 5.0 User's Guide, Department of Operations Research, Stanford University, Systems optimization Laboratory, Technical Report No. SOL 83-20, 1983.
19. J. Nocedal and S. Wright: *Numerical Optimization*, Springer-Verlag, New York, 1999.
20. M. Pyrz and J. Zawidzka: Optimal discrete truss design using improved sequential and genetic algorithm, *Engineering Computations*, 18(2001), 1078-1090.
21. I. Ribeiro: Global Optimization and Applications to Structural Engineering (in Portuguese), PhD thesis, University of Porto, Porto, 2005.
22. G. Rozvany, M. Bendsoe and Kirsch: Layout optimization of structures, *Applied Mechanics Reviews*, 48(1995), 41-118.
23. E. Salajegheh and G. Vanderplaats: Optimum design of trusses with discrete sizing and shape variables, *Structural Optimization* 6(1993), 79-85.
24. H. Sherali and W. Adams: *A Reformulation-Linearization Technique for Solving Discrete and Continuous Nonconvex Problems*, Kluwer Academic Publishers, Boston, 1999.
25. A. Templeman: Heuristic methods in discrete structural optimization, In: W. Gutkowski (Ed.): Discrete Structural Optimization, CISM Courses and Lectures No. 373, Udine, Italy, 1997, 135-165.
26. O. Zienkiewicz: *The Finite Element Method*, MacGraw-Hill, Berkshire, 1997.

On the control of an evolutionary equilibrium in micromagnetics

Michal Kočvara[1], Martin Kružík[12] and Jiří V. Outrata[1]

[1] Institute of Information Theory and Automation, Academy of Sciences of the Czech Republic, Pod vodárenskou věží 4, CZ-182 08 Praha 8, Czech Republic
{kocvara,kruzik,outrata}@utia.cas.cz

[2] Fakulty of Civil Engineering, Czech Technical University, Thákurova 7, CZ-166 29 Praha 6, Czech Republic

Summary. Optimal control of magnetization in a ferromagnet is formulated as a mathematical program with evolutionary equilibrium constraints. To this purpose, we construct an evolutionary infinite-dimensional model which is discretized both in the space as well as in time variables. The evolutionary nature of this equilibrium is due to the hysteresis behavior of the respective magnetization process. To solve the problem numerically, we adapted the implicit programming technique. The adjoint equations, needed to compute subgradients of the composite objective, are derived using the generalized differential calculus of B. Mordukhovich. We solve two test examples and discuss numerical results.

Keywords: coderivatives, ferromagnetism, hysteresis, implicit programming technique, mathematical programs with equilibrium constraints

2000 Mathematical Subject Classification: 35Q60, 49J, 78M30, 90C26

1 Introduction

In connection with the study of rate-independent processes ([20, 21]), an evolutionary equilibrium model has been introduced which takes into account irreversibility of changes occurring during the process. In this model, after a suitable discretization, one gets a sequence of optimization problems, where the solution of the ith problem enters the $(i+1)$th problem as a parameter. We would like to point out that this sequence of coupled optimization problems does not amount to a discrete-time optimal control problem. Indeed, in that problem one optimizes over the whole time interval, whereas in the evolutionary equilibrium single optimization problems are associated with each (discrete) time instant. Evolutionary equilibria were studied in connection with hysteresis ([21]), with elastoplasticity ([19]), and with delamination ([9]). In

S. Dempe and V. Kalashnikov (eds.), *Optimization with Multivalued Mappings*, pp. 143-168
©2006 Springer Science + Business Media, LLC

[10], two of the authors studied an optimization problem, in which such an evolutionary equilibrium arises among the constraints. As to our knowledge, this was the first attempt to study programs with such constraints. Using the generalized differential calculus of B. Mordukhovich, first order optimality conditions were derived there. Unfortunately, due to multiplicity of equilibria, it is very difficult to propose a reasonable procedure to the numerical solution of the problem from [10]. It seems, however, that evolutionary equilibrium models can arise in other application areas, where the uniqueness of equilibria can be ensured. This is, for instance, the case of the model from [14, 36] describing the magnetization of a piece of material in an external magnetic field. This model, based on a detailed study of magnetic microstructures of the considered material, leads to a sequence of coupled optimization problems which are convex, nonsmooth and the uniqueness of their solutions can be enforced. One can consider the external magnetic field as a control variable and think about an "optimal" magnetization of the sample at the given terminal time. This problem is similar to the control of delamination studied in [10] so that the respective optimality conditions can be derived exactly in the same way. Moreover, due to the uniqueness of equilibria, an effective method to the numerical solution of this problem can be proposed.

The aim of our paper is

- to adopt the so-called implicit programming approach (ImP), analyzed thoroughly in [17, 26, 3, 11] to optimization problems with (locally) uniquely solvable evolutionary equilibria among the constraints, and
- to apply the proposed technique to a particular problem, in which we control magnetization of a specimen.

The structure of the paper is as follows. In Section 2 we develop a variant of ImP for evolutionary equilibria. One such equilibrium is thoroughly studied in Section 3. We start with the original infinite-dimensional formulation, construct a suitable discretization and investigate the relevant properties. Section 4 then deals with an application of the results obtained in Section 2 to the discretized equilibrium derived in Section 3. This section also contains results of numerical experiments.

We have not found any work dealing with this type of problems. We refer to Reimers and Della Torre [29] for an inverse hysteresis model based on a Preisach hysteresis operator. The authors consider there a hysteresis model calculating for a given external magnetic field the magnetization response. Mathematically, it amounts to the solution of a system of nonlinear first order ordinary differential equations. The inverse model is then obtained by formal inversion of system equations. Any analysis validating this approach is, however, missing.

Having a given evolution of the magnetization, the external magnetic field is calculated in [29] using the inverse model. Afterwards, the authors use the calculated external field as an input for the Preisach-based hysteresis model and compare the original and calculated magnetization. This goal can be, however, rigorously achieved by a slight generalization of the approach proposed in the sequel.

Our notation is standard. The unit matrix is denoted by $\mathbb{I}$ and lin a is the linear hull of a vector a. If f is a differentiable function of three variables, then $\nabla_i f$ means its gradient (Jacobian) with respect to the ith variable, $i = 1, 2, 3$. For a multifunction $\Phi : \mathbb{R}^p \rightsquigarrow \mathbb{R}^q$, Gph $\Phi = \{(x, y) \in \mathbb{R}^p \times \mathbb{R}^q | \ y \in \Phi(x)\}$ and for a function $f : \mathbb{R}^p \to \bar{\mathbb{R}}$, epi $f = \{(x, r) \in \mathbb{R}^p \times \mathbb{R} | \ r \geq f(x)\}$. If A is an $m \times n$ matrix and $J \subset \{1, 2, \ldots, n\}$, then A_J is the submatrix of A, whose columns are specified by J. Similarly, for a vector $d \in \mathbb{R}^n$, d_J is a subvector composed from the components specified by J. In Sections 3 and 4 we work with vectors having the structure $d = (d^1, \ldots, d^s)$, where $d^i \in \mathbb{R}^l$, $i = 1, \ldots, s$. Then $(d)^j$ is the jth component of the whole vector d, whereas d^{ij} means the jth component of the subvector d^i. Finally, $o : \mathbb{R}_+ \to \mathbb{R}$ is a function such that $\lim_{\lambda \to 0} o(\lambda)/\lambda = 0$.

For reader's convenience, we recall definitions of basic notions from the generalized differential calculus of B. Mordukhovich that will be extensively used in the sequel.

Consider a closed set $\Pi \subset \mathbb{R}^p$.

Definition 1.1 *Let $\overline{a} \in \Pi$.*
(i) The Fréchet normal cone *to Π at $\overline{a}$, denoted $\widehat{N}_\Pi(\overline{a})$, is given by*

$$\widehat{N}_\Pi(\overline{a}) = \{v \in \mathbb{R}^p \mid \langle v, a - \overline{a} \rangle \leq o(\|a - \overline{a}\|) \quad \text{for } a \in \Pi\}.$$

(ii) The limiting normal cone *to Π at $\overline{a}$, denoted $N_\Pi(\overline{a})$, is given by*

$$N_\Pi(\overline{a}) := \limsup_{a \xrightarrow{\Pi} \overline{a}} \widehat{N}_\Pi(a),$$

*where "*lim sup*" stands for the upper limit of multifunctions in the sense of Kuratowski-Painlevé ([1]) and $a \xrightarrow{\Pi} \overline{a}$ means $a \to \overline{a}$ with $a \in \Pi$.*

If $N_\Pi(\bar{a}) = \hat{N}_\Pi(\bar{a})$, we say that Π is *normally regular* at $\bar{a}$. Each convex set is normally regular at all its points and N_Π amounts then to the classic normal cone from convex analysis. In general, however, $N_\Pi(\overline{a})$ is nonconvex, but the multifunction $N_\Pi(\cdot)$ is upper semicontinuous at each point of Π (with respect to Π). The local behavior of (extended) real-valued functions and multifunctions is described by subdifferentials and coderivatives defined next.

Definition 1.2 *Let $\varphi : \mathbb{R}^p \to \overline{\mathbb{R}}$ be an arbitrary extended real-valued function and $a \in \operatorname{dom} \varphi$. The set*

$$\partial\varphi(a) := \{\alpha \in \mathbb{R}^p | (\alpha, -1) \in N_{\text{epi}\varphi}(a, \varphi(a))\}$$

is called the limiting subdifferential *of φ at a.*

Definition 1.3 *Let* $\Phi : \mathbb{R}^p \rightsquigarrow \mathbb{R}^q$ *be an arbitrary multifunction with a closed graph and* $(a, b) \in \operatorname{Gph} \Phi$.
(i) The multifunction $\widehat{D}^*\Phi(a,b) : \mathbb{R}^q \rightsquigarrow \mathbb{R}^p$ *defined by*

$$\widehat{D}^*\Phi(a,b)\,(v) := \{u \in \mathbb{R}^p | (u, -v) \in \widehat{N}_{\operatorname{Gph}\Phi}(a,b)\}, \quad v \in \mathbb{R}^q$$

is called the regular coderivative *of* Φ *at* (a, b).
(ii) The multifunction $D^*\Phi(a,b) : \mathbb{R}^q \rightsquigarrow \mathbb{R}^p$ *defined by*

$$D^*\Phi(a,b)\,(v) := \{u \in \mathbb{R}^p | (u, -v) \in N_{\operatorname{Gph}\Phi}(a,b)\}, \quad v \in \mathbb{R}^q$$

is called the coderivative *of* Φ *at* (a, b).
If Φ *is single-valued, we write simply* $\widehat{D}^*\Phi(a)\,(v)$ *and* $D^*\Phi(a)\,(v)$.

The interested reader is referred, e.g., to [22] and [33] where the properties of the above objects are studied in detail.

2 Problem formulation and implicit programming

In accordance with the literature, under mathematical program with equilibrium constraints (MPEC) we understand the optimization problem

$$\begin{aligned} \text{minimize} \quad & \varphi(x, y) \\ \text{subject to} \quad & \\ & y \in S(x) \\ & (x, y) \in \kappa\,, \end{aligned} \tag{1}$$

where $x \in \mathbb{R}^n$ is the *control* or *design* variable, $y \in \mathbb{R}^m$ is the *state* variable, the multifunction $S : \mathbb{R}^n \rightsquigarrow \mathbb{R}^m$ assigns x a (possibly empty) set of solutions to an equilibrium problem and $\kappa \subset \mathbb{R}^n \times \mathbb{R}^m$ comprises all "nonequilibrial" constraints. S is usually defined by a *generalized equation* (GE) which may attain, e.g., the form

$$0 \in F(x, y) + Q(y)\,. \tag{2}$$

In (2), $F : \mathbb{R}^n \times \mathbb{R}^m \to \mathbb{R}^m$ is a continuously differentiable operator, whereas $Q : \mathbb{R}^m \rightsquigarrow \mathbb{R}^m$ is a multifunction with a closed graph. In the formulation of an evolutionary equilibrium one usually has a finite process time interval. After the time discretization, we will thus be dealing with $T \in \mathbb{N}$ time instants uniformly distributed over this interval. Put $m = kT$ with a positive integer k and $y = (y_1, y_2, \ldots, y_T) \in (\mathbb{R}^k)^T$. The "state map" S is now given by the sequence of GEs

$$0 \in F_i(x, y_{i-1}, y_i) + Q_i(y_{i-1}, y_i), \;\; i = 1, 2, \ldots, T \tag{3}$$

with a given *initial state* y_0. In (3) the maps $F_i : \mathbb{R}^n \times \mathbb{R}^k \times \mathbb{R}^k \to \mathbb{R}^k$ are continuously differentiable and $Q_i : \mathbb{R}^k \times \mathbb{R}^k \rightsquigarrow \mathbb{R}^k$ are closed-graph multifunctions. We will call (1) with S defined by (3) *mathematical program with evolutionary equilibrium constraints* and use the acronym MPEEC. Similarly as in [10], we confine ourselves to the case, where φ depends only on y_T, the terminal component of y. In contrast to [10], however, we now concentrate on the numerical solution of MPEEC. To this purpose, we impose the following simplifying assumptions:

A1: φ is continuously differentiable.
A2: $\kappa = \omega \times \mathbb{R}^m$, where $\omega \subset \mathbb{R}^n$ is a nonempty closed set of admissible controls.
A3: The state map S (defined via GEs (3)) is single-valued and locally Lipschitz on an open set containing ω.

While assumption (A1) is not too restrictive for many applications, assumption (A2) complicates the treatment of possible state or mixed state-control constraints which have to be handled via a smooth penalty. The most restrictive is, however, assumption (A3). In fact, this assumption can be replaced by the requirement that S possesses a Lipschitz single-valued localization at each pair (x, y), $x \in \omega$, $y \in S(x)$; see [33]. Such a weakened variant can be verified by using the concept of strong regularity due to Robinson [31]. We return to the question of the verification of (A3) later.

The key idea of implicit programming (ImP) consists in the reformulation of (1) to the form

$$\begin{aligned} &\text{minimize} \quad \Theta(x) \\ &\text{subject to} \\ &\qquad\qquad x \in \omega\,, \end{aligned} \tag{4}$$

where $\Theta(x) := \varphi(x, S_T(x))$, and S_T assigns x the respective terminal state y_T. Due to (A1),(A3), Θ is locally Lipschitz on an open set containing ω and so various methods can be used to the numerical solution of (4); see [17],[26]. In this paper, we apply a bundle method of nonsmooth optimization; in particular, the BT algorithm described in [38]. This means that we must be able to compute for each admissible control x at least one arbitrary subgradient ξ from $\bar{\partial}\Theta(x)$, the Clarke subdifferential of Θ at x. If S happens to be a PC^1 map (see [37]), one can apply the classic implicit function theorem to an essentially active component of S and arrives in this way at a desired subgradient ξ. This technique has been used in [26]. In [11], a different approach is suggested based on the generalized differential calculus of Mordukhovich. This approach is not restricted to PC^1 state maps and we will apply it also in the case of our evolutionary equilibrium. To simplify the notation, let $\widetilde{Q}_1(y_1) := Q_1(y_0, y_1)$.

Theorem 2.1 *Let assumptions (A1)–(A3) hold true,*

$$\bar{x} \in \omega, \bar{y} = (\bar{y}_1, \bar{y}_2, \ldots, \bar{y}_T) = S(\bar{x}), \ \textit{and}$$

$$\bar{c}_i := -F_i(\bar{x}, \bar{y}_{i-1}, \bar{y}_i), i = 1, 2, \ldots, T, \ \textit{with } \bar{y}_0 := y_0.$$

Assume that the three sequences of adjoint vectors

$$q_2, q_3, \ldots, q_T, \; v_1, v_2, \ldots, v_T, \; w_1, w_2, \ldots, w_T$$

fulfill the relations

$$\begin{aligned} v_1 &\in \widehat{D}^* \widetilde{Q}_1(\bar{y}_1, \bar{c}_1)(w_1) \\ (q_i, v_i) &\in \widehat{D}^* Q_i(\bar{y}_{i-1}, \bar{y}_i, \bar{c}_i)(w_i), \quad i = 2, 3, \ldots, T, \end{aligned} \tag{5}$$

and satisfy the adjoint system

$$\begin{aligned} 0 &= \nabla_2 \varphi(\bar{x}, \bar{y}_T) + (\nabla_3 F_T(\bar{x}, \bar{y}_{T-1}, \bar{y}_T))^T w_T + v_T \\ 0 &= (\nabla_3 F_{T-1}(\bar{x}, \bar{y}_{T-2}, \bar{y}_{T-1}))^T w_{T-1} + v_{T-1} \\ &\quad + (\nabla_2 F_T(\bar{x}, \bar{y}_{T-1}, \bar{y}_T))^T w_T + q_T \\ &\cdots\cdots \\ 0 &= (\nabla_3 F_1(\bar{x}, \bar{y}_0, \bar{y}_1))^T w_1 + v_1 + (\nabla_2 F_2(\bar{x}, \bar{y}_1, \bar{y}_2))^T w_2 + q_2 . \end{aligned} \tag{6}$$

Then one has

$$\xi := \nabla_1 \varphi(\bar{x}, \bar{y}_T) + \sum_{i=1}^{T} (\nabla_1 F_i(\bar{x}, \bar{y}_{i-1}, \bar{y}_i))^T w_i \in \bar{\partial}\Theta(\bar{x}). \tag{7}$$

Proof. Let b denote the vector

$$(0, 0, \ldots, \nabla_2 \varphi(\bar{x}, \bar{y}_T)) \in (\mathbb{R}^k)^T.$$

From [33, Thm.10.49] it follows that

$$\begin{aligned} \bar{\partial}\Theta(\bar{x}) \supset \partial\Theta(\bar{x}) \supset \nabla_1 \varphi(\bar{x}, \bar{y}_T) &+ \widehat{D}^* S_T(\bar{x}, \bar{y}_T)(\nabla_2 \varphi(\bar{x}, \bar{y}_T)) \\ &= \nabla_1 \varphi(\bar{x}, \bar{y}_T) + \widehat{D}^* S(\bar{x}, \bar{y})(b). \end{aligned} \tag{8}$$

Further, one observes that

$$S(x) = \{y \in \mathbb{R}^{kT} | \Phi(x, y) \in \Lambda\} ,$$

where

$$\Phi(x, y) = \begin{bmatrix} y_1 \\ -F_1(x, y_0, y_1) \\ y_1 \\ y_2 \\ -F_2(x, y_1, y_2) \\ \cdots\cdots \\ y_{T-1} \\ y_T \\ -F_T(x, y_{T-1}, y_T) \end{bmatrix} \quad \text{and } \Lambda = \operatorname{Gph} \widetilde{Q}_1 \times \mathop{\mathsf{X}}_{i=2}^{T} \operatorname{Gph} Q_i.$$

By virtue of [33, Thm.6.14],

$$\widehat{N}_{\text{Gph } S}(\bar{x},\bar{y}) \supset (\nabla \Phi(\bar{x},\bar{y}))^T \widehat{N}_\Lambda(\Phi(\bar{x},\bar{y})),$$

and thus, by definition,

$$\widehat{D}^* S(\bar{x},\bar{y})(b) \supset \{(\nabla_1 \Phi(\bar{x},\bar{y}))^T p | -b = (\nabla_2 \Phi(\bar{x},\bar{y}))^T p, p \in \widehat{N}_\Lambda(\Phi(\bar{x},\bar{y}))\}. \quad (9)$$

Putting $p = (v_1, -w_1, q_2, v_2, -w_2, \ldots, q_T, v_T, -w_T)$, we can now make use of the decomposition

$$\widehat{N}_\Lambda(\Phi(\bar{x},\bar{y})) = \widehat{N}_{\text{Gph } \widetilde{Q}_1}(\bar{x},\bar{y}_1,\bar{c}_1) \times \bigtimes_{i=2}^{T} \widehat{N}_{\text{Gph } Q_i}(\bar{x},\bar{y}_{i-1},\bar{y}_i,\bar{c}_i).$$

In this way, relation $p \in \widehat{N}_{\text{Gph } \Lambda}(\Phi(\bar{x},\bar{y}))$ implies (5), the adjoint system (6) amounts exactly to

$$(\nabla_2 \Phi(\bar{x},\bar{y}))^T p + b = 0$$

and formula (7) follows from (8) and (9). □

Unfortunately, even in case of very simple equilibria, it can be rather difficult to fulfill relations (5). These relations become, however, substantially easier if we replace the regular coderivatives by (standard) coderivatives. This possibility is examined in the following statement.

Theorem 2.2 *Let all assumptions of Theorem 2.1 be fulfilled with relations (5) replaced by*

$$\begin{aligned} v_1 &\in D^* \widetilde{Q}_1(\bar{y}_1,\bar{c}_1)(w_1) \\ (q_i, v_i) &\in D^* Q_i(\bar{y}_{i-1},\bar{y}_i,\bar{c}_i)(w_i), \quad i = 2,3,\ldots,T. \end{aligned} \quad (10)$$

Further suppose that the map $\Xi : \mathbb{R}^{3kT-1} \rightsquigarrow \mathbb{R}^n \times \mathbb{R}^{kT}$, *defined by*

$$\Xi(z) = \{(x,y) \in \mathbb{R}^n \times \mathbb{R}^{kT} | \Phi(x,y) - z \in \Lambda\},$$

is either polyhedral (see [32]) or possesses the Aubin property around $(0,\bar{x},\bar{y})$ *(see [33, Def. 9.36]). Then one has*

$$\partial \Theta(\bar{h}) \subset \nabla_1 \varphi(\bar{x},\bar{y}_T) + \Bigg\{ \sum_{i=1}^{T} (\nabla_1 F_i(\bar{x},\bar{y}_{i-1},\bar{y}_i))^T w_i | (w_1, w_2, \ldots, w_T)$$
$$\text{fulfills relations (6),(10) with suitable vectors } (v_1, v_2, \ldots, v_T) \text{ and } (q_2, q_3, \ldots, q_T) \Bigg\}. \quad (11)$$

Inclusion (11) becomes equality provided $\widehat{N}_{\text{Gph } \widetilde{Q}_1}(\bar{y}_1,\bar{c}_1) = N_{\text{Gph } \widetilde{Q}_1}(\bar{y}_1,\bar{c}_1)$ *and*
$\widehat{N}_{\text{Gph } Q_i}(\bar{y}_{i-1},\bar{y}_i,\bar{c}_i) = N_{\text{Gph } Q_i}(\bar{y}_{i-1},\bar{y}_i,\bar{c}_i), i = 2,3,\ldots,T$ *(i.e. the graphs of* $\widetilde{Q}_1$ *and* $Q_i, i = 2,3,\ldots,T$, *are normally regular at the respective points).*

Proof. From [33, Thm. 10.49] it follows that under our assumptions

$$\partial\Theta(\bar{x}) \subset \nabla_1\varphi(\bar{x}, \bar{y}_T) + D^*S(\bar{x}, \bar{y})(b).$$

Under the assumption imposed on Ξ, one has the inclusion

$$N_{\mathrm{Gph}\ S}(\bar{x}, \bar{y}) \subset (\nabla\Phi(\bar{x}, \bar{y}))^T N_\Lambda(\Phi(\bar{x}, \bar{y}))\,;$$

see [6, Thm. 4.1],[22, Cor. 5.5]. Thus we can decompose Λ in the same way as in the proof of Theorem 2.1 and arrive directly at formula (11). Concerning the second assertion, it suffices to compare (7) and (11) under the imposed regularity of $\mathrm{Gph}\,\widetilde{Q}_1$ and $\mathrm{Gph}\ Q_i, i = 2, 3. \ldots, T$. □

On the basis of the above analysis, we can now return to the assumption (A3). The single-valuedness of S can be ensured in different ways and in many cases does not represent a serious problem. To enforce the local Lipschitz continuity of S, one can require that all functions F_i are affine and all multifunctions Q_i are polyhedral. The result then follows from [32]. Alternatively, by virtue of Theorem 2.3, we can impose the assumption (A4) below at all pairs $(\bar{x}, \bar{y})$, where $\bar{x} \in \omega$ and $\bar{y} = S(\bar{x})$.

Let us call the adjoint system (6) *homogeneous*, provided $\nabla_2\varphi(\bar{x}, \bar{y})$ is replaced by the zero vector.

A4: The only vectors $q_2, q_3, \ldots, q_T$, $v_2, v_3, \ldots, v_T$ and $w_2, w_3, \ldots, w_T$ satisfying (10) and the homogeneous adjoint system are the zero vectors.

The role of (A4) is explained in the following statement.

Theorem 2.3 *Let $\bar{x} \in \omega$, S be single-valued around $\bar{x}$ and $\bar{y} = S(\bar{x})$. Assume that (A4) is fulfilled. Then Ξ possesses the Aubin property around $(0, \bar{x}, \bar{y})$ and S is Lipschitz around $\bar{x}$.*

Proof. Following [23, Cor. 4.4], we easily infer that the Aubin property of Ξ around $(0, \bar{x}, \bar{y})$ is implied by the requirement

$$\left.\begin{array}{r}(\nabla\Phi(\bar{x}, \bar{y}))^T p = 0 \\ p \in N_\Lambda(\Phi(\bar{x}, \bar{y}))\end{array}\right\} \Rightarrow p = 0.$$

To ensure the Aubin property of S around $\bar{x}$, we can apply the same result and arrive at the stronger condition

$$\left.\begin{array}{r}(\nabla_2\Phi(\bar{x}, \bar{y}))^T p = 0 \\ p \in N_\Lambda(\Phi(\bar{x}, \bar{y}))\end{array}\right\} \Rightarrow p = 0. \tag{12}$$

It thus suffices to decompose Λ as in the previous statements and observe that (12) amounts exactly to (A4). Since S is single-valued on a neighborhood of $\bar{x}$ and possesses the Aubin property around $(\bar{x}, \bar{y})$, it is in fact Lipschitz near $\bar{x}$ and we are done. □

Inclusion (11) provides us with a desired element $\xi \in \bar{\partial}\Theta(\bar{x})$ only if it becomes equality. This happens in the regular case, mentioned in Theorem 2.2, and also under another additional conditions which will not be discussed here. Fortunately, the used bundle methods converge mostly to a Clarke stationary point even if we replace a vector from $\bar{\partial}\Theta(\bar{x})$ by a vector from the right-hand side of (11). Moreover, if some difficulties occur, one can attempt to modify suitably the rules for the selection of vectors $q_2, q_3, \dots, q_T$, $v_1, v_2, \dots, v_T$ and $w_1, w_2, \dots, w_T$ in (10). This remedy will be explained in detail in the last section. To summarize, for reasons of computational complexity, we will supply our bundle methods by vectors coming from the right-hand side of (11). It turns out that this way is sufficient for a successful numerical solution of various difficult MPEECs.

Remark 2.1 In [3] one finds a theory investigating the behavior of a general bundle method, if $\bar{\partial}\Theta$ is replaced by a larger set, satisfying a few reasonable assumptions.

The construction of relations (10) and the adjoint system (6) will be illustrated in Section 4 by means of an equilibrium model (22) derived in the next section.

3 State problem – hysteresis in micromagnetics

In this section we describe a hysteresis model in micromagnetics which will be further used for the formulation of an MPEEC. This model was developed in [13, 14, 15, 35, 36] and is based on Brown's theory [2] for static ferromagnetism tailored to large specimens by DeSimone [4]. In order to get hysteresis behavior, we enrich the static model by a suitable rate-independent dissipative mechanism. The main difficulty is that the formulation leads to a minimization problem with a nonconvex feasible set. As, in general, the minimum is not attained, one must seek a generalization of the notion "solution". This is done here by means of Young measures [34, 40]. A formulation of the continuum model is the content of Subsection 3.1. Subsection 3.2 is devoted to a discretization of the problem. We show that the problem leads to a finite sequence of problems having the structure (3).

Readers not interested in the physical model and/or in its mathematical treatment may skip this section up to formula (25). The main message of this section is that the equilibrium constraint, after a suitable spatial discretization, has the structure (3). Also the respective maps F_i, Q_i are computed.

3.1 Model

The theory of rigid ferromagnetic bodies [2, 16, 18] assumes that a *magnetization* $M : \Omega \to \mathbb{R}^N$, describing the state of a body $\Omega \subset \mathbb{R}^N$, $N = 2, 3$, is subjected to the *Heisenberg-Weiss constraint*, i.e., has a given (in general, temperature dependent) magnitude

$$|M(z)| = m_s \text{ for almost all } z \in \Omega ,$$

where $m_s > 0$ is the *saturation magnetization*, considered here constant.

In the no-exchange formulation, which is valid for large bodies [4], the Helmholtz free energy of a rigid ferromagnetic body $\Omega \subset \mathbb{R}^N$ consists of two parts. The first part is the *anisotropy energy* $\int_\Omega \varphi(M(z))\,dz$ related to crystallographic properties of the ferromagnet. Denoting $S^{N-1} := \{s \in \mathbb{R}^N | \, |s| = m_s\}$, a typical $\varphi : S^{N-1} \to \mathbb{R}$ is a nonnegative function vanishing only at a few isolated points on S^{N-1} determining *directions of easy magnetization.* We are especially interested in uniaxial materials (e.g. cobalt), where φ vanishes exactly at two points. From now on we will assume that the easy axis of the material coincides with the Nth coordinate axis.

The second part of the Helmholtz energy, $\frac{1}{2}\int_{\mathbb{R}^N} |\nabla u_M(z)|^2\,dz$, is the *energy of the demagnetizing field* ∇u_M self-induced by the magnetization M; its potential u_M is governed by

$$\operatorname{div}\big(-\mu_0 \nabla u_M + M\chi_\Omega\big) = 0 \quad \text{in } \mathbb{R}^N , \tag{13}$$

where $\chi_\Omega : \mathbb{R}^N \to \{0,1\}$ is the characteristic function of Ω and μ_0 is the vacuum permeability. The demagnetizing-field energy thus penalizes non-divergence-free magnetization vectors. Standardly, we will understand (13) in the weak sense, i.e. $u_M \in W^{1,2}(\mathbb{R}^N)$ will be called a weak solution to (13) if the integral identity $\int_{\mathbb{R}^N} \big(M\chi_\Omega - \nabla u_m(z)\big) \cdot \nabla v(z)\,dz = 0$ holds for all $v \in W^{1,2}(\mathbb{R}^N)$, where $W^{1,2}(\mathbb{R}^N)$ denotes the Sobolev space of functions in $L^2(\mathbb{R}^N)$ with all first derivatives (in the distributional sense) also in $L^2(\mathbb{R}^N)$. Altogether, the Helmholtz energy $E(M)$, has the form

$$E(M) = \int_\Omega \varphi(M(z))\,dz + \frac{1}{2}\int_{\mathbb{R}^N} |\nabla u_M(z)|^2\,dz . \tag{14}$$

If the ferromagnetic specimen is exposed to some external magnetic field $h = h(z)$, the so-called *Zeeman's energy* of interactions between this field and magnetization vectors equals to $H(M) := \int_\Omega h(z)\cdot M(z)\,dz$. Finally, the following variational principle governs equilibrium configurations:

$$\begin{aligned} \text{minimize} \quad & \mathcal{G}(M) := E(M) - H(M) \qquad (15)\\ & = \int_\Omega (\varphi(M(z)) - h(z)\cdot M(z))\,dz + \frac{1}{2}\int_{\mathbb{R}^N} |\nabla u_M(z)|^2\,dz \end{aligned}$$

subject to

$$(13),\ (M, u_M) \in \mathcal{A} \times W^{1,2}(\mathbb{R}^N) ,$$

where the introduced notation $\mathcal{G}$ stands for *Gibbs' energy* and $\mathcal{A}$ is the set of admissible magnetizations

$$\mathcal{A} := \{M \in L^\infty(\Omega; \mathbb{R}^N) |\; |M(z)| = m_s \text{ for almost all } z \in \Omega\} .$$

As $\mathcal{A}$ is not convex, we cannot rely on direct methods in proving the existence of a solution. In fact, the solution to (15) need not exist in $\mathcal{A} \times W^{1,2}(\mathbb{R}^N)$; see [8] for the uniaxial case. There is a competition of the anisotropy energy in $\mathcal{G}$ preferring the magnetization of the constant length and the demagnetizing field energy preferring it to be zero, which is just what explains quite generic occurrence of the domain microstructure. Mathematically, this is expressed by nonexistence of an exact minimizer of $\mathcal{G}$ and by finer and finer self-similar spatial oscillations necessarily developed in any minimizing sequence of $\mathcal{G}$.

To pursue evolution in an efficient manner, it is important to collect some information about the fine structure "around" a current point $x \in \Omega$ in the form of a *probability measure*, denoted by ν_x, supported on the sphere S^{N-1}. Hence, one has $\nu_x \in \mathfrak{M}(S^{N-1})$, the set of all probability measures on S^{N-1}. Let us furthermore denote by $\mathbb{B}$ the ball $\{M \in \mathbb{R}^N |\; |M| \leq m_s\}$ of the radius m_s. The collection $\nu = \{\nu_x\}_{x\in\Omega}$ is often called a Young measure [40, 34] and can be considered a certain "mesoscopic" description of the magnetization. The average, let us call it *macroscopic magnetization*, $M = M(x)$ at a material point $x \in \Omega$ still remains a worthwhile quantity; it is just the first momentum of the Young measure $\nu = \{\nu_x\}_{x\in\Omega}$, i.e.

$$M(x) = \int_{S^{N-1}} s\, \nu_x(\mathrm{d}s) . \tag{16}$$

Note that the macroscopic magnetization $M : \Omega \to \mathbb{B}$ "forgets" detailed information about the microstructure in contrast with the mesoscopic magnetization $\nu : \Omega \to \mathfrak{M}(S^{N-1})$ which can capture *volume fractions* related to particular directions of the magnetization. It should be emphasized that, though we speak about (collections of) probability measures, our approach is fully deterministic.

On the mesoscopic level we write the Gibbs' energy as [28, 34]:

$$\bar{G}(\nu) = \int_\Omega \Big(\varphi \bullet \nu - h(z) \cdot M(z)\Big)\, \mathrm{d}z + \frac{1}{2}\int_{\mathbb{R}^N} |\nabla u_M(z)|^2\, \mathrm{d}z , \tag{17}$$

where $[v \bullet \nu](z) := \int_{\mathbb{R}^N} v(s)\nu_z(\mathrm{d}s)$, $\mathrm{id} : \mathbb{R}^N \to \mathbb{R}^N$ is the identity, and

$$M(z) = \int_{S^{N-1}} s\nu_z(\mathrm{d}z) . \tag{18}$$

An important property is that $\bar{G}$ is convex with respect to the natural geometry of probability measures $\nu = \{\nu_z\}_{z\in\Omega}$. Let us denote by $\mathcal{Y}(\Omega; S^{N-1})$ the convex set of families of probability measures in $\mathfrak{M}(S^{N-1})$ parameterized by $x \in \Omega$. It is well-known that the minimum of $\bar{G}$ over $\mathcal{Y}(\Omega; S^{N-1})$ is truly attained. We refer to [4, 27, 28] for a mathematically rigorous reasoning.

We can now model the behavior of low-hysteresis materials with reasonable accuracy by minimizing of $\bar{G}$. Varying the external magnetic field h in time produces, however, only a functional graph in an h/M diagram, but no hysteresis loop. On the other hand, magnetically hard materials, as e.g. CoZrDy, display significant hysteresis and cannot be modeled by mere minimization of $\bar{G}$. Therefore, we must enrich our model and focus thereby on hysteresis losses which are independent of the time frequency of h. We refer to [7] for an exposition of various kinds of energy losses in ferromagnets.

Inspired by [39], we describe energetic losses during the magnetization by a phenomenological dissipation potential ϱ depending on the time derivative $\dot{M}$ of the form

$$\varrho(\dot{M}) = \int_\Omega H_c \left| \frac{\mathrm{d}M_N}{\mathrm{d}t} \right| \mathrm{d}z \ ,$$

where the constant $H_c > 0$ is the so-called coercive force describing the width of the hysteresis loop and M_N is the Nth component of the magnetization M.

Following [19], we define a dissipation distance $\mathcal{D}$ by

$$\mathcal{D}(M, \tilde{M}) := \int_\Omega H_c |M_N - \tilde{M}_N| \, \mathrm{d}z \ . \tag{19}$$

Equivalently, this quantity can be written in terms of Young measures as

$$D(\nu, \tilde{\nu}) := \int_\Omega H_c \left| \int_{S^{N-1}} s_N \nu_z(\mathrm{d}s) - \int_{S^{N-1}} s_N \tilde{\nu}_z(\mathrm{d}s) \right| \mathrm{d}z \ , \tag{20}$$

where $s = (s_1, \dots, s_N) \in S^{N-1}$. Obviously, the equivalence of (19) and (20) holds if $M = \mathrm{id} \bullet \nu$ and $\tilde{M} = \mathrm{id} \bullet \tilde{\nu}$. Both the formulas evaluate how much energy is dissipated if we change the magnetization of the specimen from M to $\tilde{M}$. Analogously to [20, 21, 36] we assume that, having a sequence of discrete time instants, an optimal mesoscopic magnetization at the instant $1 \le i \le T$ minimizes $\bar{G}(\nu) + \mathcal{D}(M, M_{i-1})$ over $\mathcal{Y}(\Omega; S^{N-1})$ and subject to (18), where M_{i-1} is the solution at the $(i-1)$th time step.

To be more precise, we first define the Gibbs energy at the time $1 \le i \le T$, by

$$\bar{G}(i, \nu) = \int_\Omega \Big(\varphi \bullet \nu \ - h(i, z) \cdot M(z) \Big) \, \mathrm{d}z + \frac{1}{2} \int_{\mathbb{R}^N} |\nabla u_M(z)|^2 \, \mathrm{d}z \ , \tag{21}$$

where $h(i, \cdot)$ is an external field at this time. Then, starting with an initial condition $\nu_0 \in \mathcal{Y}(\Omega; S^{N-1})$, we find consecutively for $i = 1, \dots, T$ a solution $\nu_i \in \mathcal{Y}(\Omega; S^{N-1})$ of the minimization problem:

$$\begin{aligned} &\text{minimize} \quad I(\nu) := \bar{G}(i, \nu) + D(\nu, \nu_{i-1}) \\ &\text{subject to} \\ &\qquad\qquad \nu \in \mathcal{Y}(\Omega; S^{N-1}) \ . \end{aligned} \tag{22}$$

Next we are going to show that the solution to (22) is unique. The key observation is that the history dependence enters (22) only through the first momentum of the Young measure, i.e., through M_{i-1} and that we can rule out the Young measure ν from the definition of $\bar{G}$ by replacing the term $\int_\Omega \varphi \bullet \nu \, dz$ by $\int_\Omega \varphi^{**}(M(z)\, dz$, with φ^{**} being the convex envelope of $\hat{\varphi}$, where

$$\hat{\varphi}(m) = \begin{cases} \varphi(M) & \text{if } |M| = m_s \\ +\infty & \text{otherwise.} \end{cases}$$

Put

$$G^{**}(i, M) = \int_\Omega (\varphi^{**}(M(z)) - h(i, z) \cdot M(z))\, dz + \frac{1}{2} \int_{\mathbb{R}^N} |\nabla u_M(z)|^2 \, dz \, .$$

Let $M_0 \in L^2(\Omega; \mathbb{R}^N)$ with $M_0(z) \in \mathbb{B}$ for a.a. $z \in \Omega$ be given. We look for a solution M_i, $1 \le i \le T$ of the problem

$$\begin{aligned} &\text{minimize} \quad I^{**}(M) := G^{**}(i, M) + \mathcal{D}(M, M_{i-1}) \\ &\text{subject to} \\ &\qquad M \in L^2(\Omega; \mathbb{R}^N) \quad \text{and } M(z) \in \mathbb{B} \text{ for a.a. } z \in \Omega \, . \end{aligned} \tag{23}$$

The problems (22) and (23) are equivalent in the following sense. If $M_0 = \mathrm{id} \bullet \nu_0$ and if $\nu = (\nu_1, \ldots, \nu_T)$ solves (22), then $\int_{S^{N-1}} s\nu_z(ds)$ solves (23). Conversely, if $M = (M_0, \ldots, M_T)$ solves (23) and $\nu_i \in \mathcal{Y}(\Omega; S^{N-1})$, $1 \le i \le T$, is such that $M_i(z) = \int_{S^{N-1}} s(\nu_i)_z(ds)$ and $\varphi^{**}(M_i(z)) = \int_{S^{N-1}} \varphi(s)(\nu^i)_z(ds)$ for almost all $x \in \Omega$ and $1 \le i \le T$, then $\nu = (\nu_1, \ldots \nu_T)$ solves (22).

We have the following uniqueness result.

Proposition 3.1 *Let $\varphi(s) = \gamma \sum_{i=1}^{N-1} s_i^2$, $\gamma > 0$, $|s| = m_s$. Then the problem (23) has a unique solution.*

Proof. Under the assumptions, $\varphi^{**}(s) = \gamma \sum_{i=1}^{N-1} s_i^2$ for all $s \in \mathbb{R}^N$, $|s| \le m_s$; see [4]. We will proceed by induction.

Suppose that $M_{i-1} \in L^2(\Omega; \mathbb{R}^N)$, $M_{i-1}(z) \in \mathbb{B}$ for a.a. $z \in \Omega$, is given uniquely. Let $\hat{M}$ and $\tilde{M}$ be two different minimizers to I^{**}. Then $\nabla u_{\hat{M}} = \nabla u_{\tilde{M}}$ a.e. in $\mathbb{R}^N$. Indeed, if they were different, the convexity I^{**}, the strict convexity of the demagnetizing field energy, i.e. of $\| \cdot \|^2_{L^2(\mathbb{R}^N; \mathbb{R}^N)}$, and the linearity of the map $M \mapsto \nabla u_M$ would give us that $0.5\hat{M} + 0.5\tilde{M}$ has a strictly lower energy than $I^{**}(\tilde{M}) = I^{**}(\hat{M})$. Similarly, as φ^{**} is strictly convex in the first $(N-1)$ variables, we get that $\tilde{M}_i = \hat{M}_i$ a.e. in Ω for $i = 1, \ldots, N-1$. Put $\beta := \tilde{M}\chi_\Omega - \hat{M}\chi_\Omega$. Then div $\beta = 0$ because $u_\beta = 0$ a.e. in Ω, where u_β is calculated from (13). Moreover, the only nonzero component of β is the Nth one. Therefore $\beta = (0, \ldots, \beta_N)$ with $\beta_N = \beta_N(z_1, \ldots, z_{N-1})$. As β has a compact support we get $\beta = 0$ identically. The proposition is proved. □

Remark 3.1 If there is a unique representation of φ^{**} in terms of a probability measure μ_s on S^{N-1}, i.e., if $\varphi^{**}(s) = \int_{S^{N-1}} \sigma \, \mu_s(d\sigma)$ for all $s \in \mathbb{B}$, then also (22) has a unique solution. This in indeed the case under the assumption of Proposition 3.1. The basic advantage of (22) over (23) is that in (22) we do not work with the convex envelope φ^{**}. Consequently, the formulation (22) can be used even if we do not know an explicit formula for φ^{**}.

3.2 Spatial discretization of (22)

The aim of this subsection is to develop a suitable numerical approximation of the problem (22). Besides a discretization of the domain Ω we will also discretize the support of the Young measure, the sphere S^{N-1}. The, simplest but for our purposes sufficient, discretization is to divide $\bar{\Omega}$ into finite volumes $\{\triangle\}_{j=1}^{s}$, $s \in \mathbb{N}$, such that $\bar{\Omega} = \cup_{j=1}^{s} \triangle_j$, where $\triangle_i$ and $\triangle_j$ have disjoint interiors if $i \neq j$. Moreover, we assume that all $\triangle_j$ have the same N-dimensional Lebesgue measure denoted by $|\triangle|$. Then we will assume that $\nu \in \mathcal{Y}(\Omega; S^{N-1})$ is constant within each finite volume and consists of a finite sum of Dirac masses. Saying differently, we suppose that that for any $x \in \triangle_j$

$$\nu_x = \sum_{i=1}^{l} \lambda^{ji} \delta_{r^i} \ ,$$

with fixed points $r^i \in S^{N-1}$ and coefficients λ^{ji} satisfying the conditions $0 \leq \lambda^{ji} \leq 1$, $\sum_{i=1}^{l} \lambda^{ji} = 1$ for all $1 \leq j \leq s$ and $1 \leq i \leq l$. Thus, ν is fully characterized by coefficients $\{\lambda^{ji}\}$. We refer to [34] for convergence properties of this approximation. As the macroscopic magnetization M is the first moment of ν, we have that M is constant over each $\triangle_j$. We denote its value on $\triangle_j$ by $M^j = (M^{j1}, \ldots, M^{jN})$. Therefore,

$$M^j = \sum_{i=1}^{l} \lambda^{ji} r^i \ , \ 1 \leq j \leq s \ .$$

The demagnetization field energy $\frac{1}{2} \int_{\mathbb{R}^N} |\nabla u_M(z)|^2 \, \mathrm{d}z$ is quadratic in M and therefore it is quadratic in $\lambda = \{\lambda^{ji}\}$. We denote the matrix of the quadratic form assigning λ the energy $\int_{\mathbb{R}^N} |\nabla u_M(z)|^2 \, \mathrm{d}z$ by C. As M is constant over finite volumes, we can work only with spatial averages of the external field h. Hence, we put for $1 \leq i \leq T$ and $1 \leq j \leq s$

$$h_i^j = \frac{1}{|\triangle|} \int_{\triangle_j} h(i, z) \, \mathrm{d}z \ .$$

As a result of this, the discrete anisotropy and external field energies at the time i equal to

$$\sum_{j=1}^{s} |\triangle| \sum_{t=1}^{l} \lambda_i^{jt} (\varphi(r^k) - r^k \cdot h_i^k)$$

and

$$D(\nu, \nu_{i-1}) = \sum_{j=1}^{s} H_c|\triangle| \left| \sum_{k=1}^{l} (\lambda^{jk} - \lambda_{i-1}^{jk}) r^{kN} \right| .$$

If we denote

$$\lambda_i = (\lambda_i^{11}, \dots, \lambda_i^{1l}, \lambda_i^{21}, \dots, \lambda_i^{2l}, \dots, \lambda_i^{s1} \dots \lambda_i^{sl}) ,$$

$$a = H_c|\triangle|(r^{1N}, \dots, r^{lN}) ,$$

and

$$x = |\triangle|(\varphi(r^1) - r^1 \cdot h_i^1, \dots, \varphi(r^l) - r^l \cdot h_i^1, \varphi(r^1) - r^1 \cdot h_i^2, \dots, \varphi(r^l) - r^l \cdot h_i^2, \dots, \varphi(r^1) - r^1 \cdot h_i^s, \varphi(r^l) - r^l \cdot h_i^s) , \quad (24)$$

we see that the discretized version of (22) reads: starting with λ_0, find consecutively for $i = 1, \dots, T$ a solution λ_i of the optimization problem

$$\text{minimize} \quad \frac{1}{2} \langle C\lambda, \lambda \rangle + \langle x, \lambda \rangle + \sum_{j=1}^{s} \left| \left\langle a, \lambda^j - \lambda_{i-1}^j \right\rangle \right| \quad (25)$$

subject to

$$B\lambda = b ,$$
$$\lambda \geq 0 ,$$

where the constraint $B\lambda = b$ expresses the condition $\sum_{i=1}^{l} \lambda^{ji} = 1$ for all j.

Although the spatially continuous problem (22) has a unique solution for the anisotropy energy considered in Remark 3.1, this is not necessarily the case in (25). We can, however, consider the so-called constrained theory proposed in [5], where φ is considered finite only at two points in S^{N-1} which define the easy-axis of the material. Physically this means that φ steeply grows from its zero value at magnetic poles. A direct simple remedy consists in a modification of C in (25): one adds to C an arbitrarily small multiple of the identity matrix (Prop. 4.1). This way has been used in our numerical tests.

Each optimization problem (25) can be equivalently written down as a GE of the type (3) which is given by the respective KKT conditions. The state variable y_i amounts to the triple $(\lambda_i, \tau_i, \mu_i)$, where τ_i and μ_i are the multipliers associated with the equality and inequality constraints in (25). Hence, one has $(\lambda_i, \tau_i, \mu_i) \in \mathbb{R}^n \times \mathbb{R}^s \times \mathbb{R}^n$ (so that $k = 2n + s$), $\lambda_i = (\lambda_i^1, \lambda_i^2, \dots, \lambda_i^s) \in (\mathbb{R}^l)^s$ (so that $n = ls$),

$$F_i(x, y_{i-1}, y_i) = F(x, \lambda_i, \tau_i, \mu_i) = \begin{bmatrix} C\lambda_i + x + B^T\tau_i - \mu_i \\ B\lambda_i - b \\ \lambda_i \end{bmatrix} \quad (26)$$

and

$$Q_i(y_{i-1}, y_i) = Q(\lambda_{i-1}, \lambda_i, \mu_i) = \begin{bmatrix} \partial_{\lambda_i} \sum_{j=1}^{s} |\langle a, \lambda_i^j - \lambda_{i-1}^j \rangle| \\ 0 \\ N_{\mathbb{R}_+^n}(\mu_i) \end{bmatrix}, \quad i = 1, 2, \ldots, T. \tag{27}$$

In fact, we do not control our equilibrium model directly via x. From (24) it follows that $x = \tilde{x} - \hat{x}$, where

$$\tilde{x} = |\triangle|(\varphi(r^1), \ldots, \varphi(r^l), \varphi(r^1), \ldots, \varphi(r^l), \ldots, \varphi(r^1), \ldots, \varphi(r^l)) \in (\mathbb{R}^l)^s$$

is a constant vector and

$$\hat{x} = |\triangle|(r^1 \cdot h_i^1, \ldots, r^l \cdot h_i^1, r^1 \cdot h_i^2, \ldots, r^l \cdot h_i^2, \ldots, r^1 \cdot h_i^s, \ldots, r^l \cdot h_i^s) . \tag{28}$$

Considering an external field $h \in \mathbb{R}^N$ spatially constant and depending on time through two smooth functions $\alpha_1, \alpha_2 : \mathbb{R} \to \mathbb{R}$, we infer that $\hat{x}$ depends only on α_1, α_2 and consequently $x = G(\alpha_1, \alpha_2)$, where $G : \mathbb{R}^2 \to \mathbb{R}^n$ is a continuously differentiable function.

Our goal is now to minimize a cost function φ depending on α and λ_T. As an example we may suppose that φ penalizes a deviation of volume fractions of the resulting magnetization at the time T from a desired value $\overline{\lambda} = (\overline{\lambda}^1, \ldots, \overline{\lambda}^s)$. More specifically,

$$\varphi(G(\alpha_1, \alpha_2), \lambda_T) = \tilde{\varphi}(\lambda_T) := \sum_{j=1}^{s} \beta^j \|\lambda_T^j - \overline{\lambda}^j\|^2 , \tag{29}$$

where $\beta^j \geq 0$, $1 \leq j \leq s$, represent weight coefficients.

In this way we obtain the MPEEC

$$\text{minimize} \quad \sum_{j=1}^{s} \beta^j \|\lambda_T^j - \overline{\lambda}^j\|^2 \tag{30}$$

subject to

$$0 \in F(G(\alpha_1, \alpha_2), \lambda_i, \tau_i, \mu_i) + Q(\lambda_{i-1}, \lambda_i) \text{ with } \lambda_0 \text{ given}, \quad i = 1, \ldots, T$$

$$\alpha_1, \alpha_2 \in \omega ,$$

where F, Q are given in (26), (27), and $\omega \subset \mathbb{R}^2$ is the set of admissible controls α_1, α_2. The next section is devoted to its numerical solution by the implicit programming approach (ImP) developed in Section 2.

4 Numerical solution

We start with the verification of assumption (A3).

Proposition 4.1 *Let C be positive definite and ω' be an open convex set containing ω. Then the state map $\widehat{S}_T : (\alpha_1, \alpha_2) \mapsto (\lambda_T, \tau_T, \mu_T)$ defined via the GEs in (30) is single-valued and locally Lipschitz for all $(\alpha_1, \alpha_2) \in \omega'$.*

Proof. The uniqueness of the solution λ_i to (22) has been already mentioned. The uniqueness of the multipliers τ_i, μ_i follows from the fact that, due to the special structure of B, the constraint system $B\lambda = b$ and $\lambda \geq 0$ satisfies the linear independence constraint qualification (cf. [26]) at each feasible point.

Let $\widetilde{S}$ be the mapping which assigns x to the unique solution $(\lambda_1, \tau_1, \mu_1, \lambda_2, \tau_2, \mu_2, \ldots, \lambda_T, \tau_T, \mu_T)$ of the system of GEs

$$0 \in F(x, \lambda_i, \tau_i, \mu_i) + Q(\lambda_{i-1}, \lambda_i) \text{ with } \lambda_0 \text{ given, } i = 1, \ldots, T .$$

We observe that $\widetilde{S}$ is polyhedral and hence locally Lipschitz over $\mathbb{R}^n$ by virtue of [32], [26, Cor. 2.5]. For the investigated map $\widehat{S}_T$ one has

$$\widehat{S}_T = D \circ \widetilde{S} \circ G ,$$

where the matrix D realizes the appropriate canonical projection. The Lipschitz continuity of $\widehat{S}_T$ on ω' thus follows from the Lipschitz continuity of G. □

We conclude that all assumptions, needed for the application of ImP to (30), are fulfilled and concentrate on the computation of subgradients of the composite objective.

4.1 Adjoint equation

The next step consists in the evaluation of D^*Q which maps $\mathbb{R}^n \times \mathbb{R}^s \times \mathbb{R}^n$ in subsets of $\mathbb{R}^n \times \mathbb{R}^n \times \mathbb{R}^n$. By elementary coderivative calculus, at a fixed point $(\bar{\alpha}, \bar{\lambda}_{i-1}, \bar{\lambda}_i, \bar{\tau}_i, \bar{\mu}_i)$ and for $\bar{c}_i = ({}^1\bar{c}_i, 0, {}^3\bar{c}_i) = -F(\bar{\alpha}, \bar{\lambda}_i, \bar{\tau}_i, \bar{\mu}_i)$, one has

$$D^*Q(\bar{\lambda}_{i-1}, \bar{\lambda}_i, \bar{\mu}_i, {}^1\bar{c}_i, 0, {}^3\bar{c}_i)({}^1w, {}^2w, {}^3w) = \begin{bmatrix} D^*P(\bar{\lambda}_{i-1}, \bar{\lambda}_i, {}^1\bar{c}_i)({}^1w) \\ 0 \\ D^*N_{\mathbb{R}^n_+}(\bar{\mu}_i, {}^3\bar{c}_i)({}^3w) \end{bmatrix} \tag{31}$$

for any $({}^1w, {}^2w, {}^3w) \in \mathbb{R}^n \times \mathbb{R}^s \times \mathbb{R}^n$. In (31), P denotes the partial subdifferential mapping in the first line of (27). The coderivative of $N_{\mathbb{R}^n_+}$ can easily be computed, e.g., on the basis of [25, Lemma 2.2]. The computation of D^*P is, however, substantially more demanding. First we observe that, due to a separation of variables,

$$D^*P(\bar{\lambda}_{i-1}, \bar{\lambda}_i, {}^1c_i)({}^1w) = \begin{bmatrix} D^*P_1(\bar{\lambda}^1_{i-1}, \bar{\lambda}^1_i, {}^1\bar{c}^1_i)({}^1w^1) \\ \vdots \\ D^*P_s(\bar{\lambda}^s_{i-1}, \bar{\lambda}^s_i, {}^1\bar{c}^s_i)({}^1w^s) \end{bmatrix}, \tag{32}$$

where ${}^1\bar{c}_i = ({}^1\bar{c}_i^1, {}^1\bar{c}_i^2, \ldots, {}^1\bar{c}_i^s) \in (\mathbb{R}^l)^s$, ${}^1w = ({}^1w^1, {}^1w^2, \ldots, {}^1w^s) \in (\mathbb{R}^l)^s$ and

$$P_j(\lambda_{i-1}^j, \lambda_i^j) = \partial_{\lambda_i} |\langle a, \lambda_i^j - \lambda_{i-1}^j \rangle|, \; j = 1, 2, \ldots, s.$$

Let $\mathcal{M}_i(\bar{\lambda}_{i-1}, \bar{\lambda}_i)$ denote the subset of the index set $\{1, 2, \ldots, s\}$ such that for $j \in \mathcal{M}_i(\bar{\lambda}_{i-1}, \bar{\lambda}_i)$

$$\langle a, \bar{\lambda}_i^j - \bar{\lambda}_{i-1}^j \rangle = 0.$$

From Definition 1.3 it easily follows that $D^*P_j(\bar{\lambda}_{i-1}^j, \bar{\lambda}_i^j, {}^1\bar{c}_i^j)({}^1w^j) \equiv 0$ whenever $j \notin \mathcal{M}_i(\bar{\lambda}_{i-1}, \bar{\lambda}_i)$. If $j \in \mathcal{M}_i(\bar{\lambda}_{i-1}, \bar{\lambda}_i)$, then we can make use of [33, Cor.10.11] and perform a slight modification of [24, Thm. 3.4]. In this way we arrive at the formula

$$D^*P_j(\bar{\lambda}_{i-1}^j, \bar{\lambda}_i^j, {}^1\bar{c}_i^j)({}^1w^j) = \begin{bmatrix} -a \\ a \end{bmatrix} D^*\mathfrak{Z}(0, \eta)(\langle a, {}^1w^j \rangle), \tag{33}$$

where $\mathfrak{Z}(\cdot) = \partial|\cdot|$, and $\eta \in \partial\mathfrak{Z}(0) = [-1, 1]$ is uniquely determined by the equation $\bar{c}_i^j = a\eta$. One easily verifies that

$$D^*\mathfrak{Z}(0, \eta)(0) = \mathbb{R}.$$

From this it follows that for $j \in \mathcal{M}_i(\bar{\lambda}_{i-1}, \bar{\lambda}_i)$

$$(q_i^j, {}^1v_i^j) \in D^*P_j(\bar{\lambda}_{i-1}^j, \bar{\lambda}_i^j, {}^1\bar{c}_i^j)({}^1w_i^j), \tag{34}$$

whenever

$$q_i^j \in \text{lin}\, a, \; {}^1v_i^j = -q_i^j \text{ and } \langle a, {}^1w_i^j \rangle = 0. \tag{35}$$

These relations can be used together with

$$q_i^j = {}^1v_i^j = 0 \text{ for } j \notin \mathcal{M}_i(\bar{\lambda}_{i-1}, \bar{\lambda}_i)\;, \tag{36}$$

$${}^3v_i \in D^*N_{\mathbb{R}^n_+}(\bar{\mu}_i, {}^3\bar{c}_i)({}^3w_i) \tag{37}$$

in the adjoint system

$$\begin{aligned}
0 &= \nabla\tilde{\varphi}(\bar{y}_T) + \begin{bmatrix} C & B^T & -\mathbb{I} \\ B & 0 & 0 \\ E & 0 & 0 \end{bmatrix} \begin{bmatrix} {}^1w_T \\ {}^2w_T \\ {}^3w_T \end{bmatrix} + \begin{bmatrix} {}^1v_T \\ 0 \\ {}^3v_T \end{bmatrix} \\
0 &= \begin{bmatrix} q_T \\ 0 \\ 0 \end{bmatrix} + \begin{bmatrix} C & B^T & -\mathbb{I} \\ B & 0 & 0 \\ E & 0 & 0 \end{bmatrix} \begin{bmatrix} {}^1w_{T-1} \\ {}^2w_{T-1} \\ {}^3w_{T-1} \end{bmatrix} + \begin{bmatrix} {}^1v_{T-1} \\ 0 \\ {}^3v_{T-1} \end{bmatrix} \\
&\cdots\cdots\cdots \\
0 &= \begin{bmatrix} q_2 \\ 0 \\ 0 \end{bmatrix} + \begin{bmatrix} C & B^T & -\mathbb{I} \\ B & 0 & 0 \\ E & 0 & 0 \end{bmatrix} \begin{bmatrix} {}^1w_1 \\ {}^2w_1 \\ {}^3w_1 \end{bmatrix} + \begin{bmatrix} {}^1v_1 \\ 0 \\ {}^3v_1 \end{bmatrix}.
\end{aligned} \tag{38}$$

As a subgradient of the respective function Θ one can now provide the used bundle method with the vector

$$\xi = \sum_{i=1}^{T} (\nabla G(\bar{\alpha}))^T w_i^1 \, . \tag{39}$$

To connect relations (34), (35), (36), (37) and the adjoint system (38), we proceed according to [25, Lemma 2.2] and introduce for each $i = 1, 2 \ldots, T$ the index sets

$$\begin{aligned} I_i^+(\bar{\lambda}_i, \bar{\mu}_i) &:= \{j \in \{1, 2 \ldots, n\} \mid (\bar{\lambda}_i)^j = 0, (\bar{\mu}_i)^j > 0\} \\ I_i^0(\bar{\lambda}_i, \bar{\mu}_i) &:= \{j \in \{1, 2 \ldots, n\} \mid (\bar{\lambda}_i)^j = 0, (\bar{\mu}_i)^j = 0\} \\ I_i^f(\bar{\lambda}_i, \bar{\mu}_i) &:= \{j \in \{1, 2 \ldots, n\} \mid (\bar{\lambda}_i)^j > 0, (\bar{\mu}_i)^j = 0\} \end{aligned}$$

related to the inequality constraints $(\lambda_i)^j \geq 0$, $j = 1, 2, \ldots, n$. We take any partitioning of $I_i^0(\bar{\lambda}_i, \bar{\mu}_i)$:

$$I_i^0(\bar{\lambda}_i, \bar{\mu}_i) = {}^1\beta_i \cup {}^2\beta_i \tag{40}$$

and define

$$\widetilde{I}_i^+ := I_i^+(\bar{\lambda}_i, \bar{\mu}_i) \cup {}^1\beta_i \qquad \widetilde{I}_i^f := I_i^f(\bar{\lambda}_i, \bar{\mu}_i) \cup {}^2\beta_i \, .$$

Consider now the equation system

$$\begin{bmatrix} C & B^T & -\mathbb{I}_{\widetilde{I}_i^+} & A_{\mathcal{M}_i} \\ B & 0 & 0 & 0 \\ (\mathbb{I}_{\widetilde{I}_i^+})^T & 0 & 0 & 0 \\ (A_{\mathcal{M}_i})^T & 0 & 0 & 0 \end{bmatrix} \begin{bmatrix} {}^1w_i \\ {}^2w_i \\ {}^3\widetilde{w}_i \\ \widetilde{v}_i \end{bmatrix} = \begin{bmatrix} \tilde{q}_i \\ 0 \\ 0 \\ 0 \end{bmatrix} , \tag{41}$$

where A is the $(n \times s)$ matrix, defined by

$$A := \begin{bmatrix} a & 0_{l\times 1} & \cdots & 0_{l\times 1} \\ 0_{l\times 1} & a & \cdots & 0_{l\times 1} \\ & & \ddots & \\ 0_{l\times 1} & 0_{l\times 1} & \cdots & a \end{bmatrix} ,$$

and ${}^3\widetilde{w}_i$ is a subvector of ${}^3w_i$, whose components belong to $\widetilde{I}_i^+(\bar{\lambda}_i, \bar{\mu}_i)$. For notational simplicity, we occasionally omit the arguments of $\mathcal{M}_i$, I_i^+, I_i^0 and I_i^f.

This system has to be solved backwards for $i = T, T-1, \ldots, 1$ with the terminal condition $\tilde{q}_T = -\nabla\tilde{\varphi}(\bar{\lambda}_T)$ by using the updates

$$(\tilde{q}_i)_{\mathcal{M}_i} = \tilde{v}_i$$

$$(\tilde{q}_i)^j = 0 \text{ for } j \notin \mathcal{M}_i(\bar{\lambda}_{i-1}, \bar{\lambda}_i) \, ,$$

$i = T-1, T-2, \ldots, 1$. One can easily verify that the component ${}^1w_i$ to each solution of (41) is feasible also with respect to (34), (35), (36), (37) and the adjoint system (38). It may thus be used in formula (39). Note that the sizes of $\tilde{v}_i$ and ${}^3\tilde{w}_i$ vary, depending on the cardinality of $\mathcal{M}_i(\bar{\lambda}_{i-1}, \bar{\lambda})$ and $\widetilde{I}_i^+$, respectively.

Remark 4.1 The choice of the partitioning (40) has an essential influence on which subgradient we actually compute. As already mentioned, we may even compute a vector that does not belong to $\partial\Theta$. This was indeed the case in one of our numerical examples. We considered a problem with only one design variable. The data were set so that the composite objective was constant in a neighborhood of α' and that $I_i^0 \neq \emptyset$ at α'. Still, when choosing ${}^1\beta_i = \emptyset$, i.e., $\widetilde{I}_i^+ = I_i^+$, we have obtained a nonzero subgradient at α'. A simple remedy to this unpleasant situation is to change the partitioning (40); we know that there *is* a partitioning that leads to a true subgradient. In our particular case, we have taken ${}^2\beta_i = \emptyset$, i.e., $\widetilde{I}_i^+ = I_i^+ \cup I_i^0$. For this choice, we got a subgradient equal to zero, a correct one.

4.2 Computational procedures

Numerical solution of the MPEEC (30) requires solution of several subproblems. First, for a given control variable, we have to solve the state problem: a series of nonsmooth convex optimization problems (25). We transform each of them by a simple trick to a convex quadratic program. By introducing an auxiliary variable $z \in \mathbb{R}^s$ we can write (25) equivalently as

$$\begin{aligned}
\min_{\lambda, z} \quad & \frac{1}{2}\langle C\lambda, \lambda\rangle + \langle x, \lambda\rangle + \sum_{j=1}^{s} z^j \qquad (42)\\
\text{subject to} \quad & \lambda \geq 0\\
& B\lambda = b\\
& z^j \geq \left\langle a, \lambda^j - \lambda_{i-1}^j \right\rangle, \quad j = 1, \ldots, s\\
& z^j \geq -\left\langle a, \lambda^j - \lambda_{i-1}^j \right\rangle, \quad j = 1, \ldots, s\,.
\end{aligned}$$

This problem can be solved by any QP solver; we have opted for the code PENNON that proved to be very efficient for general nonlinear programming problems [12].

Second, we must solve the adjoint systems (41). These are medium-size systems of linear equations with nonsymmetric matrices. The matrices, however, may be singular in case $\mathcal{M}_i \neq \emptyset$. When we know that the matrix is nonsingular (i.e., when $\mathcal{M}_i = \emptyset$), we use the LAPACK subroutine DGESV based on LU decomposition with partial pivoting; in the general case, we use the least-square subroutine DGELSD. Note that even for a singular matrix, the first component of the solution ${}^1w_i$ is always unique—and this is the only component we need.

Finally, we have to minimize the respective composite objective Θ over ω. It was already mentioned that this is a nonsmooth, nonconvex and locally Lipschitz function. We have used one of the few suitable codes for minimization of such functions, the code BT [38].

4.3 Examples

The examples presented in this section are purely academic. Their purpose is to verify that the proposed technique can be successfully used for solving MPEECs. On the other hand, it is not difficult to modify these problems to real-world applications, by changing the sample geometry, level of discretization and, in particular, the cost function and the design variables.

We consider $N = 2$, $\varphi(r_1, r_2) = 30r_1^2$, a sample of dimensions 2×1, spatial discretization by 8×8 finite volumes and 10 time-steps. We further set $l = 4$ (the discretization of the sphere S^{N-1}; see Section 3.2).

The control variables are the amplitude and frequency of the second component of the external magnetic field h_i (independent of spatial variable). We consider $h_i = (0, f_i)$ with

$$f_i = \alpha_1 \sin(\frac{2\pi}{\alpha_2}\frac{i}{T}), \quad i = 1, 2, \ldots, T.$$

This, by (28), defines the function G.

The admissible control set ω is $[0.1, 10^5] \times [0.1, 10^5]$.

Example 1 In the first example we try to magnetize fully the specimen in $\mathbb{R}^2$ in the second direction, so that the desired magnetization M is $(0, 1)$. This can be done by setting $\overline{\lambda}^{j1} = 1.1$, $j = 1, \ldots, s$ (note that the maximal possible value of λ is equal to one). The initial value of α is set to $(50, 4)$. The problem appears to be "easy" and BT finds the optimum in just a few steps. Below we show the output of the code:

```
 BT-Algorithm
 ============
niter ncomp        f                 gp                alpha
   1     1   0.43362311E+00   0.67345917E-01   0.00000000E+00
   2     2   0.43362311E+00   0.53359109E-01   0.10000000E-01
   3     3   0.38786221E+00   0.56068671E-01   0.12958132E-01
   4     4   0.32000000E+00   0.17855778E-15   0.00000000E+00
 convergence
```

We can see that the exact minimum $\lambda_T^{j1} = 1$, $j = 1, \ldots, s/2$, (giving the optimal objective value $f^* = 32 \cdot (1.0 - 1.1)^2$) was found in just four steps. The reason for that is simple: from a certain value of the amplitude α_1' up (for a given frequency), the cost function is constant (and optimal) — the specimen is fully magnetized and any increase of the amplitude cannot change it. The BT code then quickly hits an arbitrary $\alpha_1 > \alpha_1'$ and finishes.

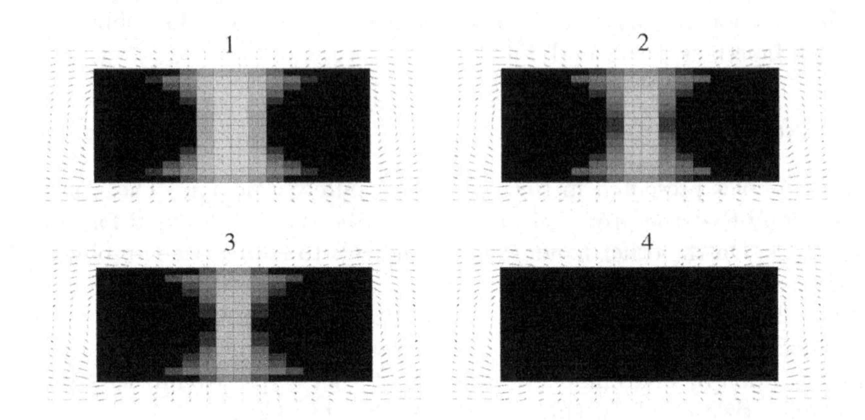

Fig. 1. Example 1—the specimen, its magnetization, and the demagnetizing field ∇u_M around the specimen in each iteration of BT. The darker is the color in a point of the specimen the closer is the magnetization to $(0, 1)$ (in the Euclidean norm).

Due to the simplicity of the example, it is not surprising that we obtain the same result when we change the time discretization, even to just one time step. Only the BT iterates will be slightly different. It is, however, interesting to note that, in each iteration (and for any time discretization), only one time step contributed to the second term of the subgradient (39); that means, $w_i^1 \neq 0$ for only one $i \in \{1, \ldots, T\}$, whereas this i was different at every BT iterate.

Example 2 In the second example, we set $\overline{\lambda}^{j1} = 0.9$, $j = 1, \ldots, s$, and try to identify this value. This time, the objective function is not constant around the optimum and the behavior of the BT code reminds more a standard behavior of a minimization method:

```
 BT-Algorithm
 ============
niter ncomp       f               gp              alpha
   1     1   0.25601552E+00   0.24693051E-01   0.00000000E+00
   2     2   0.25601552E+00   0.24693051E-01   0.00000000E+00
   3     3   0.23502032E+00   0.24693051E-01   0.10000000E-01
   4     4   0.17673858E+00   0.20989044E-01   0.00000000E+00
   5     5   0.17673858E+00   0.17128780E-01   0.69166119E-02
   6     6   0.17673858E+00   0.11764406E-01   0.91444459E-02
   7     7   0.16785966E+00   0.58038202E-02   0.00000000E+00
```

```
 8   8  0.16652221E+00  0.35283823E-02  0.23490041E-03
 9  10  0.16652221E+00  0.16319657E-02  0.12254303E-03
10  11  0.16643681E+00  0.58649082E-03  0.00000000E+00
11  12  0.16643681E+00  0.29875680E-03  0.31325080E-05
12  13  0.16643417E+00  0.10889954E-03  0.00000000E+00
13  14  0.16643417E+00  0.54258298E-04  0.10805660E-06
14  15  0.16643408E+00  0.20402428E-04  0.00000000E+00
15  16  0.16643408E+00  0.10207925E-04  0.37928982E-08
16  17  0.16643408E+00  0.38203558E-05  0.00000000E+00
17  18  0.16643408E+00  0.19099423E-05  0.13298890E-09
18  19  0.16643408E+00  0.71546057E-06  0.00000000E+00
19  20  0.16643408E+00  0.35774567E-06  0.46642216E-11
20  21  0.16643408E+00  0.13400225E-06  0.00000000E+00
21  22  0.16643408E+00  0.67024577E-07  0.16361832E-12
convergence
```

However, the minimal point $(45.17065468, 4.264825646)$ is not unique, again. When we fix the value of α_1 to 50.0 and keep only α_2 free, BT finds a point $(50.0, 5.619174535)$ with the same value of the objective function.

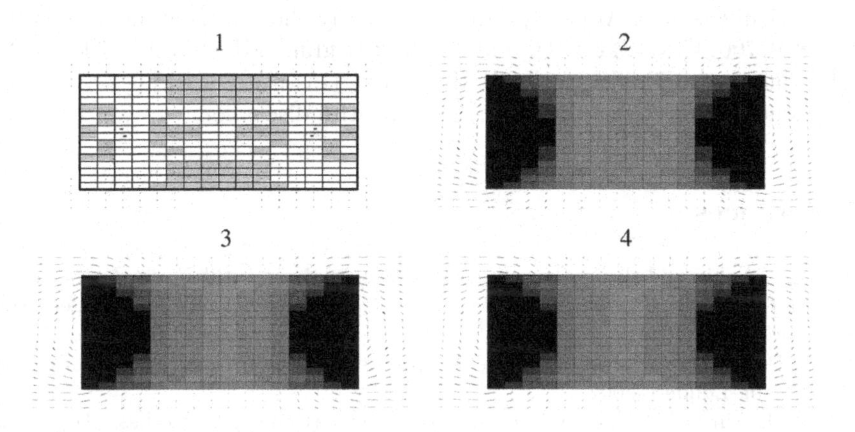

Fig. 2. Example 2—the specimen, its magnetization, and the demagnetizing field ∇u_M around the specimen in each 5th iteration of BT. The darker is the color in a point of the specimen the closer is the magnetization to $(0,1)$ (in the Euclidean norm).

5 Conclusion

We formulated a problem of optimal control of a ferromagnet in a form of an MPEEC and developed a solution approach based on an implicit programming technique. The adjoint equations, needed to compute the subgradients of the composite objective, are derived using the generalized differential calculus of B. Mordukhovich. Up to our knowledge, this is the first attempt to solve mathematical programs with evolutionary equilibria numerically. Computational tests demonstrate the applicability of this approach.

The implicit programming technique requires local uniqueness of the equilibrium problem. To enforce uniqueness, we modified slightly the matrix C in (25). If, however, the cost functions φ depends only on the spatial average of the magnetization over the specimen, i.e., on $|\Omega|^{-1}\int_\Omega M(z)\,\mathrm{d}z$ or on the self-induced magnetic field ∇u_M then, as these quantities are uniquely defined even in the discrete unconstrained case (25) [4], assumption (A3) holds automatically. Hence, the control problems may be to find an external field h such that the average magnetization is as "close" as possible to a given vector and similarly for the self-induced field. This demonstrates a wider applicability of our results than the examples solved in this contribution.

Acknowledgment

This research was supported by the Academy of Sciences of the Czech Republic through grant No. A1075402. M.K. was also supported by the IMA at the University of Minnesota during his stay there in November 2004, by VZ6840770021 (MŠMT ČR), and by the EU grant MRTN-CT-2004-505226. The authors would like to thank Miloš Jirsa (Institute of Physics, Academy of Sciences of the Czech Republic) for a fruitful discussion.

References

1. J. P. Aubin and H. Frankowska: *Set-Valued Analysis*, Birkhäuser, Boston, 1990.
2. W. F. Brown, Jr.: *Magnetostatic principles in ferromagnetism*, Springer, New York, 1966.
3. S. Dempe: *Foundations of Bilevel Programming*, Kluwer Acad. Publ., Dordrecht-Boston-London, 2002.
4. A. DeSimone: Energy minimizers for large ferromagnetic bodies, *Arch. Rat. Mech. Anal.* 125(1993), 99–143.
5. A. DeSimone and R. D. James: A constrained theory of magnetoelasticity, *J. Mech. Phys. Solids* 50(2002), 283–320.
6. R. Henrion, A. Jourani and J. Outrata: On the Calmness of a Class of Multifunctions, *SIAM Journal on Optimization* 13(2002)2, 603–618.
7. A. Hubert and R. Schäffer: *Magnetic Domains*, Springer, Berlin, 1998.
8. R. D. James and D. Kinderlehrer: Frustration in ferromagnetic materials, *Continuum Mech. Thermodyn.* 2(1990), 215–239.
9. M. Kočvara, A. Mielke and T. Roubíček: Rate-independent approach to the delamination problem, SFB 404 "Mehrfeldprobleme in der Kontinuumsmechanik", Universität Stuttgart, Preprint 2003/29, 2003 (To appear in *Mathematics and Mechanics of Solids*).

10. M. Kočvara and J. V. Outrata: On the modeling and control of delamination processes, In: J. Cagnol and J.-P. Zolesion (eds.): Control and Boundary Analysis, Marcel Dekker, New York, 2004, 171–190.
11. M. Kočvara and J. V. Outrata: Optimization Problems with Equilibrium Constraints and their Numerical Solution, *Mathematical Programming B* 101(2004), 119–150.
12. M. Kočvara and M. Stingl: PENNON—A Code for Convex Nonlinear and Semidefinite Programming, *Optimization Methods and Software* 18(2003), 317–333.
13. M. Kružík: Maximum principle based algorithm for hysteresis in micromagnetics, *Adv. Math. Sci. Appl.* 13(2003), 461–485.
14. M. Kružík: Periodic solutions to a hysteresis model in micromagnetics, IMA Preprint 1946/2003, University of Minnesota, Minneapolis, 2003 (To appear in *J. Convex Anal.*).
15. M. Kružík: Periodicity properties of solutions to a hysteresis model in micromagnetics, In: M. Feistauer et al. (eds.): Proceedings of ENUMATH 2003 the 5th European Conference on Num. Math. and Adv. Appl., Springer, Berlin, 2004, 605–614.
16. L. D. Landau and E. M. Lifshitz: On the theory of the dispersion of magnetic permeability of ferromagnetic bodies, *Physik Z. Sowjetunion* 8(1935), 153–169.
17. Z.-Q. Luo, J.–S. Pang and D. Ralph: *Mathematical Programs with Equilibrium Constraints*, Cambridge University Press, Cambridge, 1996.
18. M. Luskin and L. Ma: Analysis of the finite element approximation of microstructure in micromagnetics, *SIAM J. Num. Anal.* 29(1992), 320–331.
19. A. Mielke: Energetic formulation of multiplicative elastoplasticity using dissipation distances, *Continuum Mech. Thermodyn.* 15(2003), 351–382.
20. A. Mielke and F. Theil: Mathematical model for rate-independent phase transformations, In: H.-D. Alber, R. Balean and R.Farwig (eds.): Models of Cont. Mechanics in Analysis and Engineering, Shaker-Verlag, Aachen, 1999, 117–129.
21. A. Mielke and F. Theil: On rate-independent hysteresis models, *Nonlin. Diff. Eq. Appl.* 11(2004), 151–189.
22. B. S. Mordukhovich: Generalized Differential Calculus for Nonsmooth and Set-Valued Mappings, *Journal of Mathematical Analysis and Applications* 183(1994), 250–288.
23. B. S. Mordukhovich: Lipschitzian stability of constraint systems and generalized equations, *Nonlinear Anal.- Th. Meth. Appl.*, 22(1994), 173–206.
24. B. S. Mordukhovich and J. V. Outrata: On second-order subdifferentials and their applications, *SIAM J. Optimization* 12(2001), 139–169.
25. J. V. Outrata: Optimality conditions for a class of mathematical programs with equilibrium constraints, *Mathematics of Operations Research* 24(1999), 627–644.
26. J. V. Outrata, M. Kočvara and J. Zowe: *Nonsmooth Approach to Optimization Problems with Equilibrium Constraints: Theory, Applications and Numerical Results*, Kluwer Acad. Publ., Dordrecht-Boston-London, 1998.
27. P. Pedregal: Relaxation in ferromagnetism: the rigid case, *J. Nonlin. Sci.* 4(1994), 105–125.
28. P. Pedregal: *Parametrized Measures and Variational Principles*, Birkhäuser, Basel, 1997.
29. A. Reimers and E. Della Torre: Fast Preisach-based magnetization model and fast inverse hysteresis model, *IEEE Trans. on Magnetics* 34(1998), 3857–3866.

30. S. M. Robinson: Generalized equations and their solutions, Part I: Basic theory, *Mathematical Programming Study*, 10(1979), 128–141.
31. S. M. Robinson: Strongly regular generalized equations, *Mathematics of Operations Research* 5(1980), 43–62.
32. S. M. Robinson: Some continuity properties of polyhedral multifunctions, *Mathematical Programming Study*, 14(1981), 206–214.
33. R. T. Rockafellar and R. Wets: *Variational Analysis*, Springer Verlag, Berlin, 1998.
34. T. Roubíček: *Relaxation in Optimization Theory and Variational Calculus*, W. de Gruyter, Berlin, 1997.
35. T. Roubíček and M. Kružík: Microstructure evolution model in micromagnetics: *Zeitschrift f. Angew. Math. Phys.* 55(2004), 159–182.
36. T. Roubíček and M. Kružík: Mesoscopic model for ferromagnets with isotropic hardening, *Zeitschrift f. Angew. Math. Phys.* 56(2005), 107–135.
37. S. Scholtes: Introduction to Piecewise Differential Equations, Habilitation Thesis, Institut für Statistik und Mathematische Wirtschaftstheorie, Universität Karlsruhe, Germany, 1994.
38. H. Schramm and J. Zowe: A version of the bundle idea for minimizing a nonsmooth function: conceptual idea, convergence analysis, numerical results, *SIAM J. Optimization* 2(1992), 121–152.
39. A. Visintin: A Weiss-type model of ferromagnetism, *Physica B* 275(2000), 87–91.
40. L. C. Young: Generalized curves and existence of an attained absolute minimum in the calculus of variations, *Comptes Rendus de la Société et des Lettres de Varsovie, Classe III*, 30(1937), 212–234.

Complementarity constraints as nonlinear equations: Theory and numerical experience

Sven Leyffer

Mathematics and Computer Science Division, Argonne National Laboratory, Argonne, IL 60439, USA leyffer@mcs.anl.gov

Summary. Recently, it has been shown that mathematical programs with complementarity constraints (MPCCs) can be solved efficiently and reliably as nonlinear programs. This paper examines various nonlinear formulations of the complementarity constraints. Several nonlinear complementarity functions are considered for use in MPCC. Unlike standard smoothing techniques, however, the reformulations do not require the control of a smoothing parameter. Thus they have the advantage that the smoothing is *exact* in the sense that Karush-Kuhn-Tucker points of the reformulation correspond to strongly stationary points of the MPCC. A new *exact smoothing* of the well-known min function is also introduced and shown to possess desirable theoretical properties. It is shown how the new formulations can be integrated into a sequential quadratic programming solver, and their practical performance is compared on a range of test problems.

Keywords: Nonlinear programming, SQP, MPCC, complementarity constraints, NCP functions.

2000 Mathematics Subject Classification: 90C30, 90C33, 90C55, 49M37, 65K10.

1 Introduction

Equilibrium constraints in the form of complementarity conditions often appear as constraints in optimization problems, giving rise to mathematical programs with complementarity constraints (MPCCs). Problems of this type arise in many engineering and economic applications; see the survey by Ferris and Pang [FP97] and the monographs by Luo et al. [LPR96] and Outrata et al. [OKZ98]. The growing collections of test problems by Dirkse [Dir01], and our MacMPEC [Ley00] indicate that this an important area. MPCCs can be expressed in general as

S. Dempe and V. Kalashnikov (eds.), *Optimization with Multivalued Mappings*, pp. 169-208
©2006 Springer Science + Business Media, LLC

$$\begin{aligned} &\text{minimize} && f(x) && \text{(1a)} \\ &\text{subject to} && c_{\mathcal{E}}(x) = 0 && \text{(1b)} \\ &&& c_{\mathcal{I}}(x) \geq 0 && \text{(1c)} \\ &&& 0 \leq x_1 \perp x_2 \geq 0, && \text{(1d)} \end{aligned}$$

where $x = (x_0, x_1, x_2)$ is a decomposition of the problem variables into controls $x_0 \in \mathbb{R}^n$ and states $x_1, x_2 \in \mathbb{R}^p$. The equality constraints $c_i(x) = 0,\ i \in \mathcal{E}$ are abbreviated as $c_{\mathcal{E}}(x) = 0$, and similarly $c_{\mathcal{I}}(x) \geq 0$ represents the inequality constraints. The notation $\perp$ represents complementarity and means that either a component $x_{1i} = 0$ or the corresponding component $x_{2i} = 0$.

Clearly, more general complementarity constraints can be included in (1) by adding slack variables. Adding slacks does not destroy any properties of the MPCC such as constraint qualification or second-order condition. One convenient way of solving (1) is to replace the complementarity conditions (1d) by

$$x_1,\ x_2 \geq 0, \quad \text{and} \quad X_1 x_2 \leq 0, \tag{2}$$

where X_1 is a diagonal matrix with x_1 along its diagonal. This transforms the MPCC into an equivalent nonlinear program (NLP) and is appealing because it appears to allow standard large-scale NLP solvers to be used to solve (1).

Unfortunately, Chen and Florian [CF95] have shown that (2) violates the Mangasarian-Fromovitz constraint qualification (MFCQ) at *any* feasible point. This failure of MFCQ has a number of unpleasant consequences: The multiplier set is unbounded, the central path fails to exist, the active constraint normals are linearly dependent, and linearizations of the NLP can be inconsistent *arbitrarily close* to a solution. In addition, early numerical experience with (2) has been disappointing; see Bard [Bar88]. As a consequence, solving MPCCs as NLPs has been commonly regarded as numerically unsafe.

Recently, exciting new developments have demonstrated that the gloomy prognosis about the use of (2) may have been premature. We have used standard sequential quadratic programming (SQP) solvers to solve a large class of MPCCs, written as NLPs, reliably and efficiently [FL04]. This numerical success has motivated a closer investigation of the (local) convergence properties of SQP methods for MPCCs. Fletcher et al. [FLRS02] show that an SQP method converges locally to strongly stationary points. Anitescu [Ani05] establishes that an SQP method with *elastic mode* converges locally for MPCCs with (2). The key idea is to penalize $X_1 x_2 \leq 0$ and consider the resulting NLP, which satisfies MFCQ. Near a strongly stationary point, a sufficiently large penalty parameter can be found, and standard SQP methods converge.

The convergence properties of interior point methods (IPMs) have also received renewed attention. Numerical experiments by Benson et al. [BSSV03] and by Raghunathan and Biegler [RB02b] have shown that IPMs with minor modifications can be applied successfully to solve MPCCs. This practical success has encouraged theoretical studies of the convergence properties of IPMs for MPCCs. Raghunathan and Biegler [RB02a] relax $x_1^T x_2 \leq 0$ by a quantity proportional to the barrier parameter, which is driven to zero. Liu and Sun [LS04] propose a primal-dual IPM that also relaxes the complementarity constraint.

In this paper, we extend our results of [FLRS02] by considering NLP formulations of (1) in which the complementarity constraint (1d) is replaced by a nonlinear complementarity problem (NCP) function. This gives rise to the following NLP:

$$\begin{aligned} &\text{minimize} && f(x) && \text{(3a)}\\ &\text{subject to} && c_{\mathcal{E}}(x) = 0 && \text{(3b)}\\ & && c_{\mathcal{I}}(x) \geq 0 && \text{(3c)}\\ & && x_1,\ x_2 \geq 0,\ \Phi(x_{1i}, x_{2i}) \leq 0, && \text{(3d)} \end{aligned}$$

where $\Phi(x_1, x_2)$ is the vector of NCP functions, $\Phi(x_1, x_2) = (\varphi(x_{11}, x_{21}), \ldots, \varphi(x_{1p}, x_{2p}))^T$, and φ is any NCP function introduced in the next section. Problem (3) is in general nonsmooth because the NCP functions used in (3d) are nonsmooth at the origin. We will show that this nonsmoothness does not affect the local convergence properties of the SQP method.

The use of NCP functions for the solution of MPCCs has been considered by Dirkse et al. [DFM02] and by Facchinei et al. [FJQ99], where a sequence of smoothed NCP reformulations is solved. Our contribution is to show that this smoothing is not required. Thus we avoid the need to control the smoothing parameter that may be problematic in practice. Moreover, the direct use of NCP functions makes our approach exact, in the sense that first-order points of the resulting NLP coincide with strongly stationary points of the MPCC. As a consequence we can prove superlinear convergence under reasonable assumptions.

The paper is organized as follows. The next section reviews the NCP functions that will be used in (3d) and their pertinent properties. We also introduce new NCP functions shown to possess certain desirable properties. Section 3 shows the equivalence of first-order points of (1) and (3). Section 4 formally introduces the SQP algorithm for solving MPCCs. The equivalence of the first-order conditions forms the basis of the convergence proof of the SQP method, presented in Section 5. In Section 6, we examine the practical performance of the different NCP functions on the MacMPEC test set [Ley00]. In Section 7 we summarize our work and briefly discuss open questions.

2 NCP functions for MPCCs

An NCP function is a function $\varphi : \mathbb{R}^2 \to \mathbb{R}$ such that $\varphi(a, b) = 0$ if and only if $a, b \geq 0$, and $ab \leq 0$. Several NCP functions can be used in the reformulation (3). Here, we review some existing NCP functions and introduce new ones that have certain desirable properties.

1. The Fischer-Burmeister function [Fis95] is given by
$$\varphi_{FB}(a,b) = a + b - \sqrt{a^2 + b^2}. \tag{4}$$
It is nondifferentiable at the origin, and its Hessian is unbounded at the origin.
2. The min-function due to Chen and Harker [CH97] is the nonsmooth function
$$\varphi_{\min}(a,b) = \min(a,b). \tag{5}$$
It can be written equivalently in terms of the natural residual function [CH97]:
$$\varphi_{NR}(a,b) = \frac{1}{2}\left(a + b - \sqrt{(a-b)^2}\right). \tag{6}$$
This function is again nondifferentiable at the origin and along the line $a = b$.
3. The Chen-Chen-Kanzow function [CCK00] is a convex combination of the Fischer-Burmeister function and the bilinear function. For a fixed parameter $\lambda \in (0,1)$, it is defined as
$$\varphi_{CCK}(a,b) = \lambda \varphi_{FB}(a,b) + (1-\lambda) a_+ b_+,$$
where $a_+ = \max(0,a)$. Note that for $a \geq 0$, $a_+ = a$; hence, for any method that remains feasible with respect to the simple bounds,
$$\varphi_{CCK}(a,b) = \lambda \varphi_{FB}(a,b) + (1-\lambda)\varphi_{BL}(a,b) \tag{7}$$
holds.

In addition, we consider the bilinear form
$$\varphi_{BL}(a,b) = ab, \tag{8}$$
which is analytic and has the appealing property that its gradient vanishes at the origin (this makes it consistent with strong stationarity, as will be shown later). We observe, however, that it is not an NCP function, since $\varphi_{BL}(a,b) = 0$ does not imply nonnegativity of a, b.

We note that all functions (except for (8)) are nondifferentiable at the origin. In addition, the Hessian of the Fischer-Burmeister function is unbounded at the origin. This has to be taken into account in the design of robust SQP methods for MPCCs.

The min-function has the appealing property that linearizations of the resulting NLP (3) are consistent sufficiently close to a strongly stationary point (see Proposition 3.6). This property motivates the derivation of smooth approximations of the min-function. The first approximation is obtained by smoothing the equivalent natural residual function (6) by adding a term to the square root (which causes the discontinuity along $a = b$). For a *fixed* parameter $\sigma_{NR} > 1/2$, let

$$\varphi_{NR_s}(a,b) = \frac{1}{2}\left(a + b - \sqrt{(a-b)^2 + \frac{ab}{\sigma_{NR}}}\right). \tag{9}$$

This smoothing is similar to [CH97, FJQ99], where a positive parameter $4\mu^2 > 0$ is added to the discriminant. This has the effect that complementarity is satisfied only up to μ^2 at the solution. In contrast, adding the term ab/σ_{NR}, implies that the NCP function remains *exact* in the sense that $\varphi_{NR_s}(a,b) = 0$ if and only if $a, b \geq 0$ and $ab = 0$ for *any* $\sigma_{NR} > 1/2$. Figure 2 shows the contours of $\varphi_{NR_s}(a,b)$ for $\sigma_{NR} = 32$ and for the min-function ($\sigma_{NR} = \infty$). An interesting observation is that as $\sigma_{NR} \to \frac{1}{2}$, the smoothed min-function $\varphi_{NR_s}(a,b)$ becomes the Fischer-Burmeister function (up to a scaling factor).

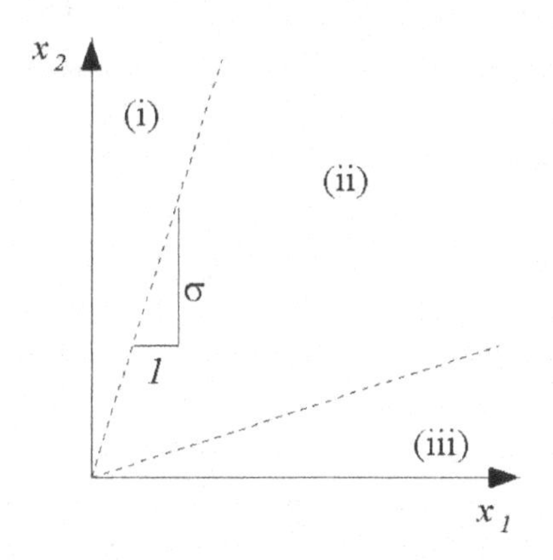

Fig. 1. Piecewise regions for smoothing the min-function

An alternative way to smooth the natural residual function is to work directly on smoothing the contours of the min-function, which are parallel to either the x_1, or the x_2 axis. The contours can be smoothed by dividing the positive orthant into (for example) three regions as shown in Figure 1. The dashed lines separate the three regions (i) to (iii), and their slope is $\sigma > 1$ and σ^{-1}, respectively. In regions (i) and (iii), the contours are identical to the min-function. This feature ensures consistency of the linearization. In region (ii), different degrees of smoothing can be applied.

The first smoothed min-function is based on a piecewise linear approximation, given by

$$\varphi_{lin}(a,b) = \begin{cases} b & b \leq a/\sigma_l \\ (a+b)/(1+\sigma_l) & a/\sigma_l < b < \sigma_l a \\ a & b \geq \sigma_l a, \end{cases} \tag{10}$$

where $\sigma = \sigma_l > 1$ is the parameter that defines the three regions in Figure 1. The idea is that close to the axis, the min-function is used, while for values of a, b that are in the center, the decision as to which should be zero is delayed.

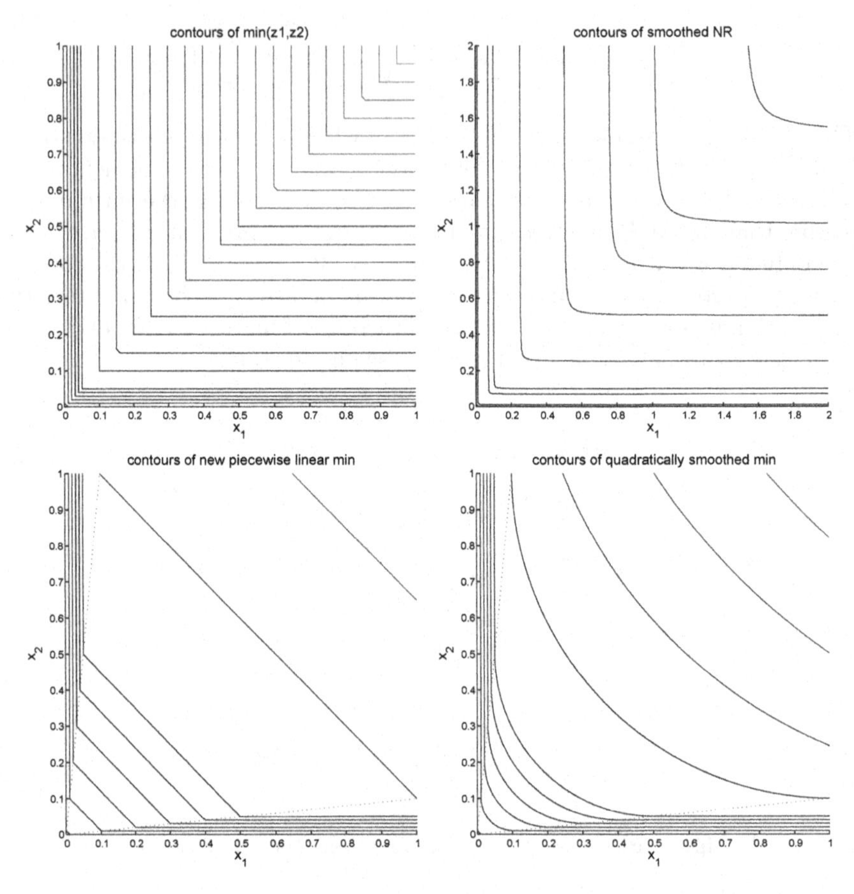

Fig. 2. Contours of the min-function, the smoothed natural residual function, the piecewise linear min-function, and the piecewise quadratic with $\sigma_l = \sigma_q = 3$

The second smoothed min-function is based on the idea of joining the linear parts in sectors (i) and (iii) with circle segments. This gives rise to the following function,

$$\varphi_{qua}(a,b) = \begin{cases} b & b \leq a/\sigma_q \\ \sqrt{\dfrac{(a-\theta)^2+(b-\theta)^2}{(\sigma_q-1)^2}} & a/\sigma_q < b < \sigma_q a \\ a & b \geq \sigma_q a, \end{cases} \tag{11}$$

where θ is the center of the circle, depending on a, b, and σ_q and is given by

$$\theta = \frac{a+b}{2-\frac{(\sigma_q-1)^2}{\sigma_q^2}} + \sqrt{\left(\frac{a+b}{2-\frac{(\sigma_q-1)^2}{\sigma_q^2}}\right)^2 - \frac{a^2+b^2}{2-\frac{(\sigma_q-1)^2}{\sigma_q^2}}}.$$

The contours of both smoothing functions are given in Figure 2. Note that the contours are parallel to the axis in regions (i) and (iii). This fact will be exploited to show that linearizations of the min-function and its two variants remain consistent arbitrarily close to a strongly stationary point. This observation, in effect, establishes a constraint qualification for the equivalent NLP (3).

The smoothing also avoids another undesirable property of the min-function: It projects iterates that are far from complementary onto the nearest axis. Close to the axis $a = b$, this projection results in an arbitrary step. Consider, for example, a point $a = 99, b = 101$. Linearizing the min-function about this point results in a first-order approximation in which $a = 0, b \geq 0$. In contrast, other NCP functions "delay" this decision and can be viewed as smoothing methods.

3 Equivalence of first-order conditions

This section shows that there exists a one-to-one correspondence between strongly stationary points of the MPCC (1) and the first-order stationary points of the equivalent NLP (3). We start by reviewing MPCC stationarity concepts. Next, we derive some properties of the linearizations of (3d) that play a crucial role in the equivalence of first-order conditions.

3.1 Strong stationarity for MPCCs

The pertinent condition for stationarity for analyzing NLP solvers applied to (3) is strong stationarity. The reason is that there exists a relationship between strong stationarity defined by Scheel and Scholtes [SS00] and the Karush-Kuhn-Tucker (KKT) points of (3). This relationship has been exploited in [FLRS02] to establish convergence of SQP methods for MPCCs formulated as NLPs. Strong stationarity is defined as follows.

Definition 3.1 *A point x is called* strongly stationary *if and only if there exist multipliers λ, $\hat{\nu}_1$, and $\hat{\nu}_2$ such that*

$$\begin{aligned} \nabla f(x) - \nabla c^T(x)\lambda - \begin{pmatrix} 0 \\ \hat{\nu}_1 \\ \hat{\nu}_2 \end{pmatrix} &= 0 \\ c_{\mathcal{E}}(x) &= 0 \\ c_{\mathcal{I}}(x) &\geq 0 \\ x_1,\ x_2 &\geq 0 \\ x_{1j} = 0 \quad or \quad x_{2j} &= 0 \\ \lambda_{\mathcal{I}} &\geq 0 \\ c_i\lambda_i = x_{1j}\hat{\nu}_{1j} = x_{2j}\hat{\nu}_{2j} &= 0 \\ if\ x_{1j} = x_{2j} = 0 \quad then \quad \hat{\nu}_{1j} \geq 0 \quad and \quad \hat{\nu}_{2j} &\geq 0. \end{aligned} \tag{12}$$

Strong stationarity can be interpreted as the KKT conditions of the relaxed NLP (13) at a feasible point x. Given two index sets $\mathcal{X}_1,\ \mathcal{X}_2 \subset \{1,\ldots,p\}$ with

$$\mathcal{X}_1 \cup \mathcal{X}_2 \ = \ \{1,\ldots,p\},$$

denote their respective complements in $\{1,\ldots,p\}$ by $\mathcal{X}_1^{\perp}$ and $\mathcal{X}_2^{\perp}$. For any such pair of index sets, define the *relaxed NLP corresponding to the MPCC (1)* as

$$\begin{array}{lllll}
\underset{x}{\text{minimize}} & f(x) & & & \\
\text{subject to} & c_{\mathcal{E}}(x) = 0 & & & \\
& c_{\mathcal{I}}(x) \geq 0 & & & \\
& x_{1j} = 0 \quad \forall j \in \mathcal{X}_2^{\perp} & \text{and} & x_{1j} \geq 0 \quad \forall j \in \mathcal{X}_2 & \\
& x_{2j} = 0 \quad \forall j \in \mathcal{X}_1^{\perp} & \text{and} & x_{2j} \geq 0 \quad \forall j \in \mathcal{X}_1. &
\end{array} \tag{13}$$

Concepts such as MPCC constraint qualifications (CQs) and second-order conditions are defined in terms of this relaxed NLP (see, e.g., [FLRS02]). Formally, the linear independence constraint qualification (LICQ) is extended to MPCCs as follows:

Definition 3.2 *The MPCC (1) is said to satisfy an* MPCC-LICQ *at x if the corresponding relaxed NLP (13) satisfies an LICQ.*

Next, a second-order sufficient condition (SOSC) for MPCCs is given. Like strong stationarity, it is related to the relaxed NLP (13). Let $\mathcal{A}^*$ denote the set of active constraints of (13) and $\mathcal{A}_+^* \subset \mathcal{A}^*$ the set of active constraints with nonzero multipliers (some could be negative). Let A denote the matrix of active constraint normals, that is,

$$A = \left[A_{\mathcal{E}}^* \ : \ A_{\mathcal{I}\cap\mathcal{A}^*}^* \ : \ \begin{matrix} 0 \\ I_1^* \\ 0 \end{matrix} \ : \ \begin{matrix} 0 \\ 0 \\ I_2^* \end{matrix} \right] =: \ [a_i^*]_{i\in\mathcal{A}^*},$$

where $A_{\mathcal{I}\cap\mathcal{A}^*}^*$ are the active inequality constraint normals and

$$I_1^* := [e_i]_{i\in\mathcal{X}_1^*} \ \text{ and } \ I_2^* := [e_i]_{i\in\mathcal{X}_2^*}$$

are parts of the $p \times p$ identity matrices corresponding to active bounds. Define the set of feasible directions of zero slope of the relaxed NLP (13) as

$$S^* = \left\{ s \mid s \neq 0\,,\ g^{*^T} s = 0\,,\ a_i^{*^T} s = 0\,,\ i \in \mathcal{A}_+^*\,,\ a_i^{*^T} s \geq 0\,,\ i \in \mathcal{A}^* \backslash \mathcal{A}_+^* \right\}.$$

The MPCC-SOSC is defined as follows.

Definition 3.3 *A strongly stationary point z^* with multipliers $(\lambda^*, \hat{\nu}_1^*, \hat{\nu}_2^*)$ satisfies the* MPCC-SOSC *if for every direction $s \in S^*$ it follows that $s^T \nabla^2 \mathcal{L}^* s > 0$, where $\nabla^2 \mathcal{L}^*$ is the Hessian of the Lagrangian of (13) evaluated at $(z^*, \lambda^*, \hat{\nu}_1^*, \hat{\nu}_2^*)$.*

3.2 Linearizations of the NCP functions

All NCP functions with the exception of the bilinear form are nonsmooth at the origin. In addition, the min-function is also nonsmooth along $a = b$, and the linearized min-function is nonsmooth along $a = \sigma^{-1}b$ and $a = \sigma b$. Luckily, SQP methods converge for a simple choice of subgradient.

We start by summarizing some well-known properties of the gradients of the Fischer-Burmeister function (4) for $(a, b) \neq (0, 0)$:

$$\nabla\varphi_{FB}(a,b) = \begin{pmatrix} 1 - \dfrac{a}{\sqrt{a^2+b^2}} \\ 1 - \dfrac{b}{\sqrt{a^2+b^2}} \end{pmatrix}.$$

It can be shown that $0 < 1 - \frac{a}{\sqrt{a^2+b^2}} < 2$ for all $(a, b) \neq (0, 0)$. In addition, if $a > 0$ and $b > 0$, it can be shown that

$$\nabla\varphi_{FB}(a,0) = \begin{pmatrix} 0 \\ 1 \end{pmatrix} \text{ and } \nabla\varphi_{FB}(0,b) = \begin{pmatrix} 1 \\ 0 \end{pmatrix}.$$

Similarly, the gradient of the smoothed natural residual function is

$$\nabla\varphi_{NRs}(a,b) = \frac{1}{2}\begin{pmatrix} 1 - \dfrac{a - b + \frac{b}{2\sigma}}{\sqrt{(a-b)^2 + \frac{ab}{\sigma}}} \\ 1 - \dfrac{b - a + \frac{a}{2\sigma}}{\sqrt{(a-b)^2 + \frac{ab}{\sigma}}} \end{pmatrix}.$$

For $a > 0$ and $b > 0$, it follows that

$$\nabla\varphi_{NRs}(a,0) = \begin{pmatrix} 0 \\ 1 - \frac{1}{4\sigma} \end{pmatrix} \text{ and } \nabla\varphi_{NRs}(0,b) = \begin{pmatrix} 1 - \frac{1}{4\sigma} \\ 0 \end{pmatrix}.$$

Despite the fact that the NCP functions are not differentiable everywhere, it turns out that a particular choice of subgradient gives fast convergence for SQP methods. To show equivalence of the first-order conditions in [FLRS02], we exploit the fact that $\nabla\varphi_{BL}(0,0) = 0$. Fortunately, 0 is a generalized gradient of the other NCP functions, that is, $0 \in \partial\varphi(0,0)$. Similarly, we will choose a suitable subgradient for the min-function along $a = b$. With a slight abuse of notation, we summarize the subgradient convention:

Convention 3.4 *The following convention is used for subgradients of the nonsmooth NCP functions:*

1. $\nabla\varphi(0,0) = 0$ *for any NCP function.*
2. $\nabla\varphi_{\min}(a,a) = (\frac{1}{2}, \frac{1}{2})^T$ *for the min-function for* $a > 0$.
3. $\nabla\varphi_{lin}(a,\sigma a) = (0,1)$ *and* $\nabla\varphi_{lin}(a,\sigma^{-1}a) = (1,0)$ *for the linearized min-function, for* $a > 0$.

This convention is consistent with the subgradients of the NCP functions and is readily implemented. The most important convention is to ensure that $\nabla\varphi(0,0) = 0$ because, otherwise, we would not be able to establish equivalence of first-order conditions. The other conventions could be relaxed to allow other subgradients. The convention on the subgradients also has an important practical implication. We have observed convergence to M-stationary, or even C-stationary points that are not strongly stationary for other choices of $0 \neq v \in \partial\varphi(0,0)$. Setting $v = 0 \in \partial\varphi(0,0)$ prevents convergence to such spurious stationary points.

It turns out that a straightforward application of SQP to (3) is not very efficient in practice. The reason is that the linearization of the complementarity constraint (2) together with the lower bounds has no strict interior. Therefore, we relax the linearization of (2). Let $0 < \delta < 1$, and $0 < \kappa \leq 1$ be constants, and consider

$$a \geq 0,\ b \geq 0,\ \varphi(\hat{a}, \hat{b}) + \nabla\varphi(\hat{a}, \hat{b})^T \begin{pmatrix} a - \hat{a} \\ b - \hat{b} \end{pmatrix} \leq \delta \left(\min(1, \varphi(\hat{a}, \hat{b}))\right)^{1+\kappa}. \quad (14)$$

Clearly, this is a relaxation of the linearization of (2). The following proposition summarizes some useful properties of the linearizations of the NCP functions.

Proposition 3.5 *Let $\varphi(a,b)$ be one of the functions (8)–(11). Then it follows that*

1. *$a, b \geq 0$ and $\varphi(a,b) \leq 0$ is equivalent to $0 \leq a \perp b \geq 0$.*
2. *If $\hat{a}, \hat{b} \geq 0$ and $\hat{a} + \hat{b} > 0$, then it follows that the perturbed system of inequalities (14) is consistent for any $0 \leq \delta < 1$, and $0 < \kappa \leq 1$. In addition, if $\delta > 0$ and $\hat{a}, \hat{b} > 0$, then (14) has a nonempty interior for the Fischer-Burmeister function, the bilinear function, and the smoothed natural residual function.*

Proof. Part 1 is obvious. For Part 2, consider each NCP function in turn. For the bilinear function (8), it readily follows that $(a,b) = (0,0)$ is feasible in (14) because for $\hat{a}, \hat{b} \geq 0$, we get

$$\hat{a}\hat{b} + \nabla\varphi_{BL}^T \begin{pmatrix} -\hat{a} \\ -\hat{b} \end{pmatrix} = -\hat{a}\hat{b} \leq 0,$$

and clearly, if $\delta > 0$, there exists a nonempty interior.

Next consider the Fischer-Burmeister function (4), for which (14) with $\delta = 0$ becomes

$$\left(1 - \frac{\hat{a}}{\sqrt{\hat{a}^2 + \hat{b}^2}}\right) a + \left(1 - \frac{\hat{b}}{\sqrt{\hat{a}^2 + \hat{b}^2}}\right) b \leq 0.$$

Since the terms in the parentheses are positive, it follows that $(a,b) = 0$ is the only point satisfying $a, b \geq 0$ and (14). On the other hand, if $\delta > 0$, then the right-hand side of (14) is positive, and there exists a nonempty interior of $a, b \geq 0$ and (14).

For (5) and (6), it follows for $\hat{a} < \hat{b}$ that (14) becomes $a = 0, b \geq 0$. The conclusion for $\hat{a} > \hat{b}$ follows similarly. If $\hat{a} = \hat{b}$, then (14) becomes $\frac{1}{2}a + \frac{1}{2}b \leq \delta \left(\min(1, \varphi(\hat{a}, \hat{b}))\right)^{1+\kappa}$, and the results follow.

The result for (7) follows from the fact that (7) is a linear combination of the Fischer-Burmeister function and (8).

To show the result for the smoothed min functions, we observe that for $b \leq a/\sigma$ and $b \geq \sigma a$ the functions are identical to the min-function and the result follows. For $a/\sigma < b < \sigma a$, we consider (10) and (11) in turn. The linearization of (10) is equivalent to $a + b \leq 0$, which implies feasibility. It can also be shown that linearization of (11) about any point is feasible at the origin $(a, b) = (0, 0)$.

The smoothed natural residual function also has feasible linearizations. For (9), (14) is equivalent to (using $\sigma = \sigma_{NR}$ to simplify the notation)

$$\varphi_{NRs}(\hat{a}, \hat{b}) + \left(1 - \frac{\hat{a} - \hat{b} + \frac{\hat{b}}{2\sigma}}{\sqrt{(\hat{a} - \hat{b})^2 + \frac{\hat{a}\hat{b}}{\sigma}}}\right)(a - \hat{a}) + \left(1 - \frac{\hat{b} - \hat{a} + \frac{\hat{a}}{2\sigma}}{\sqrt{(\hat{a} - \hat{b})^2 + \frac{\hat{a}\hat{b}}{\sigma}}}\right)(b - \hat{b}) \leq 0.$$

Rearranging, we have

$$-\varphi_{NRs}(\hat{a}, \hat{b}) + \left(1 - \frac{\hat{a} - \hat{b} + \frac{\hat{b}}{2\sigma}}{\sqrt{(\hat{a} - \hat{b})^2 + \frac{\hat{a}\hat{b}}{\sigma}}}\right) a + \left(1 - \frac{\hat{b} - \hat{a} + \frac{\hat{a}}{2\sigma}}{\sqrt{(\hat{a} - \hat{b})^2 + \frac{\hat{a}\hat{b}}{\sigma}}}\right) b \leq 0.$$

The first term is clearly nonpositive, and it can be shown that the terms multiplying a and b are nonnegative, thus implying consistency and a nonempty interior, even when $\delta = 0$. □

A disadvantage of the functions (8), (7), and (9) is that arbitrarily close to a strongly stationary point, the linearizations may be inconsistent [FLRS02]. The next proposition shows that the min-function and its smoothed versions (10) and (11) do not have this disadvantage.

Proposition 3.6 *Consider (3) using any of the min-functions, (5), (10), or (11), and assume that the MPCC-MFCQ holds at a strongly stationary point. Then it follows that the linearization of (3) is consistent for all $x_1, x_2 \geq 0$ sufficiently close to this strongly stationary point.*

Proof. Under MPCC-MFCQ, it follows that the linearization of the relaxed NLP (13) is consistent in a neighborhood of a strongly stationary point. Now consider the linearization of the min-function near a strongly stationary point, x^* say. For components i, such that $x^*_{1i} = 0 < x^*_{2i}$, it follows for any point x^k sufficiently close to x^* that $0 \leq x^k_{1i} < x^k_{2i}$. Thus, the linearization of the corresponding min-function gives $d_{1i} \leq -x^k_{1i}$. Together with the lower bound

$d_{1i} \geq -x_{1i}^k$, this is equivalent to $d_{1i} = -x_{1i}^k$, the linearization of the same component in the relaxed NLP. A similar conclusion holds for components with $x_{1i}^* > 0 = x_{2i}^*$.

Finally, for components i, such that $x_{1i}^* = 0 = x_{2i}^*$, it follows that the origin $x_{1i}^{k+1} = x_{2i}^{k+1} = 0$ is feasible (Proposition 3.5). This point is also feasible for the relaxed NLP.

A similar argument can be made for the smoothed min-functions (10) and (11) by observing that for $x_{1i}^* = 0 < x_{2i}^*$, there exists a neighborhood where these functions agree with the min-function and for $x_{1i}^* = 0 = x_{2i}^*$, feasibility follows from Proposition 3.5. □

An important consequence of this proposition is that the quadratic convergence proof for MPCCs in [FLRS02] can now be applied *without* the assumption that all QP subproblems are consistent. In this sense, Proposition 3.6 implies that the equivalent NLP (3) using the min-functions satisfies a constraint qualification.

3.3 NCP functions and strong stationarity

A consequence of the gradient convention is that the gradients of all NCP functions have the same structure. In particular, it follows that for $a, b > 0$

$$\nabla\varphi(a,0) = \begin{pmatrix} 0 \\ \tau_a \end{pmatrix}, \nabla\varphi(0,b) = \begin{pmatrix} \tau_b \\ 0 \end{pmatrix}, \nabla\varphi(a,b) = \begin{pmatrix} \tau_b \\ \tau_a \end{pmatrix}, \nabla\varphi(0,0) = \begin{pmatrix} 0 \\ 0 \end{pmatrix}$$

for some parameters $\tau_a, \tau_b > 0$ that depend on a, b and the NCP function. As a consequence, we can generalize the proof of equivalence of first-order conditions from [FLRS02] to all NCP functions from Section 2. Let $\Phi(x_1, x_2)$ denote the vector of functions $\varphi(x_{1i}, x_{2i})$. The KKT conditions of (3) are that there exist multipliers $\mu := (\lambda, \nu_1, \nu_2, \xi)$ such that

$$\begin{aligned} \nabla f(x) - \nabla c(x)^T\lambda - \begin{pmatrix} 0 \\ \nu_1 \\ \nu_2 \end{pmatrix} + \begin{pmatrix} 0 \\ \nabla_{x_1}\Phi(x_1,x_2)\xi \\ \nabla_{x_2}\Phi(x_1,x_2)\xi \end{pmatrix} &= 0 \\ c_{\mathcal{E}}(x) &= 0 \\ c_{\mathcal{I}}(x) &\geq 0 \\ x_1, x_2 &\geq 0 \\ \Phi(x_1,x_2) &\leq 0 \\ \lambda_{\mathcal{I}} &\geq 0 \\ \nu_1, \nu_2 &\geq 0 \\ \xi &\geq 0 \\ c_i(x)\lambda_i = x_{1j}\nu_{1j} = x_{2j}\nu_{2j} &= 0\,. \end{aligned} \qquad (15)$$

There is also a complementarity condition $\xi^T\Phi(x_1, x_2) = 0$, which is implied by feasibility of x_1, x_2 and has been omitted. Note that the choice $\nabla\varphi(0,0) = 0$ makes (15) consistent with strong stationarity, as will be shown next.

Theorem 3.7 *$(x^*, \lambda^*, \hat{\nu}_1, \hat{\nu}_2)$ is a strongly stationary point satisfying (12) if and only if there exist multipliers $(x^*, \lambda^*, \nu_1^*, \nu_2^*, \xi^*)$ satisfying the KKT conditions (15) of the equivalent NLP (3). If φ is any of the NCP functions of Section 2, then*

$$\hat{\nu}_1 = \nu_1^* - \boldsymbol{\tau}_1 \xi^* \tag{16a}$$
$$\hat{\nu}_2 = \nu_2^* - \boldsymbol{\tau}_2 \xi^*, \tag{16b}$$

where $\boldsymbol{\tau}_1$ and $\boldsymbol{\tau}_2$ are diagonal matrices with $\tau_j, j = 1, 2$ along their diagonals. Moreover, $\tau_{ji} = 0$, if $x_{1i} = x_{2i} = 0$ and otherwise satisfies the relationship

$$\tau_{1i} = \begin{cases} 1 \;\; \text{if } x_{2i} > 0 & \text{for } (4), (5), (6), (10), (11) \\ 1 - \frac{1}{4\sigma} \;\; \text{if } x_{2i} > 0 & \text{for } (9) \\ x_{2i} & \text{for } (8) \\ \lambda + (1-\lambda) x_{2i} \;\; \text{if } x_{2i} > 0 & \text{for } (7) \end{cases} \tag{17}$$

and

$$\tau_{2i} = \begin{cases} 1 \;\; \text{if } x_{1i} > 0 & \text{for } (4), (5), (6), (10), (11) \\ 1 - \frac{1}{4\sigma} \;\; \text{if } x_{1i} > 0 & \text{for } (9) \\ x_{1i} & \text{for } (8) \\ \lambda + (1-\lambda) x_{1i} \;\; \text{if } x_{1i} > 0 & \text{for } (7). \end{cases} \tag{18}$$

Proof. Note that gradients $\nabla \Phi$ have the same structure for all NCP functions used. Then (16) follows by comparing (15) and (12) and taking the gradients of the NCP functions into account. □

The failure of MFCQ for (3) implies that the multiplier set is unbounded. However, this unboundedness occurs in a special way. The multipliers of (3) form a ray, similar to [FLRS02], and there exists a multiplier of minimum norm, given by

$$\nu_{1i}^* = \max(\hat{\nu}_{1i}, 0), \tag{19a}$$
$$\nu_{2i}^* = \max(\hat{\nu}_{2i}, 0), \tag{19b}$$
$$\xi_i^* = -\min\left(\frac{\hat{\nu}_{1i}}{\tau_{1i}}, \frac{\hat{\nu}_{2i}}{\tau_{2i}}, 0\right). \tag{19c}$$

This implies the following complementarity conditions for the multipliers

$$0 \le \nu_{1i}^* \perp \xi_i^* \ge 0 \text{ and } 0 \le \nu_{2i}^* \perp \xi_i^* \ge 0. \tag{20}$$

This multiplier will be referred to as the *minimal, or basic, multiplier*. This term is justified by the observation (to be proved below) that the constraint normals corresponding to nonzero components of the basic multiplier are linearly independent, provided the MPCC satisfies an LICQ.

4 An SQP algorithm for NCP functions

This section describes an SQP algorithm for solving (3). The algorithm is an iterative procedure that solves a quadratic programming (QP) approximation of (3) around the iterate x^k for a step d at each iteration:

$$(QP^k)\begin{cases} \underset{d}{\text{minimize}} & g^{k^T} d + \frac{1}{2} d^T W^k d \\ \text{subject to} & c_{\mathcal{E}}^k + A_{\mathcal{E}}^{k^T} d = 0 \\ & c_{\mathcal{I}}^k + A_{\mathcal{I}}^{k^T} d \geq 0 \\ & x_1^k + d_1 \geq 0 \\ & x_2^k + d_2 \geq 0 \\ & \Phi^k + \nabla_{x_1} \Phi^{k^T} d_1 + \nabla_{x_2} \Phi^{k^T} d_2 \leq \delta \left(\min(1, \Phi^k)\right)^{1+\kappa}, \end{cases}$$

where $\mu^k = (\lambda^k, \nu_1^k, \nu_2^k, \xi^k)$ and $W^k = \nabla^2 \mathcal{L}(x^k, \mu^k)$ is the Hessian of the Lagrangian of (1):

$$W^k = \nabla^2 \mathcal{L}(x^k, \mu^k) = \nabla^2 f(x^k) - \sum_{i \in \mathcal{I} \cup \mathcal{E}} \lambda_i \nabla^2 c_i(x^k).$$

Note that the Hessian W^k does not include entries corresponding to $\nabla^2 \Phi$. This omission is deliberate as it avoids numerical difficulties near the origin, where $\nabla^2 \varphi_{FB}$ becomes unbounded. It will be shown that this does not affect the convergence properties of SQP methods.

The last constraint of (QP^k) is the relaxation of the linearization of the complementarity condition (14). We will show that the perturbation does not impede fast local convergence. Formally, the SQP algorithm is defined in Algorithm 1.

Let $k = 0$, x^0 given
while *not optimal* **do**
  Solve (QP^k) for a step d
  Set $x^{k+1} = x^k + d$, and $k = k + 1$

Algorithm 1: Local SQP Algorithm for MPCCs

In practice, we also include a globalization scheme to stabilize SQP. In our case, we use a filter [FL02] and a trust region to ensure convergence to stationary points [FLT02]. The convergence theory of filter methods allows for three possible outcomes [FLT02, Theorem 1]:

(A) The algorithm terminates at a point that is locally infeasible.
(B) The algorithm converges to a Kuhn-Tucker point.
(C) The algorithm converges to a feasible point at which MFCQ fails.

Clearly, **(B)** cannot happen because (3) violates MFCQ at any feasible point. Outcome **(A)** is typically associated with convergence to a local minimum of the norm of the constraint violation and cannot be avoided unless global optimization techniques are used. Therefore, we deal mainly with outcome **(C)** if we apply a filter algorithm to MPCC formulated as NLPs (3). The next section presents a local convergence analysis of the SQP algorithm applied to (3).

5 Local convergence of SQP for MPCCs

This section establishes superlinear convergence of SQP methods a strongly stationary point under mild conditions. The notation $\boldsymbol{\tau}_1, \boldsymbol{\tau}_2$ introduced in Theorem 3.7 allows the convergence analysis of all NCP functions to be unified. We note that the presence of the perturbation term $\delta\left(\min(1, \Phi^k)\right)^{1+\kappa}$, with $\kappa < 1$, implies that we cannot obtain quadratic convergence in general.

The convergence analysis is concerned with strongly stationary points. Let x^* be a strongly stationary point, and denote by $\mathcal{A}(x^*)$ the set of active general constraints:

$$\mathcal{A}(x^*) := \{i | c_i(x^*) = 0\}.$$

We also denote the set of active bounds by

$$\mathcal{X}_j(x^*) := \{i | x_{ji} = 0\} \quad \text{for } j = 1, 2$$

and let $\mathcal{D}(x^*) := \mathcal{X}_1(x^*) \cap \mathcal{X}_2(x^*)$ be the set of degenerate indices associated with the complementarity constraint.

Assumptions 5.1 *We make the following assumptions:*

[A0] The subgradients of the NCP functions are computed according to Convention 3.4.
[A1] The functions f and c are twice Lipschitz continuously differentiable.
[A2] (1) satisfies an MPCC-LICQ.
[A3] x^* is a strongly stationary point that satisfies an MPCC-SOSC.
[A4] $\lambda_i \neq 0$, $\forall i \in \mathcal{E}^*$, $\lambda_i^* > 0$, $\forall i \in \mathcal{A}^* \cap \mathcal{I}$, and either $\nu_{1j}^* > 0$ and $\nu_{2j}^* > 0$, $\forall j \in \mathcal{D}^*$.
[A5] The QP solver always chooses a linearly independent basis.

We note that **[A0]** is readily implemented and that assumption **[A5]** holds for the QP solvers used within `snopt` due to Gill et al. [GMS02] and `filter` [FL02]. The most restrictive assumptions are **[A2]** and **[A3]** because they exclude B-stationary points that are not strongly stationary. This fact is not surprising because it is well known that SQP methods typically converge linearly to such B-stationary points.

It is useful to divide the convergence proof into two parts. First, we consider the case where complementarity holds for some iterate k, i.e. $\Phi(x_1^k, x_2^k) = 0$.

In this case, the SQP method applied to (3) is shown to be equivalent to SQP applied to the relaxed NLP (13). In the second part, we assume that $\Phi(x_1^k, x_2^k) > 0$ for all k. Under the additional assumption that all QP approximations remain consistent, superlinear convergence can again be established.

5.1 Local convergence for exact complementarity

In this section we make the additional assumption that

[A6] $\Phi(x_1^k, x_2^k) = 0$ and (x^k, μ^k) is sufficiently close to a strongly stationary point.

Assumption **[A6]** implies that for given index sets $\mathcal{X}_j := \mathcal{X}_j(x^k) := \{i | x_{ji}^k = 0\}, j = 1,2$, the following holds:

$$\begin{array}{ll} x_{1j}^k = 0 & \forall j \in \mathcal{X}_2^\perp \\ x_{2j}^k = 0 & \forall j \in \mathcal{X}_1^\perp \\ x_{1j}^k = 0 \text{ or } x_{2j}^k = 0 & \forall j \in \mathcal{D} = \mathcal{X}_1 \cap \mathcal{X}_2. \end{array}$$

In particular, it is not necessary to assume that both $x_{1i}^k = 0$ and $x_{2i}^k = 0$ for $i \in \mathcal{D}^*$. Thus it may be possible that $\mathcal{X}_1 \neq \mathcal{X}_1^*$ (and similarly for $\mathcal{X}_2$). An important consequence of **[A6]** is that $\mathcal{X}_1$, $\mathcal{X}_2$ satisfy

$$\begin{array}{c} \mathcal{X}_1^{*^\perp} \subset \mathcal{X}_1^\perp \subset \mathcal{X}_1^{*^\perp} \cup \mathcal{D}^* \\ \mathcal{X}_2^{*^\perp} \subset \mathcal{X}_2^\perp \subset \mathcal{X}_2^{*^\perp} \cup \mathcal{D}^* \\ \mathcal{D} \subset \mathcal{D}^*, \end{array} \tag{21}$$

that is, the indices $\mathcal{X}_1^{*^\perp}$ and $\mathcal{X}_2^{*^\perp}$ of the nondegenerate complementarity constraints have been identified correctly.

Next, it is shown that SQP applied to (3) is equivalent to SQP applied to the relaxed NLP (13). For a given partition $(\mathcal{X}_1^\perp, \mathcal{X}_2^\perp, \mathcal{D})$, an SQP step for the relaxed NLP (13) is obtained by solving the QP

$$(QP_R(x^k)) \left\{ \begin{array}{ll} \underset{d}{\text{minimize}} & g^{k^T} d + \frac{1}{2} d^T W^k d \\ \text{subject to} & c_{\mathcal{E}}^k + A_{\mathcal{E}}^{k^T} d = 0 \\ & c_{\mathcal{I}}^k + A_{\mathcal{I}}^{k^T} d \geq 0 \\ & d_{1j} = 0 \ \forall j \in \mathcal{X}_2^\perp \text{ and } x_{1j}^k + d_{1j} \geq 0 \ \forall j \in \mathcal{X}_2 \\ & d_{2j} = 0 \ \forall j \in \mathcal{X}_1^\perp \text{ and } x_{2j}^k + d_{2j} \geq 0 \ \forall j \in \mathcal{X}_1. \end{array} \right.$$

The following proposition shows that SQP applied to the relaxed NLP converges quadratically and identifies the correct index sets $\mathcal{X}_1^*$ and $\mathcal{X}_2^*$ in one step. Its proof can be found in [FLRS02, Proposition 5.2].

Proposition 5.2 *Let* **[A1]**–**[A6]** *hold, and let x^k be sufficiently close to x^*. Consider the relaxed NLP for any index sets $\mathcal{X}_1$, $\mathcal{X}_2$ (satisfying (21) by virtue of* **[A6]***). Then it follows that*

1. *there exists a neighborhood U of $(z^*, \lambda^*, \nu_1^*, \nu_2^*)$ and a sequence of iterates generated by SQP applied to the relaxed NLP (13), $\{(x^l, \lambda^l, \nu_1^l, \nu_2^l)\}_{l>k}$, that lies in U and converges Q-quadratically to $(x^*, \lambda^*, \nu_1^*, \nu_2^*)$;*
2. *the sequence $\{x^l\}_{l>k}$ converges Q-superlinearly to x^*; and*
3. *$\mathcal{X}_1^l = \mathcal{X}_1^*$ and $\mathcal{X}_2^l = \mathcal{X}_2^*$ for $l > k$.*

Next, it is shown that the QP approximation to the relaxed NLP $(QP_R(x^k))$ and the QP approximation to the NCP formulation (QP^k) generate the same sequence of steps. The next lemma shows that the solution of $(QP_R(x^k))$ is feasible in (QP^k).

Lemma 5.3 *Let Assumptions* **[A1]**–**[A6]** *hold. Then it follows that a step d is feasible in $(QP_R(x^k))$ if and only if it is feasible in (QP^k).*

Proof. $(QP_R(x^k))$ and (QP^k) differ only in the way the complementarity constraint is treated. Hence we need only to prove the equivalence of those constraints. Let $j \in \mathcal{X}_2^\perp$. Then it follows that $x_{1j} = 0$, and $\frac{\partial \Phi^k}{\partial x_{1j}} = \tau_{1j} > 0$, and $\frac{\partial \Phi^k}{\partial x_{1j}} = 0$. Hence, (QP^k) contains the constraints

$$\tau_{1j} d_{1j} \leq 0 \;\text{ and }\; d_{1j} \geq 0 \Leftrightarrow d_{1j} = 0.$$

Similarly, we can show that the constraints are equivalent for $j \in \mathcal{X}_1^\perp$. Let $j \in \mathcal{D}$. Then it follows that (QP^k) contains the constraints $d_{2j} \geq 0$ and $d_{1j} \geq 0$, which are equivalent to the constraints of $(QP_R(x^k))$. The equivalence of the feasible sets follows because $(\mathcal{X}_1^\perp, \mathcal{X}_2^\perp, \mathcal{D})$ is a partition of $\{1, \ldots, p\}$. □

The next lemma shows that the solution of the two QPs are identical and that the multipliers are related.

Lemma 5.4 *Let Assumptions* **[A1]**–**[A6]** *hold. Let $(\lambda, \hat{\nu}_1, \hat{\nu}_2)$ be the Lagrange multipliers of $(QP_R(x^k, \mathcal{X}))$ (corresponding to a step d). Then it follows that the multipliers of (QP^k), corresponding to the same step d are $\mu = (\lambda, \nu_1, \nu_2, \xi)$, where*

$$\nu_{1i} = \quad \hat{\nu}_{1i} > 0, \; \forall i \in \mathcal{D} \tag{22a}$$

$$\nu_{2i} = \quad \hat{\nu}_{2i} > 0, \; \forall i \in \mathcal{D} \tag{22b}$$

$$\xi_i = \quad -\min\left(\frac{\hat{\nu}_{1i}}{\tau_{1i}}, \frac{\hat{\nu}_{2i}}{\tau_{2i}}, 0\right) \tag{22c}$$

$$\nu_{1i} = \hat{\nu}_{1i} - \xi_i \tau_{1i}, \; \forall i \in \mathcal{X}_2^\perp \tag{22d}$$

$$\nu_{2i} = \hat{\nu}_{2i} - \xi_i \tau_{2i}, \; \forall i \in \mathcal{X}_1^\perp, \tag{22e}$$

where τ_{ji} is given in (17–18). Conversely, given a solution d and multipliers μ of (QP^k), (22) shows how to construct multipliers so that $(d, \lambda, \widehat{\nu_1}, \widehat{\nu_2})$ solves $(QP_R(x^k, \mathcal{X}))$.

Proof. We equate the first-order conditions of $(QP_R(x^k))$ and (QP^k) and obtain

$$g^k + W^k d - A^k \lambda = \begin{pmatrix} 0 \\ \hat{\nu}_1 \\ \hat{\nu}_2 \end{pmatrix} = \begin{pmatrix} 0 \\ \nu_1 - \nabla_{x_1} \Phi \xi \\ \nu_2 - \nabla_{x_2} \Phi \xi \end{pmatrix}.$$

We distinguish three cases:
Case 1 ($j \in \mathcal{D}$): It follows from (21) that $j \in \mathcal{D}^*$, which implies that $\hat{\nu}_{1j}, \hat{\nu}_{2j} > 0$ for x^k sufficiently close to x^* by assumption **[A4]**. Moreover, $\frac{\partial \Phi}{\partial x_{1j}} = \frac{\partial \Phi}{\partial x_{2j}} = 0$, and hence, $\nu_{1j} = \hat{\nu}_{1j} > 0$, $\nu_{2j} = \hat{\nu}_{2j} > 0$, and $\xi_j = 0$ are valid multipliers for (QP^k).
Case 2 ($j \in \mathcal{X}_1^\perp$): We distinguish two further cases. If $j \in \mathcal{D}^*$, then a similar argument to Case 1 shows that $\nu_{1j} = \hat{\nu}_{1j} > 0$, $\nu_{2j} = \hat{\nu}_{2j} > 0$, and $\xi_j = 0$. On the other hand, if $j \in \mathcal{X}_1^{*\perp}$, then it follows that $\frac{\partial \Phi}{\partial x_{1j}} = 0$, and $\frac{\partial \Phi}{\partial x_{2j}} = \tau_{2j} > 0$ is bounded away from zero. Thus, $\nu_{1j} = \hat{\nu}_{1j} = 0$, and $\nu_{2j} = \hat{\nu}_{2j} - \tau_{2j}\xi_j$, and we can always choose $\nu_{2j}, \xi_j \geq 0$. We will show later that the QP solver in fact chooses either $\nu_{2j} > 0$, or $\xi_j > 0$.
Case 3 ($j \in \mathcal{X}_2^\perp$) is similar to Case 2. □

Next, it is shown that both QPs have the same solution in a neighborhood of $d = 0$; its proof can be found in [FLRS02, Lemma 5.6].

Lemma 5.5 *The solution d of $(QP_R(x^k))$ is the only strict local minimizer in a neighborhood of $d = 0$ and its corresponding multipliers $(\lambda, \hat{\nu}_1, \hat{\nu}_2)$ are unique. Moreover, d is also the only strict local minimizer in a neighborhood of $d = 0$ of (QP^k).*

The next theorem summarizes the results of this section.

Theorem 5.6 *If Assumptions* **[A1]**–**[A6]** *hold, then SQP applied to (3) generates a sequence $\{(x^l, \lambda^l, \nu_1^l, \nu_2^l, \xi^l)\}_{l>k}$ that converges Q-quadratically to $\{(x^*, \lambda^*, \nu_1^*, \nu_2^*, \xi^*)\}$ of (15), satisfying strong stationarity. Moreover, the sequence $\{x^l\}_{l>k}$ converges Q-superlinearly to x^* and $\Phi(x_1^l, x_2^l) = 0$ for all $l \geq k$.*

Proof. Under Assumptions **[A1]**–**[A4]**, SQP converges quadratically when applied to the relaxed NLP (13). Lemmas 5.3–5.5 show that the sequence of iterates generated by this SQP method is equivalent to the sequence of steps generated by SQP applied to (3). This implies Q-superlinear convergence of $\{x^l\}_{l>k}$. Convergence of the multipliers follows by considering (22). Clearly, the multipliers in (22a) and (22b) converge, as they are just the multipliers of the relaxed NLP, which converge by virtue of Proposition 5.2. Now observe that (22c) becomes

$$\xi_i^{k+1} = -\min\left(\frac{\hat{\nu}_{1i}^{k+1}}{\tau_{1i}^{k+1}}, \frac{\hat{\nu}_{2i}^{k+1}}{\tau_{2i}^{k+1}}, 0\right).$$

The right-hand side of this expression converges because $\hat{\nu}_{1i}^{k+1}, \hat{\nu}_{2i}^{k+1}$ converge and the denominators τ_i^{k+1} are bounded away from zero for $i \in \mathcal{X}_1^{*^\perp}, \mathcal{X}_2^{*^\perp}$. Finally, (22d) and (22e) converge by a similar argument.

$\Phi(x_1^l, x_2^l) = 0$, $\forall l \geq k$, follows from the convergence of SQP for the relaxed NLP (13) and the fact that SQP retains feasibility with respect to linear constraints. Assumption **[A4]** ensures that $d_{1j}^k = d_{2j}^k = 0, \forall j \in \mathcal{D}^*$, since $\nu_{1j}^k, \nu_{2j}^k > 0$ for biactive complementarity constraints. Thus SQP will not move out of the corner and stay on the same face. □

5.2 Local convergence for nonzero complementarity

This section shows that SQP converges superlinearly even if complementarity does not hold at the starting point, that is, if $\Phi(x_1^k, x_2^k) > 0$. It is shown in [FLRS02] that the QP approximation to (3) with $x_{1i}x_{2i} \leq 0$ can be inconsistent arbitrarily close to a strongly stationary point. Similar examples can be constructed for the NCP functions in Section 2. Only the min-function and its piecewise smooth variations guarantee feasibility of the QP approximation near a strongly stationary point (see Proposition 3.6).

Note that by virtue of the preceding section, any component for which $\varphi(x_{1i}^k, x_{2i}^k) = 0$ can be removed from the complementarity constraints and instead be treated as part of the general constraints, as $\varphi(x_{1i}^l, x_{2i}^l) = 0$ for all $l \geq k$. Hence, it can be assumed without loss of generality that $\Phi(x_1^k, x_2^k) > 0$ for all k.

In the remainder of the proof, it is assumed without loss of generality that $\mathcal{X}_1^{*^\perp} = \emptyset$, that is, the solution can be partitioned as

$$x_2^* = \begin{pmatrix} x_{21}^* \\ x_{22}^* \end{pmatrix} = \begin{pmatrix} 0 \\ x_{22}^* \end{pmatrix}, \tag{23}$$

where $x_{22}^* > 0$, and $x_1^* = 0$ is partitioned in the same way. This simplifies the notation in the proof.

SQP methods can take arbitrary steps when encountering infeasible QP approximations. In order to avoid the issue of infeasibility, the following assumption is made that often holds in practice.

[A7] All QP approximations (QP^k) are consistent for x^k sufficiently close to x^*.

This is clearly an undesirable assumption because it is an assumption on the progress of the method. However, Proposition 3.6 shows that **[A7]** holds for the NCP reformulations involving the min-function. In addition, it is shown in [FLRS02] that **[A7]** is satisfied for MPCCs with vertical complementarity constraints that satisfy a mixed-P property. Moreover, the use of the perturbation makes it less likely that the SQP method will encounter infeasible QP subproblems.

The key idea behind our convergence result is to show convergence for any "basic" active set. To this end, we introduce the set of active complementarity constraints

$$\mathcal{C}(x) := \{i \,:\, \varphi(x_{1i}, x_{2i}) = 0\}.$$

Let $\mathcal{I}(x) := \mathcal{I} \cap \mathcal{A}(x)$, and let the basic constraints be

$$\mathcal{B}(x) := \mathcal{E} \cup \mathcal{I}(x) \cup \mathcal{X}_1(x) \cup \mathcal{X}_2(x) \cup \mathcal{C}(x).$$

The set of strictly active constraints (defined in terms of the basic multiplier, μ, see (19)) is given by

$$\mathcal{B}_+(x) \;:=\; \{i \in \mathcal{B}(x) \mid \mu_i \neq 0\}.$$

Moreover, let B_+^k denote the matrix of strictly active constraint normals at $x = x^k$, namely,

$$B_+^k \;:=\; \left[a_i^k\right]_{i \in \mathcal{B}_+(x^k)},$$

where a_i^k is the constraint normal of constraint $i \in \mathcal{B}_+(x^k)$.

The failure of any constraint qualification at a solution x^* of the equivalent NLP (3) implies that the active constraint normals at x^* are linearly dependent. However, the constraint normals corresponding to strictly active constraints are linearly independent, as shown in the following lemma.

Lemma 5.7 *Let Assumptions* **[A1]**–**[A4]** *hold, and let x^* be a solution of the MPCC (1). Let $\mathcal{I}^*$ denote the set of* active *inequalities $c_{\mathcal{I}}(x)$, and consider the matrix of active constraint normals at x^*,*

$$B^* \;=\; \left[\begin{array}{c|cc|c} & 0 & 0 & 0 \\ A_{\mathcal{E}}^* \; A_{\mathcal{I}^*}^* & I & 0 & \begin{pmatrix} 0 \\ -\nabla_{x_{12}} \Phi_2 \end{pmatrix} \\ & 0 & \begin{bmatrix} I \\ 0 \end{bmatrix} & \begin{pmatrix} 0 \\ 0 \end{pmatrix} \end{array}\right], \tag{24}$$

where we have assumed without loss of generality that $\mathcal{X}_1^{\perp^*} = \emptyset$. The last column is the gradient of the complementarity constraint. Then it follows that B is linearly dependent and that

$$span\langle\begin{bmatrix} 0 \\ I_2 \end{bmatrix}\rangle \;=\; span\langle\begin{bmatrix} 0 \\ -\nabla_{x_{12}} \Phi_2 \end{bmatrix}\rangle. \tag{25}$$

Moreover, any submatrix of columns of B has full rank provided that it contains $[A_{\mathcal{E}}^* \; A_{\mathcal{I}}^*]$ and a linearly independent set from the columns in (25).

Proof. The structure of the gradient of the NCP functions and (23) show that (25) holds. Thus B^* is linearly dependent. MPCC-LICQ shows that B^* without the columns corresponding to the NCP functions has full rank. By choosing a linearly independent subset from the columns in (25), we get a basis. □

Lemma 5.7 shows that the normals corresponding to the basic multiplier are linearly independent despite the fact that the active normals are linearly dependent. The proof shows that in order to obtain a linearly independent basis, any column of $x_{12} = 0$ can be exchanged with the corresponding normal of the complementarity constraint. This matches the observation that the basic multipliers of the simple bounds and the corresponding complementarity constraint are complementary (see (20)).

Next, it is shown that for x^k sufficiently close to x^*, if both the normals corresponding to $x_{1i} \geq 0$ and $\varphi(x_{1i}, x_{2i}) \leq 0$ are active, then at the next iteration exact complementarity holds for that component and $\varphi(x_{1i}^l, x_{2i}^l) = 0$ and for all subsequent iterations by virtue of Lemma 5.3. Thus, the QP solver cannot continue to choose a basis that is increasingly ill-conditioned.

Lemma 5.8 *Let Assumptions* [**A1**]–[**A5**] *hold, and let x^k be sufficiently close to x^*. Partition the NCP function $\Phi = (\Phi_1, \Phi_2)^T$ in the same way as x_1, x_2 in (23). Consider the matrix of active constraint normals at x^k,*

$$B = \left[\begin{array}{c|cc|cc} A_{\mathcal{E}}^k \; A_{\mathcal{I}}^k & \begin{matrix} 0 \\ \begin{bmatrix} I & 0 \\ 0 & I \end{bmatrix} \\ \\ \end{matrix} & \begin{matrix} 0 \\ \\ \begin{bmatrix} I \\ 0 \end{bmatrix} \end{matrix} & \begin{matrix} 0 \\ \begin{bmatrix} -\nabla_{x_{11}} \Phi_1 \\ 0 \\ -\nabla_{x_{21}} \Phi_1 \\ 0 \end{bmatrix} \end{matrix} & \begin{matrix} 0 \\ \begin{bmatrix} 0 \\ -\nabla_{x_{12}} \Phi_2 \\ 0 \\ -\nabla_{x_{22}} \Phi_2 \end{bmatrix} \end{matrix} \end{array} \right].$$

Then it follows that the columns corresponding to the matrix $\nabla_x \Phi_2$ have the structure $(0, 0, -\tau, 0, -\varepsilon)^T$, where $\tau = \mathcal{O}(1)$ and $\varepsilon > 0$ is small. If the optimal basis of (QP^k) contains both a column i of $x_{1i} \geq 0$ and $\varphi(x_{1i}, x_{2i}) \leq 0$, then it follows that

$$x_{1i}^k > 0 \quad \text{and} \quad x_{1i}^{k+1} x_{2i}^{k+1} = 0.$$

Moreover, there exists $c > 0$ such that

$$\| (x^{k+1}, \mu^{k+1}) - (x^*, \mu^*) \| \leq c \| (x^k, \mu^k) - (x^*, \mu^*) \|. \tag{26}$$

Proof. The first part follows by observing that for x^k close to x^*, $x_{12} \geq 0$ is small and $x_{22} = \mathcal{O}(1)$, which implies the form of the columns. Exchanging them with the corresponding columns of $x_{12} \geq 0$ results in a nonsingular matrix by Lemma 5.7. The second part follows from the nonsingularity assumption [**A5**] (if $x_{1i}^k = 0$, then the basis would be singular) and the fact that if the column corresponding to $x_{1i} \geq 0$ is basic, then $x_{1i}^{k+1} = x_{1i}^k + d_{1i} = 0$ holds.

The third part follows by observing that Assumptions [**A2**] and [**A3**] imply that the relaxed NLP satisfies an LICQ and a SOSC. Hence, the basis B without the final column gives a feasible point close to x^k. Denote this solution by $(\hat{x}, \hat{\mu})$, and let the corresponding step be denoted by $\hat{d}$. Clearly, if this step also satisfies the linearization of the complementarity constraint, that is, if

$$\Phi^k + \nabla_{x_1} \Phi^{k^T} \hat{d}_1 + \nabla_{x_2} \Phi^{k^T} \hat{d}_2 \leq 0,$$

then (26) follows by second-order convergence of SQP for the relaxed NLP. If, on the other hand,

$$\Phi^k + \nabla_{x_1} \Phi^{k^T} \hat{d}_1 + \nabla_{x_2} \Phi^{k^T} \hat{d}_2 > 0,$$

then the SQP step of the relaxed NLP is not feasible in (QP^k). In this case consider the following decomposition of the SQP step. Let

$$\hat{d}^n = \begin{pmatrix} 0 \\ \hat{d}_1 \\ \begin{pmatrix} \hat{d}_{21} \\ 0 \end{pmatrix} \end{pmatrix} = \begin{pmatrix} 0 \\ -x_1^k \\ \begin{pmatrix} -x_{21}^k \\ 0 \end{pmatrix} \end{pmatrix}$$

be the normal component, and let $\hat{d}^t := \hat{d} - \hat{d}^n$ be the tangential component. Then it follows that the step of (QP^k) satisfies $d^k = \hat{d}^n + \sigma \hat{d}^t$ for some $\sigma \in [0, 1]$, and the desired bound on the distance follows from the convergence of $\hat{d}$. □

Thus, if both the normals corresponding to $\varphi(x_{1i}, x_{2i}) \leq 0$ and $x_{1i} \geq 0$ are basic, then $x_{1i}^{k+1} x_{2i}^{k+1} = 0$ for a point close to x^*. This component can then be removed from the complementarity constraint, as Lemma 5.3 shows that $x_{1i}^{k+l} x_{2i}^{k+l} = 0$ for all $l \geq 1$. In the remainder we can therefore concentrate on the case that $x_{1i}^k x_{2i}^k > 0$ for all iterates k. Next, it is shown that for x^k sufficiently close to x^*, the basis at x^k contains the equality constraints $\mathcal{E}$ and the active inequality constraints $\mathcal{I}^*$.

Lemma 5.9 *Let x^k be sufficiently close to x^*, and let Assumptions* **[A1]**–**[A5]** *and* **[A7]** *hold. Then it follows that the optimal basis B of (QP^k) contains the normals $A_{\mathcal{E}}^k$ and $A_{\mathcal{I}^*}^k$.*

Proof. This follows by considering the gradient of (QP^k),

$$0 = \nabla f^k + W^k d^k - \nabla c^{k^T} \lambda^{k+1} - \begin{pmatrix} 0 \\ \nu_1^{k+1} - \xi^{k+1} \nabla_{x_1} \Phi^k \\ \nu_2^{k+1} - \xi^{k+1} \nabla_{x_2} \Phi^k \end{pmatrix},$$

where W^k is the Hessian of the Lagrangian. For x^k sufficiently close to x^*, it follows from **[A4]** that $\lambda_i^{k+1} \neq 0$ for all $i \in \mathcal{E} \cup \mathcal{I}^*$. □

Thus, as long as the QP approximations remain consistent, the optimal basis of (QP^k) will be a subset of B satisfying the conditions in Lemma 5.8. The key idea is to show that for any such basis, there exists an equality constraint problem for which SQP converges quadratically. Since there is only a finite number of basis, this implies convergence for SQP.

We now introduce the *reduced NLP*, which is an equality constraint NLP corresponding to a basis with properties as in Lemma 5.8. Assume that x^* can be partitioned as in (23), and define the reduced NLP as

$$\begin{array}{ll}\underset{x}{\text{minimize}} & f(x)\\ \text{subject to} & c_{\mathcal{E}}(x)=0\\ & c_{\mathcal{I}^*}(x)=0\\ & x_{11}=0\\ & x_{21}=0\\ & x_{1i}=0 \quad \text{or} \quad \Phi(x_{1i},x_{2i})=0 \quad \forall i\in\mathcal{X}_2^{\perp},\end{array}$$

where the last constraint means that either $x_{1i}=0$ or $\Phi(x_{1i},x_{2i})=0$ but not both are present in the reduced NLP. Note that according to (23), $\mathcal{X}_1^{\perp}=\emptyset$. The key idea will be to relate the reduced NLP to a basis satisfying the conditions of Lemma 5.8. Next, it is shown that any reduced NLP satisfies an LICQ and an SOCS.

Lemma 5.10 *Let Assumptions* **[A1]**-**[A4]** *and* **[A7]** *hold. Then it follows that any reduced NLP satisfies LICQ and SOSC.*

Proof. Lemma 5.8 and the fact that either $x_{1i}=0$ or $\Phi(x_{1i},x_{2i})=0$ are active shows that the normals of the equality constraints of each reduced NLP are linearly independent. The SOSC follows from the MPCC-SOSC and the observation that the MPCC and the reduced NLP have the same null-space. □

Thus, applying SQP to the reduced NLP results in second-order convergence. Next, we observe that any nonsingular basis B corresponds to a reduced NLP. Unfortunately, relaxing the complementarity constraints in (QP^k) means that second-order convergence does not follow directly. However, the particular form of perturbation allows a superlinear convergence result to be established.

Proposition 5.11 *Let Assumptions* **[A1]**-**[A4]** *and* **[A7]** *hold. Then it follows that an SQP method that relaxes the complementarity as in (QP^k) converges superlinearly to x^* for any reduced NLP.*

Proof. Assume that $\delta=0$, so that no perturbation is used. Lemma 5.10 shows that the reduced NLP satisfy LICQ and SOSC and, therefore, convergence of SQP follows. In particular, it follows that for a given reduced NLP corresponding to a basis $\mathcal{B}$, there exists a constant $c_{\mathcal{B}}>0$ such that

$$\|\left(x^{k+1},\mu^{k+1}\right)-\left(x^*,\mu^*\right)\| \leq c_{\mathcal{B}}\|\left(x^k,\mu^k\right)-\left(x^*,\mu^*\right)\|^2. \qquad (27)$$

If the right-hand side of the complementarity constraint is perturbed (i.e., $\delta > 0$), then consider the Newton step corresponding to the QP approximation of the relaxed NLP about x^k. In particular, this step satisfies $d_N^k = -x_1^k$, and it follows that the perturbation is $o(\|d_N\|)$, where d_N is the Newton step. Hence, superlinear convergence follows using the Dennis-Moré characterization theorem (e.g., Fletcher, [Fle87, Theorem 6.2.3]). □

We note that the SQP method based on (QP^k) ignores the curvature corresponding to $\varphi(x_{12}, x_{22}) = 0$. However, it is easy to extend the proof of Proposition 5.11 to allow $\nabla^2 \Phi$ to be included. The key idea is to show that the limit of the projected Hessian of $\nabla \Phi^*$ is zero. Letting Z^k be a basis of the nullspace of (QP^k), we need to show that $\lim_{k \to \infty} Z^k \nabla^2 \Phi^* = 0$, which implies superlinear convergence (see, e.g., [Fle87, Chapter 12.4]). It can be shown that the Hessian of the NCP functions is unbounded in the nullspace of the active constraints of (QP^k).

Summarizing the results of this section, we obtain the following theorem.

Theorem 5.12 *Let Assumptions* **[A1]**–**[A5]** *and* **[A7]** *hold. Then it follows that SQP applied to the NLP formulation (3) of the MPCC (1) converges superlinearly near a solution* (x^*, μ^*).

Proof. Proposition 5.11 shows that SQP converges superlinearly for any possible choice of basis $\mathcal{B}$, and Assumption **[A7]** shows that (QP^k) is consistent and remains consistent. Therefore, there exists a basis for which superlinear convergence follows. Thus for each basis,

$$\lim_{k \to \infty} \frac{\|(x^{k+1}, \mu^{k+1}) - (x^*, \mu^*)\|}{\|(x^k, \mu^k) - (x^*, \mu^*)\|} = 0$$

follows. Since there are a finite number of bases, this condition holds independent of the basis and SQP converges superlinearly. □

5.3 Discussion of proofs

Several interesting observations arise from the convergence proofs of the preceding two sections. The curvature of the complementarity constraint $\Phi(x_1, x_2)$ can be ignored without losing fast local convergence. This fact is not surprising because the complementarity constraint

$$0 \leq x_1 \perp x_2 \geq 0$$

has zero curvature at any feasible point with $x_{1i} + x_{2i} > 0$. At the origin, on the other hand, the curvature is infinite. However, in this case the curvature does not affect convergence, as the reduced Hessian is zero.

If the min-function (5) or its piecewise smooth variants (10) or (11) are used, then the proof simplifies, as near a strongly stationary point, $\nabla \Phi_{x_2} = 0$. In addition, the linearizations are consistent even without the perturbation (14) and convergence follows from the convergence of the relaxed NLP. This fact can be interpreted as a constraint qualification for the NCP formulations using (5) or (10) or (11) at strongly stationary points.

The conclusions and proofs presented in this section also carry through for linear complementarity constraints but *not* for general nonlinear complementarity constraints. The reason is that the implication

$$x_{1i}^k x_{2i}^k = 0 \Rightarrow x_{1i}^{k+1} x_{2i}^{k+1} = 0 \tag{28}$$

holds for linear complementarity problems but *not* for nonlinear complementarity problems because in general, an SQP method would move off a nonlinear constraint. This is one reason for the introduction of slacks to deal with more general complementarity constraints. In addition, (28) can be made to hold in *inexact arithmetic* by taking care of handling simple bounds appropriately. The same is not true if one expression is a linear equation.

6 Numerical results

This section describes our experience with an implementation of the different NCP formulation of the MPCC (1) in our sequential quadratic programming solver. Our SQP method promotes global convergence through the use of a *filter*. The filter accepts a trial point whenever the objective or the constraint violation is improved compared with all previous iterates, Fletcher et al. [FGL+02, FL02, FLT02].

6.1 Preliminaries

The solver is interfaced to the modeling language AMPL, due to Fourer et al. [FGK03]. Our interface introduces slacks to formulate general complementarity constraints in the form (1) and handles the reformulation to the NLP (3) automatically. The interface also computes the derivatives of the NCP functions and relaxes the linearizations according to (14). A user can choose between the various formulations and set parameters such as δ, κ by passing options to the solver.

The test problems come from MacMPEC [Ley00], a collection of some 150 MPCC test problems [FL04] from a variety of backgrounds and sizes. The numerical tests are performed on a PC with an Intel Pentium 4 processor with 2.5 GHz and 512 KB RAM running Red Hat Linux version 7.3. The AMPL solver interface is compiled with the Intel C++ compiler version 6.0, and the SQP/MPCC solver is compiled with the Intel Fortran Compiler version 6.0.

Not all 150 problems in MacMPEC are included in this experiment. We have deliberately left out a number of 32×32 discretizations of the incidence set identification and packaging problems. These problems are similar to one another (a small number of them are included) but take a long time to run. This is especially true for the formulations that do not lump the complementarity constraint. In this sense, the results would have been even better for the formulation using the scalar product form.

To determine reasonable values for the various parameters introduced in the definition of the NCP functions, we run a small representative selection of MPCC problems. The overall performance is not very sensitive to a particular parameter choice. No attempt was made to "optimize" the parameter values; rather, we were interested in determining default values that would work well. Table 1 displays the default parameter values.

Table 1. Default parameter values for numerical experiments.

Parameter	Description	Default
δ	relaxation of linearization in (14)	0.1
κ	relaxation of linearization in (14)	1.0
σ_{NR}	smoothing of natural residual (9)	32.0
λ	Chen-Chen-Kanzow parameter (7)	0.7
σ_l	slope of linearized min-function (10)	4.0
σ_q	slope of quadratic min-function (11)	2.0

While the number of parameters may appear unreasonably large, each formulation requires only three parameters to be set. The choice of $\lambda = 0.7$ also agrees with [MFF+01], where $\lambda = 0.8$ is suggested. Note that since $\delta = 0.1$, the Chen-Chen-Kanzow function is relaxed further.

Care has to be taken when computing the smoothed natural residual function (9); it can be affected by cancellation error, as the following example illustrates. Suppose $a = 10^4$ and $b = 10^{-4}$ and that single-precision arithmetic is used. Then it follows that

$$2\varphi_{NR_s}(a,b) = (10^4 + 10^{-4}) - \sqrt{(10^4 - 10^{-4})^2 + \frac{1}{\sigma_{NR}}} \overset{\text{float}}{\simeq} 10^4 - \sqrt{10^8} = 0,$$

that is cancellation errors causes (9) to declare an infeasible point complementary. This situation can be avoided by employing the same trick used by Munson et al. [MFF+01] in reformulating the Fischer-Burmeister function giving rise to

$$\varphi_{NR_s}(a,b) = \frac{1}{2} \frac{\left(\frac{4\sigma_{NR}-1}{\sigma_{NR}}\right)}{a + b + \sqrt{(a-b)^2 + \frac{ab}{\sigma_{NR}}}}. \tag{29}$$

Derivative values can be computed in a similarly stable fashion.

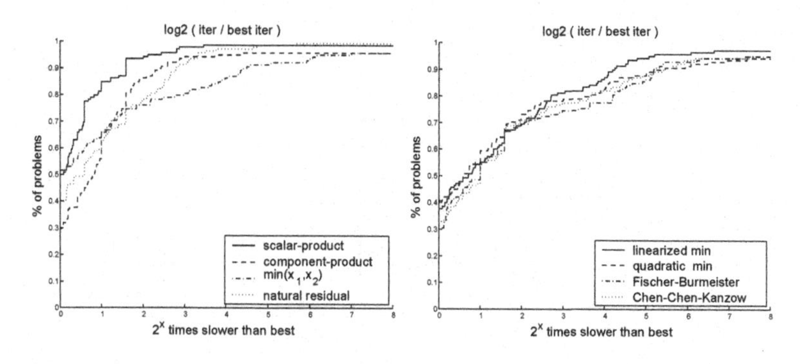

Fig. 3. Performance (iterations) plots for different NCP formulations

6.2 Performance plots and results

Results are provided in two forms. The performance plots of Dolan and Moré [DM00] in Figures 3 and 4 show the relative performance of each formulation in terms of iteration count and CPU time. These plots can be interpreted as follows. For every solver s and every problem p, the ratio of the number of iterations (or CPU time) of solver p on problem s over the fastest solve for problem s is computed and the base 2 logarithm is taken,

$$\log_2\left(\frac{\#\ \mathrm{iter}(s,p)}{\mathrm{best\_iter}(p)}\right).$$

By sorting these ratios in ascending order for every solver, the resulting plots can be interpreted as the probability that a given solver solves a problem within a certain multiple of the fastest solver.

Failures (see next section) are handled by setting the iteration count and the CPU time to a large number. This strategy ensures that the robustness can also be obtained from the performance plots. The percentage of MPCC problems solved is equivalent to the right asymptote of the performance line for each solver.

6.3 Failures of the NCP formulations

Solving MPCCs as NLPs is surprisingly robust. We observe very few failures, even though many problems are known to violate the assumptions made in this paper. Even the worst NCP formulation failed only on eight problems. Below, we list the problems that failed together with the reason for the failure.

The NLP solver can fail in three ways. The first failure mode occurs when the trust-region radius becomes smaller than the solver tolerance (`1E-6`) and no further progress can be made. This is referred to in the table below as "TR too small." Such a failure often happens at a solution where the KKT error cannot be reduced to sufficient accuracy. The second failure mode occurs if

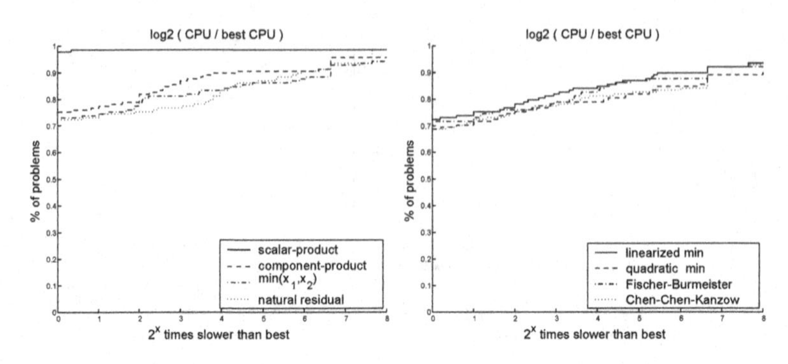

Fig. 4. Performance (CPU time) plots for different NCP formulations

the QP solver detects inconsistent linearizations near a feasible point. This is referred to as "infeasible QP." Note that the fact that MPCCs violate MFCQ implies that linearizations can become inconsistent arbitrarily close to a feasible point. Third, "iter. limit" refers to failures in which the solver reached its iteration limit (1000) without confirming optimality. The following failures were reported:

1. Scalar product form $x_1^T x_2 \leq 0$ — 2 failures
   TR too small : tollmpec1
   infeasible QP : design-cent-3
2. Bilinear form $x_{1i} x_{2i} \leq 0$ — 5 failures
   infeasible QP : design-cent-3, incid-set1c-32, pack-rig2c-32, pack-rig2p-16
   iter. limit : bem-milanc30-s
3. min-function $\min(x_{1i}, x_{2i}) \leq 0$ — 6 failures
   TR too small : ex9.2.2
   infeasible QP : pack-comp1p-8, pack-comp1p-16
   iter. limit : pack-comp2p-8, pack-comp2p-16, qpec-200-2
4. Linearized min-function (10) — 4 failures
   TR too small : jr2, qpec-200-3
   infeasible QP : bem-milanc30-s
   iter. limit : qpec-200-2
5. Quadratically smoothed min-function (11) — 8 failures
   TR too small : jr2
   infeasible QP : incid-set2c-32
   iter. limit : ex9.2.2, gauvin, incid-set1c-32, qpec-100-4, qpec-200-1,
   : qpec-200-3
6. Fischer-Burmeister function (4) — 7 failures
   infeasible QP : design-cent-3, ralphmod
   iter. limit : pack-comp1c-8, pack-rig1-16, pack-rig1c-16, pack-rig2-16,
   : pack-rig2c-16
7. Smoothed natural residual function (9) — 1 failures

TR too small : bem-milanc30-s

8. Chen-Chen-Kanzow function (7) 5 failures
   infeasible QP : pack-comp1p-8, qpec-200-3, pack-comp1c-8, pack-rig2p-16
   iter. limit : bem-milanc30-s

This list contains some problems known not to have strongly stationary limit points. For instance, ex9.2.2, ralph1, and scholtes4 have B-stationary solutions that are not strongly stationary. Problem gauvin has a global minimum at a point where the lower-level problem fails a constraint violation, so the formulation as an MPCC is not appropriate.

In the tests, two problems also gave rise to IEEE errors in the AMPL function evaluations, specifically the Chen-Chen-Kanzow function on `pack-rig1-16` and `pack-rig1c-32`. Since this type of error is caused not by the method but by the model, they are not counted in the errors.

6.4 Interpretation of the results

The results confirm that solving MPCCs as NLPs is very robust. In particular, the scalar product and the smoothed natural residual function are very robust, solving all but two problems and one problem, respectively.

The results for the min-function, on the other hand, are disappointing. Recall that these functions are theoretically attractive because they do not require an additional assumption to be made on the feasibility of QP approximations. This property makes the number of failures (6/4/8) for the min-function and its smoothed variants disappointing.

The best results in terms of performance and robustness were obtained for the scalar product formulation and the smoothed natural residual function. The performance plots in Figures 3 and 4 clearly show that these formulations are superior. In particular, the scalar product function is significantly faster than any other approach.

The formulation using $x_1^T x_2$ has two main advantages that may explain its superiority. First, it introduces only a single additional constraint, which reduces the size of the NLP to be solved. Moreover, this formulation requires less storage for the QP basis factors. Second, by lumping the complementarity conditions, the formulation allows a certain degree of nonmonotonicity in the complementarity error of each individual $x_{1i}x_{2i}$ and reduces the overall complementarity error, $x_1^T x_2$, only.

The worst results in terms of both robustness and efficiency are obtained for the Fischer-Burmeister function and the quadratically smoothed min-function. These formulations fail on seven and eight problems, respectively and are significantly slower than the other formulations. The Chen-Chen-Kanzow function improves on the Fischer-Burmeister function. This observation is not surprising because φ_{CCK} is a convex combination of the Fischer-Burmeister function and the more successful bilinear formulation. The worse behavior of φ_{FB} might be due to the fact that its linearized feasible region is smaller than for the bilinear form. This is also supported by the type of failures that can be observed for the Fischer-Burmeister function, which has many infeasible QP terminations.

Analyzing the solution characteristics of the scalar product form, we observe that only four problems have a large value of ξ. This fact shows that the SQP method converges to strongly stationary points for the remaining problems, as a bounded complementarity multiplier is equivalent to strong-stationarity (Theorem 3.7). The four problems for which ξ is unbounded are ex9.2.2, ralph1, ralphmod, and scholtes4. The last problem is known to violate an MPCC-MFCQ at its only stationary point, and the limit is B-stationary but not strongly stationary, and SQP converges linearly for this problem [FLRS02].

In addition, it can be observed that the complementarity error is exactly zero at most solutions. The reasons for this behavior are as follows:

1. Complementarity occurs only between variables. Thus, if a lower bound is in the active set, then the corresponding residual can be set to zero even in inexact arithmetic.
2. Many problems in the test set have a solution where $\xi = 0$. This indicates that the complementarity constraint $x_1^T x_2 \leq 0$ is locally redundant. Hence, exact complementarity is achieved as soon as the SQP method identifies the correct active set.
3. Our QP solver resolves degeneracy by making nearly degenerate constraints exactly degenerate and then employing a recursive procedure to remove degeneracy. This process of making nearly degenerate constraints exactly degenerate forces exact complementarity. Consider any nondegenerate index for which $x_{2i}^* > 0 = x_{1i}^*$, and assume that $x_{1i}^k > 0$ is small. The QP solver resolves the "near" degeneracy between the lower bound $x_{1i} \geq 0$ and the complementarity constraint by perturbing x_{1i} to zero. Thus exact complementarity is achieved.

This behavior is reassuring and makes the NLP approach to MPCCs attractive from a numerical standpoint.

7 Conclusions

Mathematical programs with complementarity constraints (MPCCs) are an emerging area of nonlinear optimization. Until recently researchers had assumed that the inherent degeneracy of MPCCs makes the application of standard NLP solvers unsafe. In this paper we show how MPCCs can be formulated as NLPs using a range of so-called NCP functions. Two new smoothed min-functions are introduced that exhibit desirable theoretical properties comparable to a constraint qualification.

In contrast to other smoothing approaches, the present formulations are exact in the sense that KKT points of the reformulated NLP correspond to strongly stationary points of the MPCC. Thus there is no need to control a smoothing parameter, which may be problematic.

It is shown that SQP methods exhibit fast local convergence near strongly stationary points under reasonable assumptions. This behavior is observed in practice on a large range of MPCC problems. The numerical results favor a lumped formulation in which all complementarity constraints are lumped into a single constraint. A new smoothed version of the min-function is also shown to be very robust and efficient. On the other hand, results for other standard NCP functions such as the Fischer-Burmeister function are disappointing.

The use of the simple bounds in the reformulation of complementarity (2) allows an alternative NLP formulation of the MPCC (1). This formulation lumps the nonlinear NCP functions into a single constraint, similar to $x_1^T x_2 \le 0$. Thus, an alternative NLP is given by

$$\begin{array}{ll} \text{minimize} & f(x) \\ \text{subject to} & c_{\mathcal{E}}(x) = 0 \\ & c_{\mathcal{I}}(x) \ge 0 \\ & x_1,\ x_2 \ge 0, \\ & e^T \Phi(x_1, x_2) \le 0. \end{array} \tag{30}$$

It is straightforward to see, that (30) is equivalent to (1). The convergence proof is readily extended to this formulation. We note that (30) has several advantages over (3). It reduces the number of constraints in the NLP. Moreover, our experience indicates that the lumped version of the bilinear form, $x_1^T x_2 \le 0$, often performs better than the separate version using $x_{1i}x_{2i} \le 0$. One reason may be that the lumped version allows nonmonotone changes in the complementarity residual in individual variable pairs as long as the overall complementarity is reduced.

Some open questions remain. One question concerns the global convergence of SQP methods from arbitrary starting points. Any approach to this question must take into account the globalization scheme and, in addition, provide powerful feasibility restoration. A related question is whether SQP methods can avoid convergence to spurious stationary points. Such points are sometimes referred to as C-stationary points even though they allow the existence of trivial first-order descent direction. At present, we believe that current SQP methods cannot avoid convergence to C-stationary points. Any attempt to avoid C-stationarity is likely to require algorithmic modifications.

Acknowledgments

We are grateful to an anonymous referee for many helpful comments and suggestions. This work was supported by the Mathematical, Information, and Computational Sciences Division subprogram of the Office of Advanced Scientific Computing Research, Office of Science, U.S. Department of Energy, under Contract W-31-109-ENG-38.

A Problem characteristics

This appendix lists the problem characteristics obtained with the scalar product formulation. The headings of each column are explained next. n, m, and p are the number of variables, constraints (excluding complementarity), and complementarity constraints, respectively. n_{NLP} is the number of variables after slacks were added, and k is the dimension of the nullspace at the solution. The definition of the degree of degeneracy d_1, d_2, d_m is taken from Jiang and Ralph [JR99] and refer to first-level degeneracy, d_1, second-level degeneracy, d_2, and mixed-degeneracy, d_m. The complementarity error $(x_1^T x_2)$ is in the column headed by compl, and ξ is the multiplier of the complementarity constraint $x_1^T x_2 \leq 0$.

name	n	m	p	n_{NLP}	k	d_1	d_2	d_m	compl	ξ
bard1	5	4	3	8	0	3	0	0	0.00	0.762
bard1m	6	4	3	9	0	4	0	0	0.00	0.762
bard2	12	9	3	15	0	2	1	0	0.00	0.00
bard2m	12	9	3	15	0	2	1	0	0.00	0.00
bard3	6	5	1	7	0	2	0	0	0.00	0.00
bard3m	6	5	3	9	0	2	0	0	0.00	1.09
bar-truss-3	35	34	6	35	0	13	0	0	0.00	1.45
bem-milanc30-s	3436	3433	1464	3436	1	1745	1	0	0.00	954.
bilevel1	10	9	6	12	0	6	0	0	0.00	0.150
bilevel2	16	13	8	20	1	5	0	0	0.294E-10	0.00
bilevel3	11	10	3	11	0	5	0	0	0.00	1.09
bilin	8	7	6	14	0	4	0	0	0.00	22.0
dempe	3	2	1	4	0	0	0	0	0.00	0.571E-05
design-cent-1	12	11	3	15	0	6	0	0	0.00	2.17
design-cent-2	13	15	3	16	0	11	0	0	0.00	0.00
design-cent-3	15	11	3	18	0	1	0	1	0.00	0.313E-01
design-cent-4	22	20	8	30	1	12	0	0	0.00	0.845
desilva	6	4	2	8	0	2	0	2	0.00	0.00
df1	2	3	1	3	1	1	0	1	0.00	0.00
ex9.1.1	13	12	5	13	0	4	0	0	0.00	0.00
ex9.1.10	11	9	3	11	0	5	0	2	0.00	0.00
ex9.1.2	8	7	2	8	0	4	0	0	0.00	0.00
ex9.1.3	23	21	6	23	0	14	0	1	0.00	3.20
ex9.1.4	8	7	2	8	0	3	0	1	0.00	0.00
ex9.1.5	13	12	5	13	0	8	0	2	0.00	10.0
ex9.1.6	14	13	6	14	0	6	0	1	0.00	1.56
ex9.1.7	17	15	6	17	0	8	0	1	0.00	5.00
ex9.1.8	11	9	3	11	0	5	0	2	0.00	0.00
ex9.1.9	12	11	5	12	0	5	0	1	0.00	0.444
ex9.2.1	10	9	4	10	0	6	0	1	0.00	0.762
ex9.2.2	9	8	3	9	0	3	0	1	0.183E-12	0.386E+07
ex9.2.3	14	13	4	14	0	5	1	0	0.00	0.00
ex9.2.4	8	7	2	8	0	3	0	0	0.00	1.00

name	n	m	p	n_{NLP}	k	d_1	d_2	d_m	compl	ξ
ex9.2.5	8	7	3	8	0	3	0	0	0.00	6.00
ex9.2.6	16	12	6	16	2	4	0	2	0.168E-10	0.500
ex9.2.7	10	9	4	10	0	6	0	1	0.00	0.762
ex9.2.8	6	5	2	6	0	3	0	1	0.00	0.500
ex9.2.9	9	8	3	9	0	7	0	0	0.100E-06	0.00
flp2	4	2	2	6	1	2	0	1	0.00	0.987
flp4-1	80	60	30	110	0	30	0	0	0.00	0.00
flp4-2	110	110	60	170	0	60	0	0	0.00	0.00
flp4-3	140	170	70	210	0	70	0	0	0.00	0.00
flp4-4	200	250	100	300	0	100	0	0	0.00	0.00
gauvin	3	2	2	5	0	1	0	0	0.00	0.250
gnash10	13	12	8	13	1	0	0	0	0.00	0.142
gnash11	13	12	8	13	1	0	0	0	0.00	0.918E-01
gnash12	13	12	8	13	1	0	0	0	0.00	0.397E-01
gnash13	13	12	8	13	1	0	0	0	0.00	0.149E-01
gnash14	13	12	8	13	1	0	0	0	0.00	0.199E-02
gnash15	13	12	8	13	0	3	0	0	0.00	7.65
gnash16	13	12	8	13	0	3	0	0	0.00	1.95
gnash17	13	12	8	13	1	4	0	0	0.00	1.67
gnash18	13	12	8	13	1	4	0	0	0.00	12.7
gnash19	13	12	8	13	0	2	0	0	0.00	2.80
hakonsen	9	8	4	9	0	2	0	0	0.00	0.390
hs044-i	20	14	10	26	0	7	0	1	0.00	5.69
incid-set1-16	485	491	225	485	0	232	0	5	0.00	0.00
incid-set1-8	117	119	49	117	0	54	0	4	0.00	0.00
incid-set1c-16	485	506	225	485	1	233	1	5	0.00	0.00
incid-set1c-32	1989	2034	961	1989	4	165	20	0	0.00	0.00
incid-set1c-8	117	126	49	117	0	59	0	4	0.00	0.00
incid-set2-16	485	491	225	710	3	212	13	0	0.00	0.00
incid-set2-8	117	119	49	166	5	42	7	0	0.00	0.00
incid-set2c-16	485	506	225	710	0	218	12	0	0.00	0.00
incid-set2c-32	1989	2034	961	2950	2	937	24	0	0.00	0.00
incid-set2c-8	117	126	49	166	2	46	6	0	0.00	0.00
jr1	2	1	1	3	1	0	0	0	0.00	0.00
jr2	2	1	1	3	0	0	0	0	0.00	2.00
kth1	2	1	1	2	0	0	1	0	0.00	0.00
kth2	2	1	1	2	1	0	0	0	0.00	0.00
kth3	2	1	1	2	0	0	0	0	0.00	1.00
liswet1-050	152	103	50	202	1	52	0	0	0.00	0.00
liswet1-100	302	203	100	402	1	102	0	0	0.00	0.00
liswet1-200	602	403	200	802	1	202	0	0	0.00	0.00
nash1	6	4	2	8	0	4	0	0	0.00	0.00
outrata31	5	4	4	9	0	0	1	0	0.00	0.164
outrata32	5	4	4	9	1	0	0	0	0.00	0.168
outrata33	5	4	4	9	1	1	0	0	0.00	0.714
outrata34	5	4	4	9	1	1	0	0	0.00	2.07

name	n	m	p	n_{NLP}	k	d_1	d_2	d_m	compl	ξ
pack-comp1-16	467	511	225	692	3	268	0	2	0.00	0.00
pack-comp1-8	107	121	49	156	0	113	0	0	0.414E-06	0.00
pack-comp1c-16	467	526	225	692	1	269	0	1	0.00	0.00
pack-comp1c-32	1955	2138	961	2916	3	1108	0	2	0.00	0.00
pack-comp1c-8	107	128	49	156	0	120	0	0	0.414E-06	0.00
pack-comp1p-16	467	466	225	692	5	223	2	0	0.00	0.00
pack-comp1p-8	107	106	49	156	0	83	0	0	0.00	0.00
pack-comp2-16	467	511	225	692	5	268	0	2	0.00	0.00
pack-comp2-8	107	121	49	156	5	62	0	2	0.00	0.00
pack-comp2c-16	467	526	225	692	4	268	0	2	0.00	0.00
pack-comp2c-32	1955	2138	961	2916	16	1058	0	2	0.00	0.00
pack-comp2c-8	107	128	49	156	1	62	0	2	0.00	0.00
pack-comp2p-16	467	466	225	692	13	223	2	0	0.00	0.00
pack-comp2p-8	107	106	49	156	1	47	2	0	0.00	0.00
pack-rig1-16	380	379	158	485	7	208	0	0	0.00	0.00
pack-rig1-8	87	86	32	108	6	47	0	0	0.00	0.00
pack-rig1c-16	380	394	158	485	4	206	0	0	0.00	0.00
pack-rig1c-32	1622	1652	708	2087	2	928	0	0	0.763E-06	0.00
pack-rig1c-8	87	93	32	108	5	47	0	0	0.00	0.00
pack-rig1p-16	445	444	203	550	3	229	2	0	0.00	0.00
pack-rig1p-8	105	104	47	126	5	50	2	0	0.00	0.00
pack-rig2-16	375	374	149	480	1	203	0	0	0.622E-06	0.00
pack-rig2-8	85	84	30	106	5	43	0	0	0.00	0.00
pack-rig2c-16	375	389	149	480	1	219	0	0	0.484E-06	0.00
pack-rig2c-32	1580	1610	661	2045	0	912	0	0	0.240E-06	0.00
pack-rig2c-8	85	91	30	106	2	45	0	0	0.00	0.00
pack-rig2p-16	436	435	194	541	0	215	1	0	0.00	0.00
pack-rig2p-8	103	102	45	124	6	49	2	0	0.00	0.00
portfl1	87	25	12	87	6	6	0	0	0.00	0.897
portfl2	87	25	12	87	0	7	0	0	0.00	0.682
portfl3	87	25	12	87	13	6	0	0	0.00	31.1
portfl4	87	25	12	87	16	5	0	0	0.00	114.
portfl6	87	25	12	87	13	5	0	0	0.00	55.0
qpec1	30	20	20	40	0	10	10	0	0.00	0.00
qpec-100-1	105	102	100	205	0	74	3	0	0.00	10.1
qpec-100-2	110	102	100	210	0	58	4	0	0.00	191.
qpec-100-3	110	104	100	210	0	35	2	0	0.00	4.45
qpec-100-4	120	104	100	220	0	61	4	0	0.00	15.3
qpec2	30	20	20	40	0	0	10	0	0.00	0.667
qpec-200-1	210	204	200	410	0	153	2	0	0.00	158.
qpec-200-2	220	204	200	420	0	118	2	0	0.00	3.42
qpec-200-3	220	208	200	420	0	48	6	0	0.00	35.5
qpec-200-4	240	208	200	440	0	133	7	0	0.00	7.95
ralph1	2	1	1	3	0	0	0	0	0.471E-12	0.486E+06
ralph2	2	1	1	2	1	0	0	0	0.313E-06	2.00
ralphmod	104	100	100	204	0	79	2	0	0.00	0.357E+08

name	n	m	p	n_{NLP}	k	d_1	d_2	d_m	compl	ξ
scholtes1	3	1	1	4	0	1	0	0	0.00	0.00
scholtes2	3	1	1	4	0	0	1	0	0.00	0.00
scholtes3	2	1	1	2	0	0	0	0	0.00	1.00
scholtes4	3	3	1	3	0	0	0	0	0.161E-13	0.525E+07
scholtes5	3	2	2	3	2	0	0	0	0.00	0.00
sl1	8	5	3	11	0	5	0	1	0.00	0.00
stackelberg1	3	2	1	3	1	0	0	0	0.00	0.00
tap-09	86	68	32	118	8	29	0	2	0.00	0.687E-07
tap-15	194	167	83	277	0	169	0	27	0.00	57.1
tollmpec	2403	2376	1748	4151	1	1489	1	88	0.00	2.35
tollmpec1	2403	2376	1748	4151	0	2402	0	86	0.00	0.00
water-FL	213	160	44	213	3	46	0	0	0.00	0.163E+04
water-net	66	50	14	66	2	15	0	0	0.00	0.00

B Detailed results: Iteration counts

Name	$x_1^T x_2$	(8)	(5)	(10)	(11)	(4)	(9)	(7)
bard1	3	4	9	13	2	25	3	8
bard1m	3	4	9	13	2	4	3	7
bard2	1	1	1	1	1	1	1	1
bard2m	1	1	1	1	1	1	1	1
bard3	4	4	4	4	4	4	4	4
bard3m	4	4	4	4	4	4	4	4
bar-truss-3	10	9	9	9	9	9	9	9
bem-milanc30-s	62	1000	655	111	245	144	410	1000
bilevel1	2	3	2	3	3	4	3	4
bilevel2	7	2	1	2	5	3	1	2
bilevel3	7	6	6	6	6	6	6	6
bilin	2	6	1	3	3	3	5	3
dempe	58	58	58	58	58	94	58	58
design-cent-1	4	4	4	4	4	4	4	4
design-cent-2	31	21	37	37	29	32	32	60
design-cent-3	191	164	173	173	173	217	185	163
design-cent-4	3	4	3	3	3	4	3	4
desilva	2	2	2	2	2	2	2	2
df1	2	2	2	2	2	2	2	2
ex9.1.1	1	2	1	1	1	3	2	3
ex9.1.10	1	1	1	1	1	1	1	1
ex9.1.2	2	3	1	3	3	3	3	3
ex9.1.3	3	3	1	3	4	3	3	3
ex9.1.4	2	2	2	2	2	2	2	2
ex9.1.5	3	3	1	3	3	3	3	3
ex9.1.6	3	5	2	2	2	4	4	6
ex9.1.7	3	3	1	3	3	3	3	3
ex9.1.8	1	1	1	1	1	1	1	1
ex9.1.9	3	3	2	3	8	3	3	3

Name	$x_1^T x_2$	(8)	(5)	(10)	(11)	(4)	(9)	(7)
ex9.2.1	3	4	6	13	6	8	3	8
ex9.2.2	22	22	76	71	1000	238	3	180
ex9.2.3	1	1	1	1	1	1	1	1
ex9.2.4	3	2	2	2	2	2	2	2
ex9.2.5	7	7	1	17	32	35	4	7
ex9.2.6	3	2	1	1	2	2	1	2
ex9.2.7	3	4	6	13	6	8	3	8
ex9.2.8	3	3	1	1	1	4	3	3
ex9.2.9	3	3	1	3	3	3	3	3
flp2	3	3	1	1	1	3	3	1
flp4-1	3	2	2	2	2	2	2	2
flp4-2	3	2	2	2	2	2	2	2
flp4-3	3	2	2	2	2	2	2	2
flp4-4	3	2	2	2	2	2	2	2
gauvin	3	9	71	71	1000	54	7	6
gnash10	8	8	7	7	7	8	7	8
gnash11	8	8	7	7	7	8	7	8
gnash12	9	8	8	8	8	8	8	8
gnash13	13	9	10	10	9	10	10	11
gnash14	10	10	9	9	9	13	10	11
gnash15	18	18	41	11	11	9	10	27
gnash16	16	14	26	12	10	45	11	14
gnash17	17	17	10	10	9	11	10	15
gnash18	15	19	55	73	10	184	11	128
gnash19	10	19	10	8	8	18	14	25
hakonsen	10	10	12	12	10	10	10	10
hs044-i	6	4	2	2	4	4	2	4
incid-set1-16	33	139	78	120	493	85	175	66
incid-set1-8	34	35	56	56	51	42	73	65
incid-set1c-16	34	89	89	93	168	69	109	86
incid-set1c-32	37	309	102	155	1000	127	304	161
incid-set1c-8	39	32	43	38	48	35	67	43
incid-set2-16	19	37	35	35	35	33	24	33
incid-set2-8	48	19	18	18	18	18	18	18
incid-set2c-16	37	36	40	35	305	27	71	32
incid-set2c-32	31	87	71	122	489	71	308	88
incid-set2c-8	24	20	27	23	52	29	25	27
jr1	1	1	1	1	1	1	1	1
jr2	7	7	61	66	114	22	3	18
kth1	1	1	1	1	1	1	1	1
kth2	2	2	2	2	2	2	2	2
kth3	4	5	67	67	67	3	2	4
liswet1-050	1	1	1	1	1	1	1	1
liswet1-100	1	1	1	1	1	1	1	1
liswet1-200	1	1	1	1	1	1	1	1
nash1	3	2	1	1	1	2	1	2

Name	$x_1^T x_2$	(8)	(5)	(10)	(11)	(4)	(9)	(7)
outrata31	8	8	7	7	7	8	7	7
outrata32	8	9	8	8	8	9	8	8
outrata33	7	8	7	7	7	8	7	8
outrata34	6	7	6	6	6	7	6	7
pack-comp1-16	20	39	751	64	37	12	68	12
pack-comp1-8	8	30	152	66	24	36	16	36
pack-comp1c-16	5	38	358	76	40	15	50	15
pack-comp1c-32	13	2	787	238	217	50	344	50
pack-comp1c-8	8	19	68	40	14	40	18	41
pack-comp1p-16	45	72	895	344	81	52	197	31
pack-comp1p-8	53	64	274	219	87	36	200	172
pack-comp2-16	43	49	442	42	38	44	81	35
pack-comp2-8	8	26	10	18	11	8	10	8
pack-comp2c-16	15	23	336	76	30	15	17	15
pack-comp2c-32	7	34	901	193	178	45	175	42
pack-comp2c-8	6	11	18	15	14	6	13	6
pack-comp2p-16	32	64	1000	190	142	36	232	48
pack-comp2p-8	60	57	1000	104	77	34	171	58
pack-rig1-16	64	56	81	120	178	1000	90	206
pack-rig1-8	7	10	25	13	17	145	13	148
pack-rig1c-16	11	43	15	57	53	458	19	548
pack-rig1c-32	18	238	42	302	369	99	181	107
pack-rig1c-8	6	8	13	13	10	139	9	142
pack-rig1p-16	28	48	56	164	490	97	118	59
pack-rig1p-8	14	16	22	25	60	144	29	147
pack-rig2-16	7	11	21	42	119	1000	30	421
pack-rig2-8	10	16	10	36	38	254	62	253
pack-rig2c-16	6	11	13	67	96	1000	34	421
pack-rig2c-32	11	71	31	187	222	57	551	55
pack-rig2c-8	6	12	6	15	33	254	23	253
pack-rig2p-16	10	38	79	367	436	301	86	309
pack-rig2p-8	20	16	18	46	89	197	20	196
portfl1	5	7	4	21	6	76	6	84
portfl2	4	6	3	43	8	108	5	162
portfl3	4	6	3	8	5	6	10	6
portfl4	4	4	5	7	5	50	8	48
portfl6	4	6	3	4	5	68	8	66
qpec1	3	2	2	2	2	2	2	2
qpec-100-1	7	34	114	112	251	253	43	300
qpec-100-2	7	24	47	137	427	219	44	43
qpec-100-3	6	20	121	137	713	256	27	105
qpec-100-4	5	9	103	497	1000	176	42	78
qpec2	2	2	1	1	1	2	1	2
qpec-200-1	10	24	87	25	1000	363	38	343
qpec-200-2	10	33	1000	1000	888	182	114	79
qpec-200-3	11	20	160	267	1000	377	62	357
qpec-200-4	5	13	78	95	862	92	34	89

Name	$x_1^T x_2$	(8)	(5)	(10)	(11)	(4)	(9)	(7)
ralph1	27	27	70	70	70	368	5	181
ralph2	11	21	1	1	1	168	3	179
ralphmod	7	37	25	46	114	178	48	21
scholtes1	4	3	3	3	3	3	3	3
scholtes2	2	2	2	2	2	2	2	2
scholtes3	4	6	67	67	67	1	1	1
scholtes4	26	28	71	74	74	239	6	181
scholtes5	1	1	1	1	1	1	1	1
sl1	1	1	1	1	1	1	1	1
stackelberg1	4	4	4	4	4	4	4	4
tap-09	21	23	17	18	18	12	11	23
tap-15	28	19	18	12	18	20	19	20
tollmpec	10	36	22	24	20	79	135	128
tollmpec1	10	50	20	28	24	379	139	108
water-FL	272	237	235	279	333	256	263	356
water-net	131	114	109	125	137	126	190	114

References

[Ani05] M. Anitescu. Global convergence of an elastic mode approach for a class of mathematical programs with complementarity constraints. *SIAM J. Optimization*, 16(1):120–145, 2005.

[Bar88] J. F. Bard. Convex two-level optimization. *Mathematical Programming*, 40(1):15–27, 1988.

[BSSV03] H. Benson, A. Sen, D. F. Shanno, and R. V. D. Vanderbei. Interior-point algorithms, penalty methods and equilibrium problems. Technical Report ORFE-03-02, Princeton University, Operations Research and Financial Engineering, October 2003. To appear in Computational Optimization and Applications.

[CCK00] B. Chen, X. Chen, and C. Kanzow. A penalized Fischer-Burmeister NCP-function. *Mathematical Programming*, 88(1):211–216, June 2000.

[CF95] Y. Chen and M. Florian. The nonlinear bilevel programming problem: Formulations, regularity and optimality conditions. *Optimization*, 32:193–209, 1995.

[CH97] B. Chen and P. T. Harker. Smooth approximations to nonlinear complementarity problems. *SIAM J. Optimization*, 7(2):403–420, 1997.

[DFM02] S. P. Dirkse, M. C. Ferris, and A. Meeraus. Mathematical programs with equilibrium constraints: Automatic reformulation and solution via constraint optimization. Technical Report NA-02/11, Oxford University Computing Laboratory, July 2002.

[Dir01] S. P. Dirkse. MPEC world. Webpage, GAMS Development Corp., `www.gamsworld.org/mpec/`, 2001.

[DM00] Elizabeth D. Dolan and Jorge Moré. Benchmarking optimization software with cops. Technical Report MCS-TM-246, MCS Division, Argonne National Laboratory, Argonne, IL, November 2000.

[FGK03] R. Fourer, D. M. Gay, and B. W. Kernighan. *AMPL: A Modelling Language for Mathematical Programming*. Books/Cole—Thomson Learning, 2nd edition, 2003.

[FGL+02] R. Fletcher, N. I. M. Gould, S. Leyffer, Ph. L. Toint, and A. Wächter. Global convergence of trust-region SQP-filter algorithms for general nonlinear programming. *SIAM J. Optimization*, 13(3):635–659, 2002.

[Fis95] A. Fischer. A Newton-type method for positive semi-definite linear complementarity problems. *Journal of Optimization Theory and Applications*, 86:585–608, 1995.

[FJQ99] F. Facchinei, H. Jiang, and L. Qi. A smoothing method for mathematical programs with equilibrium constraints. *Mathematical Programming*, 85:107–134, 1999.

[FL02] R. Fletcher and S. Leyffer. Nonlinear programming without a penalty function. *Mathematical Programming*, 91:239–270, 2002.

[FL04] R. Fletcher and S. Leyffer. Solving mathematical program with complementarity constraints as nonlinear programs. *Optimization Methods and Software*, 19(1):15–40, 2004.

[Fle87] R. Fletcher. *Practical Methods of Optimization, 2nd edition.* John Wiley, Chichester, 1987.

[FLRS02] R. Fletcher, S. Leyffer, D. Ralph, and S. Scholtes. Local convergence of SQP methods for mathematical programs with equilibrium constraints. Numerical Analysis Report NA/209, Department of Mathematics, University of Dundee, Dundee, UK, May 2002. To appear in SIAM J. Optimization.

[FLT02] R. Fletcher, S. Leyffer, and Ph. L. Toint. On the global convergence of a filter-SQP algorithm. *SIAM J. Optimization*, 13(1):44–59, 2002.

[FP97] M. C. Ferris and J. S. Pang. Engineering and economic applications of complementarity problems. *SIAM Review*, 39(4):669–713, 1997.

[GMS02] P.E. Gill, W. Murray, and M.A. Saunders. SNOPT: An SQP algorithm for large–scale constrained optimization. *SIAM Journal on Optimization*, 12(4):979–1006, 2002.

[JR99] H. Jiang and D. Ralph. QPECgen, a MATLAB generator for mathematical programs with quadratic objectives and affine variational inequality constraints. *Computational Optimization and Applications*, 13:25–59, 1999.

[Ley00] S. Leyffer. MacMPEC: AMPL collection of MPECs. Webpage, `www.mcs.anl.gov/~leyffer/MacMPEC/`, 2000.

[LPR96] Z.-Q. Luo, J.-S. Pang, and D. Ralph. *Mathematical Programs with Equilibrium Constraints.* Cambridge University Press, Cambridge, UK, 1996.

[LS04] X. Liu and J. Sun. Generalized stationary points and an interior point method for mathematical programs with equilibrium constraints. *Mathematical Programming B*, 101(1):231–261, 2004.

[MFF+01] T. S. Munson, F. Facchinei, M. C. Ferris, A. Fischer, and C. Kanzow. The semismooth algorithm for large scale complementarity problems. *INFORMS J. Computing*, 13(4):294–311, 2001.

[OKZ98] J. Outrata, M. Kocvara, and J. Zowe. *Nonsmooth Approach to Optimization Problems with Equilibrium Constraints.* Kluwer Academic Publishers, Dordrecht, 1998.

[RB02a] A. Raghunathan and L. T. Biegler. Barrier methods for mathematical programs with complementarity constraints (MPCCs). Technical report, Carnegie Mellon University, Department of Chemical Engineering, Pittsburgh, PA, December 2002.

[RB02b] A. Raghunathan and L. T. Biegler. MPEC formulations and algorithms in process engineering. Technical report, Carnegie Mellon University, Department of Chemical Engineering, Pittsburgh, PA, 2002.

[RB05] A. Raghunathan and L. T. Biegler. An interior point method for mathematical programs with complementarity constraints (MPCCs). *SIAM Journal on Optimization*, 15(3):720–750, 2005.

[SS00] H. Scheel and S. Scholtes. Mathematical program with complementarity constraints: Stationarity, optimality and sensitivity. *Mathematics of Operations Research*, 25:1–22, 2000.

A semi-infinite approach to design centering

Oliver Stein

RWTH Aachen University, Aachen, Germany stein@mathc.rwth-aachen.de

Summary. We consider design centering problems in their reformulation as general semi-infinite optimization problems. The main goal of the article is to show that the Reduction Ansatz of semi-infinite programming generically holds at each solution of the reformulated design centering problem. This is of fundamental importance for theory and numerical methods which base on the intrinsic bilevel structure of the problem.

For the genericity considerations we prove a new first order necessary optimality condition in design centering. Since in the course of our analysis also a certain standard semi-infinite programming problem turns out to be related to design centering, the connections to this problem are studied, too.

Key words: Optimality conditions, Reduction Ansatz, Jet transversality, Genericity.

1 Introduction

Design Centering. A design centering problem considers a container set $C \subset \mathbb{R}^m$ and a parametrized body $B(x) \subset \mathbb{R}^m$ with parameter vector $x \in \mathbb{R}^n$. The task is to inscribe $B(x)$ into C such that some functional f, e.g. the volume of $B(x)$, is maximized:

$$DC: \quad \max_{x \in \mathbb{R}^n} \; f(x) \quad \text{subject to} \quad B(x) \subset C\,.$$

In Figure 1 $B(x)$ is a disk in $\mathbb{R}^2$, parametrized by its midpoint and its radius. The parameter vector $x \in \mathbb{R}^3$ is chosen such that $B(x)$ has maximal area in the nonconvex container set C.

A straightforward extension of the model is to inscribe finitely many nonoverlapping bodies into C such that some total measure is maximized. Figure 2 shows the numerical solution of such a multi-body design centering

S. Dempe and V. Kalashnikov (eds.), *Optimization with Multivalued Mappings*, pp. 209-228
©2006 Springer Science + Business Media, LLC

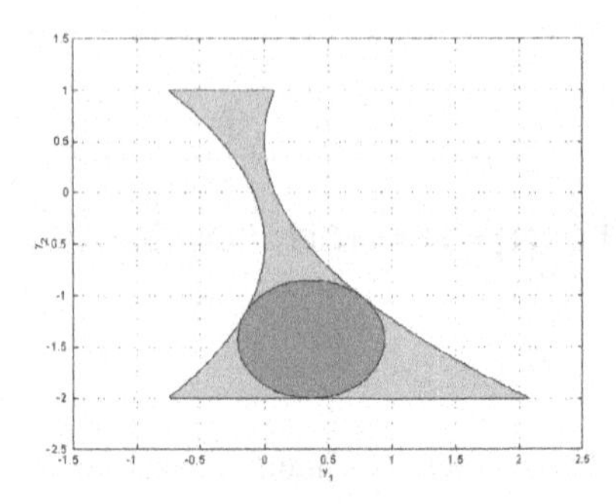

Fig. 1. A disk with maximal area in a nonconvex container

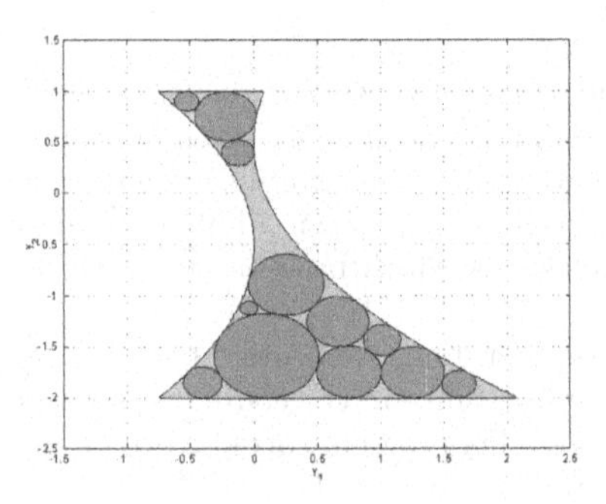

Fig. 2. Twelve disks with maximal total area in a nonconvex container

problem with the same container set as in Figure 1 and twelve nonoverlapping disks.

Single-body design centering problems with special sets $B(x)$ and C have been studied extensively, see e.g. [5] for the complexity of inscribing a convex body into a convex container, [12] for maximization of a production yield under uncertain quality parameters, and [18] for the problem of cutting a diamond with prescribed form and maximal volume from a raw diamond. The cutting stock problem ([2]) is an example of multi-body design centering.

To give an example of a design centering problem with a rather intricate container set, consider the so-called maneuverability problem of a robot from [4]:

Example 1. A robot may be viewed as a structure of connected links, where some geometrical parameters $\theta_1, ..., \theta_R$, such as lengths of the links or angles in the joints, can be controlled by drive motors (cf. Figure 3 which is taken from [8]).

The equations of motion for a robot have the form

$$F \;=\; A(\theta) \cdot \ddot{\theta} \;+\; H(\theta, \dot{\theta}) \;,$$

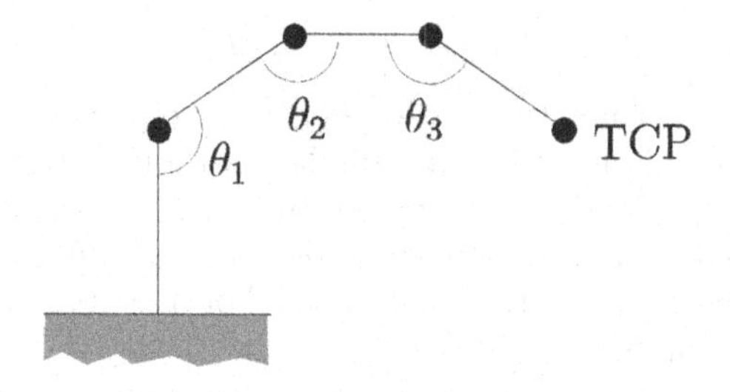

Fig. 3. A robot with connected links and a tool center point

where $F \in \mathrm{IR}^R$ denotes the vector of forces (torques), $A(\theta)$ is the inertia matrix, and $H(\theta, \dot{\theta})$ is the vector of friction, gravity, centrifugal and Coriolis forces. Given vectors $F^-, F^+ \in \mathrm{IR}^R$ of lower and upper bounds of F as well as an operating region $\Omega \subset \mathrm{IR}^R \times \mathrm{IR}^R$, the set

$$C = \{ \ddot{\theta} \in \mathrm{IR}^R | \ F^- \leq A(\theta)\ddot{\theta} + H(\theta, \dot{\theta}) \leq F^+ \ \text{ for all } \ (\theta, \dot{\theta}) \in \Omega \}$$

describes the accelerations which can be realized in every point $(\theta, \dot{\theta}) \in \Omega$. Since the size of C is a measure for the usefulness of a given robot for certain tasks, an approximation for the volume of C is sought in [4]: Find a simple body B which is parametrized by a vector x such that $B(x)$ is as large as possible and contained in C. In this way we arrive at a design centering problem DC.

The aim of this article is to use techniques from general semi-infinite programming to treat a broad class of design centering problems theoretically as well as numerically. In fact, Example 1 gave rise to one of the first formulations of a general semi-infinite optimization problem in [8].

Semi-infinite Programming. The connection of design centering to semi-infinite programming is straightforward: let C be described by the inequality constraint $c(y) \leq 0$. Then the inclusion

$$B(x) \subset C = \{ y \in \mathrm{IR}^m | \ c(y) \leq 0 \}$$

is trivially equivalent to the semi-infinite constraint

$$c(y) \leq 0 \quad \forall \, y \in B(x) \, .$$

Thus the design centering problem DC is equivalent to the general semi-infinite problem

$$GSIP_{DC}: \qquad \max_x f(x) \quad \text{subject to} \quad c(y) \leq 0 \quad \forall \, y \in B(x) \, .$$

Problems of this type are called semi-infinite as they involve a finite-dimensional decision variable x and possibly infinitely many inequality constraints

$$g(x,y) \le 0 \quad \forall\, y \in B(x)\,,$$

where in design centering the function $g(x,y) := c(y)$ does not depend on x.

On the other hand, in a so-called standard semi-infinite optimization problem there is no $x-$dependence in the set $B(x)$, i.e. the semi-infinite index set $B(x) \equiv B$ is fixed. Standard semi-infinite optimization problems have been studied systematically since the early 1960s. For an extensive survey on standard semi-infinite programming see [7].

As it turned out more recently in [16], general semi-infinite programming is intrinsically more complicated than standard semi-infinite programming, so that some basic theoretical and numerical strategies cannot be transferred from the standard to the general case. In particular, the feasible set M of *GSIP* may be nonclosed and exhibit a disjunctive structure even for defining functions in general position. An introduction to general semi-infinite programming is given in [21].

Bilevel Programming. The key to the theoretical treatment of general semi-infinite programming and to the conceptually new solution method from [23] lies in the bilevel structure of semi-infinite programming. In the following we briefly sketch the main ideas of this approach.

Consider the general semi-infinite program

$$GSIP: \qquad \max_x f(x) \quad \text{subject to} \quad g(x,y) \le 0 \quad \forall\, y \in B(x)\,,$$

where for all $x \in \mathbb{R}^n$ we have

$$B(x) = \{\, y \in \mathbb{R}^m |\; w(x,y) \le 0 \,\}\,.$$

Let the defining functions $f : \mathbb{R}^n \to \mathbb{R}$ and $g, w : \mathbb{R}^n \times \mathbb{R}^m \to \mathbb{R}$ be at least once continuously differentiable, and let $\nabla_x g$ denote the column vector of partial derivatives of g with respect to x, etc. Then the set-valued mapping $B : \mathbb{R}^n \rightrightarrows \mathbb{R}^m$ is closed. Let B also be locally bounded, i.e. for all $\bar{x} \in \mathbb{R}^n$ there exists a neighborhood U of $\bar{x}$ and a bounded set $Y \subset \mathbb{R}^m$ with $B(x) \subset Y$ for all $x \in U$. Note that then $B(x)$ is compact for each $x \in \mathbb{R}^n$. We also assume that $B(x)$ is nonempty for all $x \in \mathbb{R}^n$.

Under these assumptions it is easy to see that the semi-infinite constraint in *GSIP* is equivalent to

$$\varphi(x) := \max_{y \in B(x)} g(x,y) \le 0\,,$$

which means that the feasible set M of *GSIP* is the lower level set of some optimal value function. In fact, φ is the optimal value function of the so-called lower level problem

$$Q(x): \quad \max_{y \in \mathbb{R}^m} g(x,y) \quad \text{subject to} \quad w(x,y) \le 0\,.$$

In contrast to the upper level problem which consists in maximizing f over M, in the lower level problem x plays the role of an $n-$dimensional parameter, and y is the decision variable. The main computational problem in semi-infinite programming is that the lower level problem has to be solved to global optimality, even if only a stationary point of the upper level problem is sought.

Since under the assumptions of closedness and local boundedness of the set-valued mapping B and the continuity of g the optimal value function φ is at least upper semi-continuous, points $x \in \mathbb{R}^n$ with $\varphi(x) < 0$ belong to the topological interior of M. For investigations of the local structure of M or of local optimality conditions we are only interested in points from the boundary ∂M of M, so that it suffices to consider the zeros of φ, i.e. points $x \in \mathbb{R}^n$ for which $Q(x)$ has vanishing maximal value. We denote the corresponding globally maximal points of $Q(x)$ by

$$B_0(x) \;=\; \{\, y \in B(x)|\; g(x,y) = 0 \,\}\;.$$

The Reduction Ansatz. When studying semi-infinite problems, it is of crucial importance to control the elements of $B_0(x)$ for varying x. This can be achieved, for example, by means of the implicit function theorem. For $\bar{x} \in M$ a local maximizer $\bar{y}$ of $Q(\bar{x})$ is called nondegenerate in the sense of Jongen/Jonker/Twilt ([14]), if the linear independence constraint qualification (LICQ), strict complementary slackness (SCS) and the second order sufficiency condition $D_y^2 \Lambda(\bar{x},\bar{y},\bar{\gamma})|_{T_{\bar{y}}B(\bar{x})} \prec 0$ are satisfied. Here $\Lambda(x,y,\gamma) = g(x,y) - \gamma\, w(x,y)$ denotes the lower level Lagrangian, $T_{\bar{y}}B(\bar{x})$ is the tangent space to $B(\bar{x})$ at $\bar{y}$, and $A \prec 0$ stands for the negative definiteness of a matrix A. The *Reduction Ansatz* is said to hold at $\bar{x} \in M$ if all global maximizers of $Q(\bar{x})$ are nondegenerate. Since nondegenerate maximizers are isolated, and $B(\bar{x})$ is a compact set, the set $B_0(\bar{x})$ can only contain finitely many points. By a result from [3] the local variation of these points with x can be described by the implicit function theorem.

The Reduction Ansatz was originally formulated for standard semi-infinite problems in [6] and [24] under weaker regularity assumptions. It was transferred to general semi-infinite problems in [9]. For *standard* semi-infinite problems the Reduction Ansatz is a natural assumption in the sense that for problems with defining functions in general position it holds at each local maximizer ([19, 25]). For *GSIP* this result can be transferred to local maximizers $\bar{x}$ with $|B_0(\bar{x})| \geq n$ ([20]). Moreover, in [22] it is shown that it holds in the "completely linear" case, i.e. when the defining functions f, g and w of *GSIP* are affine linear on their respective domains. For *GSIP* without these special structures, until now it is not known whether the Reduction Ansatz generically holds at all local maximizers. Note that even if this general result was true, it would not necessarily mean that the Reduction Ansatz holds generically at local maximizers of $GSIP_{DC}$. In fact, only such specially structured perturbations of the defining functions of $GSIP_{DC}$ are allowed which leave the function c independent of x.

Under the Reduction Ansatz it was not only shown that M can locally be described by finitely many smooth inequality constraints ([9]), but it also serves as a regularity condition for the convergence proof of the numerical solution method from [23]. For completeness, we briefly sketch the main idea of this bilevel method.

A numerical method for *GSIP*. To make the global solution of the lower level problem computationally tractable, we assume that $Q(x)$ is a regular convex problem for all $x \in \mathbb{R}^n$, i.e. the functions $-g(x,\cdot)$ and $w(x,\cdot)$ are convex in y, and $B(x)$ possesses a Slater point. It is well-known that then the global solutions of the problem $Q(x)$ are exactly its Karush-Kuhn-Tucker points: y solves $Q(x)$ if and only if there exists some $\gamma \in \mathbb{R}$ such that

$$\begin{aligned} \nabla_y \Lambda(x,y,\gamma) &= 0 \\ \gamma \cdot w(x,y) &= 0 \\ \gamma,\ -w(x,y) &\geq 0 \,. \end{aligned}$$

For this reason it makes sense to replace the problem *GSIP*, in which only optimal *values* of the lower problem enter, by a problem which also uses lower level optimal *points*. In fact, we first consider the Stackelberg game

$$SG: \qquad \max_{x,y} f(x) \quad \text{subject to} \quad g(x,y) \leq 0, \quad y \text{ solves } Q(x)\,.$$

Note that the decision variable of SG resides in the higher-dimensional space $\mathbb{R}^n \times \mathbb{R}^m$, i.e. *GSIP* is lifted. In [22] it is shown that under our assumptions the orthogonal projection of the feasible set of SG to $\mathbb{R}^n$ coincides with the feasible set of *GSIP*, so that the $x-$component of any solution of SG is a solution of *GSIP*.

In a second step we replace the restriction that y solves $Q(x)$ in SG equivalently by the corresponding Karush-Kuhn-Tucker condition:

$$\begin{aligned} MPCC: \qquad \max_{x,y,\gamma} f(x) \quad \text{subject to} \quad g(x,y) &\leq 0 \\ \nabla_y \Lambda(x,y,\gamma) &= 0 \\ \gamma \cdot w(x,y) &= 0 \\ \gamma,\ -w(x,y) &\geq 0 \,. \end{aligned}$$

The resulting mathematical program with complementarity constraints lifts the problem again to a higher-dimensional space, but now $MPCC$ solution techniques may be applied. One possibility is to reformulate the complementarity conditions in $MPCC$ by means of an NCP function Φ like the Fischer-Burmeister function $\Phi(a,b) = a + b - ||(a,b)||_2$, and then to regularize the necessarily nonsmooth or degenerate NCP function by a one-dimensional parameter $\tau > 0$, e.g. to $\Phi_\tau(a,b) = a + b - ||(a,b,\tau)||_2$. An obvious idea for a numerical method is to solve the finite and regular optimization problems

$$\begin{aligned} P_\tau: \qquad \max_{x,y,\gamma} f(x) \quad \text{subject to} \quad g(x,y) &\leq 0 \\ \nabla_y \Lambda(x,y,\gamma) &= 0 \\ \Phi_\tau(\,\gamma, -w(x,y)\,) &= 0 \end{aligned}$$

for $\tau \searrow 0$. For details and for a convergence proof of this method see [21].

As mentioned before, this convergence proof relies on the Reduction Ansatz in the solution point. Although for general semi-infinite problems it is not clear yet whether the Reduction Ansatz holds generically in each local solution, in numerical tests convergence can usually be observed. The numerical examples in Figures 1 and 2 were actually generated by this algorithm, applied to the general semi-infinite reformulation $GSIP_{DC}$ of DC.

The present article will show that for the specially structured problems $GSIP_{DC}$ which stem from a reformulation of DC, the Reduction Ansatz in each local maximizer is generic. In Section 2 we derive a first order necessary optimality condition for DC which will be the basis of the genericity considerations in Section 3. Section 4 presents some connections to a *standard* semi-infinite problem that can be associated with DC, before Section 5 closes the article with some final remarks.

2 First order optimality conditions

Let us consider the slightly more general design centering problem

$$DC: \quad \max_{x \in \mathbb{R}^n} f(x) \quad \text{subject to} \quad B(x) \subset C$$

with

$$C = \{ y \in \mathbb{R}^m | \, c_j(y) \leq 0, \; j \in J \}$$

and

$$B(x) = \{ y \in \mathbb{R}^m | \, v_\ell(y) \leq 0, \; \ell \in L, \; w(x,y) \leq 0 \}$$

with finite index sets J and L, and with at least once continuously differentiable defining functions f, c_j, $j \in J$, v_ℓ, $\ell \in L$, and w. We assume that C and

$$Y = \{ y \in \mathbb{R}^m | \, v_\ell(y) \leq 0, \; \ell \in L \}$$

are nonempty and compact sets. In applications the set Y can often be chosen to contain C so that the compactness of C follows from the compactness of Y. Moreover, the local boundedness of the set-valued mapping B is a trivial consequence of the boundedness of Y.

The general semi-infinite reformulation of DC now becomes a problem with finitely many semi-infinite constraints,

$$GSIP_{DC}: \qquad \max_x f(x) \quad \text{subject to} \quad c_j(y) \leq 0 \quad \forall \, y \in B(x), \; j \in J,$$

and finitely many lower level problems $Q^j(x)$ with optimal value functions $\varphi_j(x)$ and optimal points $B_0^j(x)$, $j \in J$. For $\bar{x} \in M$ we denote by

$$J_0(\bar{x}) = \{ j \in J | \, \varphi_j(\bar{x}) = 0 \}$$

the set of active semi-infinite constraints. From the upper semi-continuity of the functions φ_j, $j \in J$, it is clear that at each feasible boundary point $\bar{x} \in M \cap \partial M$ the set $\bigcup_{j \in J_0(\bar{x})} B_0^j(\bar{x})$ is nonempty. For the problem $GSIP_{DC}$ we can show that an even smaller set is nonempty. In fact, with

$$B_{00}^j(\bar{x}) \;=\; \{\, y \in B_0^j(\bar{x}) |\; w(\bar{x}, y) = 0 \,\}$$

the following result holds.

Lemma 1. *The set* $\bigcup_{j \in J_0(\bar{x})} B_{00}^j(\bar{x})$ *is nonempty for each feasible boundary point* $\bar{x} \in M \cap \partial M$.

Proof. For $\bar{x} \in \partial M$ there exists a sequence $x^\nu \to \bar{x}$ with $x^\nu \notin M$ for all $\nu \in \mathbb{N}$. By definition of M, for all $\nu \in \mathbb{N}$ there exists some $y^\nu \in B(x^\nu)$ and some $j_\nu \in J$ with $c_{j_\nu}(y^\nu) > 0$.

As J is a finite set, the sequence $(j_\nu)_{\nu \in \mathbb{N}}$ contains some index $j_0 \in J$ infinitely many times. Taking the corresponding subsequence if necessary, we may assume $j_\nu \equiv j_0$ without loss of generality.

Moreover, as B is locally bounded at $\bar{x}$, the sequence $(y^\nu)_{\nu \in \mathbb{N}}$ is bounded and, thus, without loss of generality convergent to some $\bar{y} \in \mathbb{R}^m$. From the closedness of the set-valued mapping B and $x^\nu \to \bar{x}$ we also obtain $\bar{y} \in B(\bar{x})$. The feasibility of $\bar{x}$ means that for all $j \in J$ and all $y \in B(\bar{x})$ we have $c_j(y) \le 0$, so that we arrive at

$$0 \;\le\; \lim_{\nu \to \infty} c_{j_0}(y^\nu) \;=\; c_{j_0}(\bar{y}) \;\le\; 0\,.$$

This implies $\bar{y} \in B_0^{j_0}(\bar{x})$ as well as $j_0 \in J_0(\bar{x})$.

Next, assume that for some $\nu \in \mathbb{N}$ it is $w(\bar{x}, y^\nu) \le 0$. Since we have $y^\nu \in Y$, it follows $y^\nu \in B(\bar{x})$. From $\bar{x} \in M$ we conclude that $c_{j_0}(y^\nu) \le 0$, in contradiction to the construction of y^ν. Consequently we have

$$\text{for all } \nu \in \mathbb{N}: \quad 0 \;<\; w(\bar{x}, y^\nu)\,. \tag{1}$$

Together with $y^\nu \in B(x^\nu)$ for all $\nu \in \mathbb{N}$ it follows

$$0 \;\le\; \lim_{\nu \to \infty} w(\bar{x}, y^\nu) \;=\; w(\bar{x}, \bar{y}) \;=\; \lim_{\nu \to \infty} w(x^\nu, y^\nu) \;\le\; 0$$

and thus $\bar{y} \in B_{00}^{j_0}(\bar{x})$. □

A usual starting point for genericity considerations is a first order optimality condition which holds without any regularity assumptions. For general semi-infinite problems

$$GSIP: \qquad \max_x f(x) \quad \text{subject to} \quad g_j(x, y) \;\le\; 0 \quad \forall\, y \in B(x)\,,\; j \in J\,,$$

such a condition is given in [16]. To formulate this condition, we denote by

$$\Lambda_j(x,y,\alpha,\beta,\gamma) \;=\; \alpha\, g_j(x,y) - \beta^\top v(y) - \gamma\, w(x,y)\,,\; j\in J,$$

the Fritz-John type lower level Lagrangians, and for $\bar{x} \in M$, $j \in J_0(\bar{x})$ and $\bar{y} \in B_0^j(\bar{x})$ by

$$\begin{aligned} FJ^j(\bar{x},\bar{y}) \;=\; &\{(\alpha,\beta,\gamma)\in \mathbb{R}\times\mathbb{R}^{|L|}\times\mathbb{R}|\; (\alpha,\beta,\gamma)\geq 0,\; \|(\alpha,\beta,\gamma)\|_1 = 1,\\ &\nabla_y \Lambda_j(\bar{x},\bar{y},\alpha,\beta,\gamma) = 0,\; \Lambda_j(\bar{x},\bar{y},\alpha,\beta,\gamma) = 0\;\}\end{aligned}$$

the corresponding sets of Fritz-John multipliers.

Theorem 1 ([16]). *Let $\bar{x} \in M \cap \partial M$ be a local maximizer of GSIP. Then there exist $p_j \in \mathbb{N}$, $\bar{y}^{j,k} \in B_0^j(\bar{x})$, $(\alpha_{j,k}, \beta_{j,k}, \gamma_{j,k}) \in FJ^j(\bar{x}, \bar{y}^{j,k})$, and nontrivial multipliers $\kappa \geq 0$, $\lambda_{j,k} \geq 0$, $1 \leq k \leq p_j$, $j \in J_0(\bar{x})$, such that $\sum_{j\in J_0(\bar{x})} p_j \leq n+1$ and*

$$\kappa\nabla f(\bar{x}) \;-\; \sum_{j\in J_0(\bar{x})} \sum_{k=1}^{p_j} \lambda_{j,k}\, \nabla_x \Lambda_j(\bar{x},\bar{y}^{j,k},\alpha_{j,k},\beta_{j,k},\gamma_{j,k}) \;=\; 0\,.$$

This condition simplifies significantly for the problem $GSIP_{DC}$. In fact, in the lower level Lagrangians

$$\Lambda_j(x,y,\alpha,\beta,\gamma) \;=\; \alpha\, c_j(y) - \beta^\top v(y) - \gamma\, w(x,y)\,,\; j\in J\,,$$

only the function w depends on x, so that we obtain

$$\nabla_x \Lambda_j(x,y,\alpha,\beta,\gamma) \;=\; -\gamma\,\nabla_x w(x,y)\,.$$

The following result is thus immediate.

Corollary 1. *Let $\bar{x} \in M \cap \partial M$ be a local maximizer of DC. Then there exist $p_j \in \mathbb{N}$, $\bar{y}^{j,k} \in B_0^j(\bar{x})$, $(\alpha_{j,k}, \beta_{j,k}, \gamma_{j,k}) \in FJ^j(\bar{x}, \bar{y}^{j,k})$, and nontrivial multipliers $\kappa \geq 0$, $\lambda_{j,k} \geq 0$, $1 \leq k \leq p_j$, $j \in J_0(\bar{x})$, such that $\sum_{j\in J_0(\bar{x})} p_j \leq n+1$ and*

$$\kappa\nabla f(\bar{x}) \;+\; \sum_{j\in J_0(\bar{x})} \sum_{k=1}^{p_j} \lambda_{j,k}\,\gamma_{j,k}\, \nabla_x w(\bar{x},\bar{y}^{j,k}) = \;0\,. \tag{2}$$

A major disadvantage of condition (2) is that it does not guarantee the linear dependence of the vectors $\nabla f(\bar{x})$, $\nabla_x w(\bar{x},\bar{y}^{j,k})$, $1 \leq k \leq p_j$, $j \in J_0(\bar{x})$. In fact, it is easy to construct situations in which $\kappa = 0$ and $\gamma^{j,k} = 0$, $1 \leq k \leq p_j$, $j \in J_0(\bar{x})$. Since the linear dependence of these vectors is crucial for genericity investigations, next we will give a stronger optimality condition.

It is not surprising that this strengthening is possible if one compares the situation to that of standard semi-infinite programming: also there only one of the lower level defining functions depends on x, namely $g_j(x,y)$. The corresponding first order optimality condition deduced from Theorem 1 involves multiplier products $\lambda_{j,k}\,\alpha_{j,k}$ as coefficients of the vectors $\nabla_x g_j(\bar{x},\bar{y}^{j,k})$, whereas from John's original condition for standard semi-infinite programs ([13]) it is clear that a single coefficient $\mu_{j,k}$ would suffice.

Theorem 2. *Let $\bar{x} \in M \cap \partial M$ be a local maximizer of DC. Then there exist $p_j \in I\!N$, $\bar{y}^{j,k} \in B^j_{00}(\bar{x})$, and nontrivial multipliers $\kappa \geq 0$, $\mu_{j,k} \geq 0$, $1 \leq k \leq p_j$, $j \in J_0(\bar{x})$, such that $\sum_{j \in J_0(\bar{x})} p_j \leq n+1$ and*

$$\kappa \nabla f(\bar{x}) \; + \; \sum_{j \in J_0(\bar{x})} \; \sum_{k=1}^{p_j} \mu_{j,k} \, \nabla_x w(\bar{x}, \bar{y}^{j,k}) = \; 0 \; . \tag{3}$$

The proof of Theorem 2 needs some preparation. Recall that the outer tangent cone (contingent cone) $\Gamma^\star(\bar{x}, M)$ to a set $M \subset \mathbb{R}^n$ at $\bar{x} \in \mathbb{R}^n$ is defined by $\bar{d} \in \Gamma^\star(\bar{x}, M)$ if and only if there exist sequences $(t^\nu)_{\nu \in \mathbb{N}}$ and $(d^\nu)_{\nu \in \mathbb{N}}$ such that

$$t^\nu \searrow 0, \; d^\nu \; \to \; \bar{d} \quad \text{and} \quad \bar{x} + t^\nu d^\nu \in M \;\; \text{for all} \;\; \nu \in \mathbb{N} \, .$$

Moreover, we define the inner tangent cone $\Gamma(\bar{x}, M)$ to M at $\bar{x} \in \mathbb{R}^n$ as: $\bar{d} \in \Gamma(\bar{x}, M)$ if and only if there exist some $\bar{t} > 0$ and a neighborhood D of $\bar{d}$ such that

$$\bar{x} + t\, d \in M \;\; \text{for all} \;\; t \in (0, \bar{t}), \;\; d \in D \, .$$

It is well-known ([17]) that $\Gamma(\bar{x}, M) \subset \Gamma^\star(\bar{x}, M)$ and that $\Gamma(\bar{x}, M)^c = \Gamma^\star(\bar{x}, M^c)$, where A^c denotes the set complement of a set $A \subset \mathbb{R}^n$. Furthermore, the following primal first order necessary optimality condition holds.

Lemma 2 ([17]). *Let $\bar{x}$ be a local maximizer of f over M. Then there exists no contingent direction of first order ascent in $\bar{x}$:*

$$\{ \, d \in I\!R^n | \; \langle \nabla f(\bar{x}), d \rangle > 0 \, \} \; \cap \; \Gamma^\star(\bar{x}, M) \;\; = \;\; \emptyset \; .$$

Lemma 3. *For $\bar{x} \in M$ each solution $d^0 \in I\!R^n$ of the system*

$$\langle \nabla_x w(\bar{x}, y), d \rangle \;\; > \;\; 0 \quad \text{for all } y \in B^j_{00}(\bar{x}), \;\; j \in J_0(\bar{x}) \tag{4}$$

is an element of $\Gamma(\bar{x}, M)$.

Proof. Let d^0 be a solution of (4) and assume that $d^0 \in \Gamma(\bar{x}, M)^c$. Then we have $d^0 \in \Gamma^\star(\bar{x}, M^c)$, so that there exist sequences $(t^\nu)_{\nu \in \mathbb{N}}$ and $(d^\nu)_{\nu \in \mathbb{N}}$ such that $t^\nu \searrow 0$, $d^\nu \to d^0$ and $x^\nu := \bar{x} + t^\nu d^\nu \in M^c$ for all $\nu \in \mathbb{N}$.

Exactly like in the proof of Lemma 1 we can now construct some $j_0 \in J_0(\bar{x})$ and a sequence $y^\nu \in B(x^\nu)$ with $y^\nu \to \bar{y} \in B^{j_0}_{00}(\bar{x})$. For all $\nu \in \mathbb{N}$ the mean value theorem guarantees the existence of some $\theta^\nu \in [0,1]$ with

$$0 \; \geq \; w(\bar{x} + t^\nu d^\nu, y^\nu) \;\; = \;\; w(\bar{x}, y^\nu) + t^\nu \langle \nabla_x w(\bar{x} + \theta^\nu t^\nu d^\nu, y^\nu), d^\nu \rangle \; .$$

From (1) and $t^\nu > 0$ we conclude $0 > \langle \nabla_x w(\bar{x} + \theta^\nu t^\nu d^\nu, y^\nu), d^\nu \rangle$ for all $\nu \in \mathbb{N}$ which implies $0 \geq \langle \nabla_x w(\bar{x}, \bar{y}), d^0 \rangle$. Hence we have constructed some $j_0 \in J_0(\bar{x})$ and $\bar{y} \in B^{j_0}_{00}(\bar{x})$ with $\langle \nabla_x w(\bar{x}, \bar{y}), d^0 \rangle \leq 0$, in contradiction to the assumption. □

A combination of Lemma 2, the inclusion $\Gamma(\bar{x}, M) \subset \Gamma^\star(\bar{x}, M)$, and Lemma 3 yields that at a local maximizer $\bar{x}$ of DC the system

$$\langle \nabla f(\bar{x}), d \rangle > 0, \quad \langle \nabla_x w(\bar{x}, y), d \rangle \; > \; 0 \quad \text{for all } y \in B^j_{00}(\bar{x}),\; j \in J_0(\bar{x})$$

is not soluble in d. By a theorem of the alternative this result is equivalent to the assertion of Theorem 2. In the following $\mathrm{conv}(S)$ denotes the convex hull of a set $S \subset \mathbb{R}^n$, i.e. the set of all finite convex combinations of elements from S.

Lemma 4 (Lemma of Gordan, [1, 10]). *Let $S \subset \mathbb{R}^n$ be nonempty and compact. Then the inequality system*

$$s^\top d \; > \; 0 \;\; \textit{for all} \;\; s \in S$$

is inconsistent for $d \in \mathbb{R}^n$ if and only if $0 \in \mathrm{conv}(S)$.

Recall that in the case $0 \in \mathrm{conv}(S)$ it is possible to express the origin as the convex combination of at most $n+1$ elements from S, due to Carathéodory's theorem.

Since the set $\bigcup_{j \in J_0(\bar{x})} B^j_{00}(\bar{x})$ is compact as the finite union of closed subsets of the compact set $B(\bar{x})$, Lemma 4 implies Theorem 2. Note that if the latter union of sets was empty, we would simply obtain the condition $\nabla f(\bar{x}) = 0$ from unconstrained optimization. However, in view of Lemma 1 under the assumption $\bar{x} \in M \cap \partial M$ of Theorem 2 this is not possible.

3 Genericity of the Reduction Ansatz

Multi-jet transversality. In the following we give a short introduction to transversality theory, as far as we need it for our analysis. For details, see [11, 15]. Two smooth manifolds V, W in $\mathbb{R}^N$ are said to intersect transversally (notation: $V \pitchfork W$) if at each intersection point $u \in V \cap W$ the tangent spaces $T_u V$, $T_u W$ together span the embedding space:

$$T_u V \; + \; T_u W \; = \; \mathbb{R}^N \, . \tag{5}$$

The number $N - \dim V$ is called the codimension of V in $\mathbb{R}^N$, shortly $\mathrm{codim}\, V$, and we have

$$\mathrm{codim}\, V \; \leq \; \dim W \tag{6}$$

whenever $V \pitchfork W$ and $V \cap W \neq \emptyset$. For our purpose, the manifold W is induced by the 1-jet extension of a function $F \in C^\infty(\mathbb{R}^N, \mathbb{R}^M)$, i.e. by the mapping

$$j^1 F : \; \mathbb{R}^N \longrightarrow J(N, M, 1), \;\; z \longmapsto (z, F(z), F_z(z))$$

where $J(N, M, 1) \; = \; \mathbb{R}^{N+M+N \cdot M}$ and the partial derivatives are listed according to some order convention ([15]). Choosing W as the graph of $j^1 F$

(notation: $W = j^1F(\mathbb{R}^N)$) it is easily shown that W is a smooth manifold of dimension N in $J(N, M, 1)$. Given another smooth manifold V in $J(N, M, 1)$, we define the set

$$\pitchfork^1 V \;=\; \{F \in C^\infty(\mathbb{R}^N, \mathbb{R}^M) |\; j^1F(\mathbb{R}^N) \pitchfork V\} \,.$$

Our analysis bases on the following theorem which is originally due to R. Thom. For proofs see [11, 15].

Theorem 3 (Jet transversality). *With respect to the C_s^∞-topology, the set $\pitchfork^1 V$ is generic in $C^\infty(\mathbb{R}^N, \mathbb{R}^M)$.*

Here, C_s^∞ denotes the Whitney topology ([11, 15]). In particular, $\pitchfork^1 V$ is C_s^∞-dense in $C^\infty(\mathbb{R}^N, \mathbb{R}^M)$ and hence, C_s^d-dense in $C^d(\mathbb{R}^N, \mathbb{R}^M)$ for any $d \in \mathbb{N}_0 = \mathbb{N} \cup \{0\}$ ([11]).

Since jet transversality gives information about certain properties of the functions under investigation only at every *single* point we apply the concept of multi-jet transversality instead ([15]). Thereby we are able to study properties that have to be satisfied at all global maximizers of the lower level problem at the same time. Let D be a positive integer and define

$$\mathbb{R}_D^N \;=\; \left\{(z^1, \ldots, z^D) \in \textstyle\prod_{k=1}^D \mathbb{R}^N |\; z^i \neq z^j \text{ for } 1 \le i < j \le D\right\}$$

as well as the multi-jet space

$$J_D(N, M, 1) \;=\; \left\{(z^1, u^1, \ldots, z^D, u^D) \in \textstyle\prod_{k=1}^D J(N, M, 1) |\; (z^1, \ldots, z^D) \in \mathbb{R}_D^N\right\} \,.$$

The multi-jet extension $j_D^1 F : \mathbb{R}_D^N \longrightarrow J_D(N, M, 1)$ is the mapping

$$j_D^1 F : \; (z^1, \ldots, z^D) \;\longmapsto\; (j^1F(z^1), \ldots, j^1F(z^D)) \;,$$

and for a smooth manifold V in $J_D(N, M, 1)$ we define the set

$$\pitchfork_D^1 V \;=\; \{F \in C^\infty(\mathbb{R}^N, \mathbb{R}^M) |\; j_D^1 F(\mathbb{R}_D^N) \pitchfork V\} \,.$$

Theorem 4 (Multi-jet transversality). *With respect to the C_s^∞-topology, the set $\pitchfork_D^1 V$ is generic in $C^\infty(\mathbb{R}^N, \mathbb{R}^M)$.*

Rank conditions. For $M, N \in \mathbb{N}$ and $R \le \min(M, N)$ let us define the set of matrices of rank R,

$$\mathbb{R}_R^{M \times N} \;=\; \left\{A \in \mathbb{R}^{M \times N} \middle|\; \operatorname{rank}(A) = R\right\} \,.$$

Moreover, for $M, N \in \mathbb{N}$, $R \le \min(M, N)$, $\mathcal{I} \subset \{1, ..., M\}$ and

$$\max(R + |\mathcal{I}| - M, 0) \;\le\; S \;\le\; \min(R, |\mathcal{I}|)$$

we let

$$\mathbb{R}^{M\times N}_{R,\mathcal{I},S} = \left\{ A \in \mathbb{R}^{M\times N}_{R} \,\middle|\, A^{(\mathcal{I})} \in \mathbb{R}^{(M-|\mathcal{I}|)\times N}_{R-S} \right\},$$

where the matrix $A^{(\mathcal{I})}$ results from A by deletion of the rows with indices in $\mathcal{I}$. Observe that the above restrictions on S follow from the trivial relations $0 \le R-S \le M-|\mathcal{I}|$ and $R-|\mathcal{I}| \le R-S \le R$.

These definitions are intimately related to the Reduction Ansatz in the lower level problem. In fact, for $\bar{x} \in M$ and some $j \in J_0(\bar{x})$ let $\bar{y}$ be a maximizer of $Q^j(\bar{x})$. From the first order necessary optimality condition of Fritz John we know that then the gradient $\nabla c_j(\bar{y})$ and the gradients of the active inequality constraints are linearly dependent. To identify these constraints conveniently we put $L = \{1,...,s\}$ with $s \in \mathbb{N}$, $v_{s+1}(x,y) := w(x,y)$, $\Lambda = L \cup \{s+1\}$, $\Lambda_0(\bar{x},\bar{y}) = \{\ell \in \Lambda |\ v_\ell(\bar{x},\bar{y}) = 0\}$, and $s_0 = |\Lambda_0(\bar{x},\bar{y})|$. Let $D_y v_{\Lambda_0}(\bar{x},\bar{y})$ denote the matrix with rows $D_y v_\ell(\bar{x},\bar{y}) := \nabla_y^\top v_\ell(\bar{x},\bar{y})$, $\ell \in \Lambda_0(\bar{x},\bar{y})$. We obtain

$$\begin{pmatrix} D_y c_j(\bar{x},\bar{y}) \\ D_y v_{\Lambda_0}(\bar{x},\bar{y}) \end{pmatrix} \in \mathbb{R}^{(1+s_0)\times m}_{\rho_j}$$

with $\rho_j \le s_0$. With this notation, LICQ is equivalent to

$$\begin{pmatrix} D_y c_j(\bar{x},\bar{y}) \\ D_y v_{\Lambda_0}(\bar{x},\bar{y}) \end{pmatrix} \in \mathbb{R}^{(1+s_0)\times m}_{s_0\,,\{0\},0},$$

if we identify the first row of the matrix with the index $\ell = 0$. Moreover, SCS implies

$$\begin{pmatrix} D_y c_j(\bar{x},\bar{y}) \\ D_y v_{\Lambda_0}(\bar{x},\bar{y}) \end{pmatrix} \in \mathbb{R}^{(1+s_0)\times m}_{s_0\,,\{\ell\},0},$$

for all $\ell \in \Lambda_0(\bar{x},\bar{y})$.

For a matrix $A \in \mathbb{R}^{M\times N}$ with rows $A^1,...,A^M$ we define the function

$$\mathrm{vec}: \mathbb{R}^{M\times N} \longrightarrow \mathbb{R}^{M\cdot N},\ A \longmapsto (A^1,...,A^M).$$

Lemma 5 ([15, 20]).

(i) The set $\mathrm{vec}\left(\mathbb{R}^{M\times N}_{R}\right)$ *is a smooth manifold of codimension* $(M-R)\cdot(N-R)$ *in* $\mathbb{R}^{M\cdot N}$.

(ii) The set $\mathrm{vec}\left(\mathbb{R}^{M\times N}_{R,\mathcal{I},S}\right)$ *is a smooth manifold of codimension* $(M-R)\cdot(N-R)+S\cdot(M-R+S-|\mathcal{I}|)$ *in* $\mathbb{R}^{M\cdot N}$.

A codimension formula. Let $J = \{1,...,p\}$ as well as $p_0 = |J_0(\bar{x})|$. By Lemma 1, for $\bar{x} \in M \cap \partial M$ the set $\bigcup_{j\in J_0(\bar{x})} B^j_{00}(\bar{x})$ is nonempty. We consider the case in which it contains at least r different elements, say $\bar{y}^{j,k} \in B^j_{00}(\bar{x})$, $1 \le k \le p_j$, $j \in J_0(\bar{x})$, with $\sum_{j=1}^{p_0} p_j = r$.

As $\bar{y}^{j,k}$ is a maximizer of $Q^j(\bar{x})$ we find a unique number $\rho_{j,k} \le s_0^{j,k} := |\Lambda_0(\bar{x},\bar{y}^{j,k})|$ such that

$$\begin{pmatrix} D_y c_j(\bar{x}, \bar{y}^{j,k}) \\ D_y v_{\Lambda_0}(\bar{x}, \bar{y}^{j,k}) \end{pmatrix} \in \mathbb{R}^{(1+s_0^{j,k})\times m}_{\rho_{j,k}} ,$$

and we define the rank defect $d_{j,k} = s_0^{j,k} - \rho_{j,k}$. Moreover, we have

$$\begin{pmatrix} D_y c_j(\bar{x}, \bar{y}^{j,k}) \\ D_y v_{\Lambda_0}(\bar{x}, \bar{y}^{j,k}) \end{pmatrix} \in \mathbb{R}^{(1+s_0^{j,k})\times m}_{\rho_{j,k}\,,\, D_{j,k}\,,\, \sigma_{j,k}}$$

for several choices of $D_{j,k}$ and $\sigma_{j,k}$, where we can always choose $D_{j,k} = \emptyset$ and $\sigma_{j,k} = 0$.

Furthermore, if $\bar{x}$ is a local maximizer of DC, Theorem 2 guarantees that for some choice $\bar{y}^{j,k} \in B^j_{00}(\bar{x})$, $1 \le k \le p_j$, $j \in J_0(\bar{x})$ with $\sum_{j=1}^{p_0} p_j = r \le n+1$ we also have

$$\begin{pmatrix} Df(\bar{x}) \\ D_x w(\bar{x}, \bar{y}^{j,k})_{1\le k\le p_j\,,\,1\le j\le p_0} \end{pmatrix} \in \mathbb{R}^{(1+r)\times n}_{\rho_0\,,\, D_0\,,\, \sigma_0}$$

with $\rho_0 \le r$. We denote the corresponding rank defect by $d_0 = r - \rho_0$. Our subsequent analysis bases on the following relation:

$$\begin{aligned} 0 \;\ge\; & d_0 + d_0(n - r + d_0) + \sigma_0(1 + d_0 + \sigma_0 - |D_0|) \qquad (7) \\ + & \sum_{j=1}^{p_0}\sum_{k=1}^{p_j} \Big[\, d_{j,k} + d_{j,k}(m - s_0^{j,k} + d_{j,k}) + \sigma_{j,k}(1 + d_{j,k} + \sigma_{j,k} - |D_{j,k}|) \,\Big] . \end{aligned}$$

Put $\mathbb{Q}^d = C^d(\mathbb{R}^n, \mathbb{R}) \times \mathbb{C}^d(c) \times \mathbb{C}^d(v) \times C^d(\mathbb{R}^n \times \mathbb{R}^m, \mathbb{R})$, where $\mathbb{C}^d(c)$ and $\mathbb{C}^d(v)$ are defined to be the set of vector functions $c \in C^d(\mathbb{R}^m, \mathbb{R}^p)$ and $v \in C^d(\mathbb{R}^m, \mathbb{R}^s)$ such that C and Y are nonempty and compact, respectively. Define

$$\begin{aligned} \mathcal{F}^d = \{\, (f, c, v, w) \in \mathbb{Q}^d |\ & \text{any choice of } r \text{ elements} \\ & \text{from } \textstyle\bigcup_{j\in J_0(\bar{x})} B^j_{00}(\bar{x}) \text{ corresponding to a point} \\ & \bar{x} \in M \cap \partial M \text{ satisfies relation (7)} \,\} . \end{aligned}$$

Theorem 5. *$\mathcal{F}^\infty$ is C^∞_s-dense in $\mathbb{Q}^\infty$.*

Proof. For $r \in \mathbb{N}$ and $K := \{1, ..., r\}$ consider the reduced multi-jet

$$\begin{aligned} j^1_r(f, c, v, w)(x^1, y^1, ..., x^r, y^r) \;=\; & (x^k, y^k, Df^k, c^k_1, ..., c^k_p, Dc^k_1, ..., Dc^k_p, \\ & v^k_1, ..., v^k_s, Dv^k_1, ..., Dv^k_s, w^k, D_x w^k, D_y w^k,\ k \in K\ ) \end{aligned}$$

with $(x^1, y^1, ..., x^r, y^r) \in \mathbb{R}^{n+m}_r$ and $Df^k = Df(x^k)$, etc. In the following we call $K_j\,,\ j \in \tilde{J}_0$, a partition of K if $\bigcup_{j\in\tilde{J}_0} K_j = K$ and if the sets $K_j\,,\ j \in \tilde{J}_0$, are pairwise distinct. For

$$\left.\begin{array}{l}
r \in \mathbb{N} \\
\tilde{J}_0 \subset J \\
K_j\,,\ j \in \tilde{J}_0, \quad \text{a partition of } K = \{1, ..., r\} \\
0 \leq \rho_0 \leq \min(1+r, n) \\
D_0 \subset \{0, ..., r\} \\
\max(\rho_0 + |D_0| - 1 - r, 0) \leq \sigma_0 \leq \min(\rho_0, |D_0|) \\
\tilde{\Lambda}_0^{j,k} \subset \Lambda \\
0 \leq \rho_{j,k} \leq \min(1 + s_0^{j,k}, m) \\
D_{j,k} \subset \{0, ..., s_0^{j,k}\} \\
\max(\rho_{j,k} + |D_{j,k}| - 1 - s_0^{j,k}, 0) \leq \sigma_{j,k} \leq \min(\rho_{j,k}, |D_{j,k}|) \\
k \in K_j,\ j \in \tilde{J}_0
\end{array}\right\} \quad (8)$$

we define the C^∞-manifold $\mathcal{N}_{r,(K_j, j\in\tilde{J}_0),\rho_0,D_0,\sigma_0,\ (\tilde{\Lambda}_0^{j,k},\rho_{j,k},D_{j,k},\sigma_{j,k},\ k\in K_j, j\in\tilde{J}_0)}$ to be the set of points

$$(\tilde{x}^k, \tilde{y}^k, \tilde{F}^k, \tilde{c}_1^k, ..., \tilde{c}_p^k, \tilde{C}_1^k, ..., \tilde{C}_p^k, \tilde{v}_1^k, ..., \tilde{v}_s^k, \tilde{V}_1^k, ..., \tilde{V}_s^k, \tilde{w}^k, \tilde{X}^k, \tilde{Y}^k,\ k \in K\ )$$

with the following properties:

- dimensions:

$$\begin{aligned}
(\tilde{x}^1, \tilde{y}^1, ..., \tilde{x}^r, \tilde{y}^r) &\in \mathbb{R}_r^{n+m}, \\
\tilde{c}_j^k,\ j \in J,\ \tilde{v}_\ell^k,\ \ell \in L,\ \tilde{w}^k &\in \mathbb{R},\ k \in K \\
\tilde{F}^k,\ \tilde{X}^k &\in \mathbb{R}^n,\ k \in K \\
\tilde{C}_j^k,\ j \in J,\ \tilde{V}_\ell^k,\ \ell \in L,\ \tilde{Y}^k &\in \mathbb{R}^m,\ k \in K
\end{aligned}$$

- conditions on the independent variables:

$$\tilde{x}^1 \ = \ ... \ = \ \tilde{x}^r$$

- conditions on the functional values:

$$\tilde{c}_j^k = 0,\ k \in K_j\,,\ j \in \tilde{J}_0\,, \quad \tilde{v}_\ell^k = 0, \quad \ell \in \tilde{\Lambda}_0^{j,k},\ k \in K_j\,,\ j \in \tilde{J}_0$$

- conditions on the gradients:

$$\begin{pmatrix} \tilde{F}^1 \\ (\tilde{X}^k)_{k\in K_j,\, j\in\tilde{J}_0} \end{pmatrix} \in \mathbb{R}^{(1+r)\times n}_{\rho_0\,,\, D_0\,,\, \sigma_0}\,,$$

$$\begin{pmatrix} \tilde{C}_j^k \\ \tilde{V}^k_{\tilde{\Lambda}_0^{j,k}} \end{pmatrix} \in \mathbb{R}^{(1+s_0^{j,k})\times m}_{\rho_{j,k}\,,\, D_{j,k}\,,\, \sigma_{j,k}}\,, \quad k \in K_j\,,\ j \in \tilde{J}_0\,.$$

With the help of Lemma 5(ii) we can calculate the codimension of this manifold:

$$\begin{aligned} &\operatorname{codim} \mathcal{N}_{r,(K_j,j\in\tilde{J}_0),\rho_0,D_0,\sigma_0,\ (\tilde{A}_0^{j,k},\rho_{j,k},D_{j,k},\sigma_{j,k},\ k\in K_j,j\in\tilde{J}_0)} = \\ &= (r-1)n + r + \sum_{j\in\tilde{J}_0}\sum_{k\in K_j} s_0^{j,k} \\ &\quad +(1+r-\rho_0)(n-\rho_0) + \sigma_0(1+r-\rho_0+\sigma_0-|D_0|) \\ &\quad +\sum_{j\in\tilde{J}_0}\sum_{k\in K_j}\Big[(1+s_0^{j,k}-\rho_{j,k})(m-\rho_{j,k}) \\ &\quad +\sigma_{j,k}(1+s_0^{j,k}-\rho_{j,k}+\sigma_{j,k}-|D_{j,k}|)\Big] . \end{aligned} \tag{9}$$

Define the set

$$\mathcal{F}^\star = \bigcap_{r=1}^{\infty} \bigcap_{(K_j\cdots\tilde{J}_0)} \pitchfork_r^1 \mathcal{N}_{r,(K_j,j\in\tilde{J}_0),\rho_0,D_0,\sigma_0,\ (\tilde{A}_0^{j,k},\rho_{j,k},D_{j,k},\sigma_{j,k},\ k\in K_j,j\in\tilde{J}_0)}$$

where the inner intersection ranges over all possible choices of K_1, etc., according to (8). $\mathcal{F}^\star$ is C_s^∞-dense in $\mathbb{Q}^\infty$ by Theorem 4. It remains to be shown that $\mathcal{F}^\star \subset \mathcal{F}^\infty$. Choose a function vector $(f,c,v,w) \in \mathcal{F}^\star$ as well as a local maximizer $\bar{x}$ of DC. By Lemma 1 the set $\bigcup_{j\in J_0(\bar{x})} B_{00}^j(\bar{x})$ is non-empty. From each nonempty $B_{00}^j(\bar{x})$ choose some (pairwise distinct) $\bar{y}^{j,k}$, $k \in K_j$, and put $K_j = \emptyset$ if $B_{00}^j(\bar{x}) = \emptyset$. Denote the total number of chosen elements by r and put $K = \{1,...,r\}$. Then $K_j\,, j \in J_0(\bar{x})$, forms a partition of K, $(\bar{x},\bar{y}^1,...,\bar{x},\bar{y}^r) \in \mathbb{R}_r^{n+m}$, and $j_r^1(f,c,v,w)(\bar{x},\bar{y}^1,...,\bar{x},\bar{y}^r)$ is contained in some set $\mathcal{N}_{r,(\cdots\tilde{J}_0)}$. As the intersection of $j_r^1(f,c,v,w)(\mathbb{R}_r^{n+m})$ with $\mathcal{N}_{r,(\cdots\tilde{J}_0)}$ is transverse, (6) yields $r\,(n+m) \geq \operatorname{codim} \mathcal{N}_{r,(\cdots\tilde{J}_0)}$. Inserting (9) now yields (7) after a short calculation. □

Note that the statement of Theorem 5 is equivalent to saying that $\mathcal{F}^\infty$ is C_s^d-dense in $\mathbb{Q}^\infty$ for each $d \in \mathbb{N}_0$. Since the set $C^\infty(\mathbb{R}^N,\mathbb{R})$ is also C_s^d-dense in $C^d(\mathbb{R}^N,\mathbb{R})$ ([11]), it is no restriction to consider the space of smooth defining functions $\mathbb{Q}^\infty$ instead of the space $\mathbb{Q}^d$, $d \geq 2$.

Corollary 2. *For $(f,c,v,w) \in \mathcal{F}^\star$ let $\bar{x} \in M \cap \partial M$ be a local maximizer of DC. Then the set $\bigcup_{j\in J_0(\bar{x})} B_{00}^j(\bar{x})$ contains at most n elements $\bar{y}^1,...,\bar{y}^r$, and for each $1 \leq k \leq r$ LICQ and SCS hold at $\bar{y}^k$ in the corresponding lower level problem.*

Proof. One can easily conclude from the relations in (8) that each factor in the right hand side of (7) is nonnegative. Consequently, all summands have to vanish. In particular we find $d_0 = d_{j,k} = 0$ for all $1 \leq k \leq p_j\,,\ j \in J_0(\bar{x})$. This implies $0 \leq n - \rho_0 = n - r + d_0 = n - r$ which is the first part of the assertion.

A second consequence is $\sigma_{j,k}(1+\sigma_{j,k}-|D_{j,k}|)=0$ for all $1\le k\le p_j$, $j\in J_0(\bar{x})$. Hence, $|D_{j,k}|=1$ implies $\sigma_{j,k}=0$. This means that LICQ and SCS hold at each $\bar{y}^{j,k}$ in $Q^j(\bar{x})$. □

With a tedious evaluation of the tangent space condition (5) it is also possible to show that for $(f,c,v,w)\in\mathcal{F}^\star$ and a local maximizer $\bar{x}\in M\cap\partial M$ of DC at each $\bar{y}\in\bigcup_{j\in J_0(\bar{x})}B^j_{00}(\bar{x})$ the second order sufficiency condition holds. Altogether this means that for $(f,c,v,w)\in\mathcal{F}^\star$ the Reduction Ansatz is valid at each local maximizer of DC.

4 An associated standard semi-infinite problem

The first order necessary optimality condition in Theorem 2 has the typical structure of an optimality condition for some *standard* semi-infinite program. In fact, we can construct a certain standard semi-infinite problem which is strongly related to DC.

For the following arguments we put $C_j^{\le}=\{y\in\mathbb{R}^m|\ c_j(y)\le 0\}$, $C_j^{<}=\{y\in\mathbb{R}^m|\ c_j(y)<0\}$, etc. for $j\in J$ as well as $W^{\le}(x)=\{y\in\mathbb{R}^m|\ w(x,y)\le 0\}$ etc. The main idea is to rewrite the inclusion constraint $B(x)\subset C$ of DC in an equivalent form like $C^c\subset B(x)^c$.

Slightly modified this idea proceeds as follows. By definition we have $B(x)\subset C$ if and only $Y\cap W^{\le}(x)\subset\bigcap_{j\in J}C_j^{\le}$. The latter is equivalent to $Y\cap W^{\le}(x)\cap\bigcup_{j\in J}C_j^{>}=\emptyset$ and, thus, to $\bigcup_{j\in J}\left(Y\cap C_j^{>}\right)\subset W^{>}(x)$.

This means that an equivalent formulation of the constraint $B(x)\subset C$ is given by

$$w(x,y)\ >\ 0\quad\text{for all}\ \ y\in Y\cap C_j^{>}\,,\ j\in J\,.$$

Due to the strict inequalities these are not semi-infinite constraints in the usual sense. We can, however, formulate an *associated standard semi-infinite problem* for DC:

$$SIP_{DC}:\quad\max_x f(x)\quad\text{subject to}\quad w(x,y)\ \ge\ 0\quad\forall\, y\in Y\cap C_j^{\ge}\,,\ j\in J\,.$$

Note that the index sets $Y\cap C_j^{\ge}$, $j\in J$, of the finitely many semi-infinite constraints are compact, and certainly nonempty if $C\subset Y$. Recall that we defined the optimal value functions

$$\varphi_j(x)\ =\ \max_{y\in Y\cap W^{\le}(x)}\ c_j(y)\,,\ j\in J\,,$$

and the active index set $J_0(x)=\{j\in J|\ \varphi_j(x)=0\}$ for the problem $GSIP_{DC}$. For the problem SIP_{DC} we put analogously

$$\psi_j(x)\ =\ \min_{y\in Y\cap C_j^{\ge}}\ w(x,y)\,,\ j\in J\,,$$

$J_0^{SIP}(x) = \{j \in J |\ \psi_j(x) = 0\}$, and $Q^j_{SIP}(x)$, $j \in J$, for the corresponding lower level problems. For $j \in J_0^{SIP}(x)$ the optimal points of $Q^j_{SIP}(x)$ form the set $\{y \in Y \cap C_j^{\geq} |\ w(x,y) = 0\} = Y \cap C_j^{\geq} \cap W^{=}(x)$. Fritz John's first order optimality condition for standard semi-infinite problems thus yields the following result.

Proposition 4.1 *Let $\bar{x} \in \partial M_{SIP}$ be a local maximizer of SIP_{DC}. Then there exist $p_j \in I\!N$, $\bar{y}^{j,k} \in Y \cap C_j^{\geq} \cap W^{=}(x)$, and nontrivial multipliers $\kappa \geq 0$, $\mu_{j,k} \geq 0$, $1 \leq k \leq p_j$, $j \in J_0^{SIP}(\bar{x})$, such that $\sum_{j \in J_0^{SIP}(\bar{x})} p_j \leq n+1$ and*

$$\kappa \nabla f(\bar{x}) + \sum_{j \in J_0^{SIP}(\bar{x})} \sum_{k=1}^{p_j} \mu_{j,k} \nabla_x w(\bar{x}, \bar{y}^{j,k}) = 0 .$$

The resemblance of this result with Theorem 2 is obvious. We emphasize that we relaxed strict to nonstrict inequalities while deriving the problem SIP_{DC} from DC, so that an identical result for both problems cannot be expected. More precisely, the feasible sets

$$M = \{ x \in I\!R^n |\ \varphi_j(x) \leq 0,\ j \in J \} = \bigcap_{j \in J} \Phi_j^{\leq}$$

and

$$M_{SIP} = \{ x \in I\!R^n |\ \psi_j(x) \geq 0,\ j \in J \} = \bigcap_{j \in J} \Psi_j^{\geq}$$

do not necessarily coincide. Their relation is clarified by the next results.

Lemma 6.
(i) For all $j \in J$ we have $\Phi_j^{<} = \Psi_j^{>}$.
(ii) For all $j \in J$ and $x \in \Phi_j^{=}$ we have $x \in \Psi_j^{=}$ if and only if $w(x,\cdot)$ is active in all global solutions of $Q^j(x)$.
(iii) For all $j \in J$ and $x \in \Psi_j^{=}$ we have $x \in \Phi_j^{=}$ if and only if c_j is active in all global solutions of $Q^j_{SIP}(x)$.

Proof. For all $j \in J$ we have $x \in \Phi_j^{<}$ if and only if $Y \cap W^{\leq}(x) \subset C_j^{<}$, and we have $x \in \Psi_j^{>}$ if and only if $Y \cap C_j^{\geq} \subset W^{>}(x)$. Since both characterizations are equivalent to $Y \cap C_j^{\geq} \cap W^{\leq}(x) = \emptyset$, the assertion of part (i) follows.

From part (i) it is clear that for each $j \in J$ the set $\Phi_j^{=}$ is necessarily contained in $\Psi_j^{\leq}$. We have $x \in \Psi_j^{<}$ if and only if $Y \cap C_j^{\geq} \cap W^{<}(x) \neq \emptyset$. On the other hand, for $x \in \Phi_j^{=}$ the set $Y \cap C_j^{\geq} \cap W^{\leq}(x)$ is the set of global solutions of $Q^j(x)$. This shows the assertion of part (ii). The proof of part (iii) is analogous. □

Theorem 6.
(i) Let $x \in M$ and for each $j \in J_0(x)$ let $w(x,\cdot)$ be active in all global solutions of $Q^j(x)$. Then we have $x \in M_{SIP}$.

(ii) Let $x \in M_{SIP}$ and for each $j \in J_0^{SIP}(x)$ let c_j be active in all global solutions of $Q^j_{SIP}(x)$. Then we have $x \in M$.

Proof. Lemma 6. □

Note that under the assumption of Theorem 6(ii) the global solution set $Y \cap C_j^{\geq} \cap W^{=}(x)$ can be replaced by $Y \cap C_j^{=} \cap W^{=}(x) = B^j_{00}(x)$, so that the difference between Theorem 2 and Proposition 4.1 disappears.

5 Final remarks

A main technical assumption for the genericity proof in Section 3 is that only one of the smooth constraints in the description of $B(x)$ actually depends on x. There are, of course, design centering problems which cannot be formulated this way. These problems appear to be as difficult as the general semi-infinite optimization problem without any additional structure, so that genericity results for this case can be expected as soon as the generic validity of the Reduction Ansatz at all solutions of *GSIP* has been shown.

Under the Reduction Ansatz, locally around a local solution $\bar{x}$ the problem $GSIP_{DC}$ can be rewritten as a smooth problem with finitely many constraints. We point out that our genericity proof from Section 3 also shows that for $(f, c, v, w) \in \mathcal{F}^\star$ a local maximizer $\bar{x} \in M \cap \partial M$ of *DC* is nondegenerate for this locally reduced problem.

The results of the present article for single-body design centering problems can be transferred to the multi-body case with some additional technical effort. This and efficient numerical methods for multi-body design centering will be subject of future research.

References

1. E.W. Cheney, *Introduction to Approximation Theory,* McGraw-Hill, New York, 1966.
2. V. Chvatal, *Linear Programming,* Freeman, New York, 1983.
3. A.V. Fiacco, G.P. McCormick, *Nonlinear Programming: Sequential Unconstrained Minimization Techniques,* Wiley, New York, 1968.
4. T.J. Graettinger, B.H. Krogh, *The acceleration radius: a global performance measure for robotic manipulators,* IEEE Journal of Robotics and Automation, Vol. 4 (1988), 60-69.
5. P. Gritzmann, V. Klee, *On the complexity of some basic problems in computational convexity. I. Containment problems,* Discrete Mathematics, Vol. 136 (1994), 129-174.
6. R. Hettich, H.Th. Jongen, *Semi-infinite programming: conditions of optimality and applications,* in: J. Stoer (ed): Optimization Techniques, Part 2, Lecture Notes in Control and Information Sciences, Vol. 7, Springer, Berlin, 1978, 1-11.

7. R. Hettich, K.O. Kortanek, *Semi-infinite programming: theory, methods, and applications,* SIAM Review, Vol. 35 (1993), 380-429.
8. R. Hettich, G. Still, *Semi-infinite programming models in robotics,* in: J. Guddat, H.Th. Jongen, B. Kummer, F. Nožička (eds): Parametric Optimization and Related Topics II, Akademie Verlag, Berlin, 1991, 112-118.
9. R. Hettich, G. Still, *Second order optimality conditions for generalized semi-infinite programming problems,* Optimization, Vol. 34 (1995), 195-211.
10. R. Hettich, P. Zencke, *Numerische Methoden der Approximation und semi-infiniten Optimierung,* Teubner, Stuttgart, 1982.
11. M.W. Hirsch, *Differential Topology,* Springer, New York, 1976.
12. R. Horst, H. Tuy, *Global Optimization,* Springer, Berlin, 1990.
13. F. John, *Extremum problems with inequalities as subsidiary conditions,* in: Studies and Essays, R. Courant Anniversary Volume, Interscience, New York, 1948, 187-204.
14. H.Th. Jongen, P. Jonker, F. Twilt, *Critical sets in parametric optimization,* Mathematical Programming, Vol. 34 (1986), 333-353.
15. H.Th. Jongen, P. Jonker, F. Twilt, *Nonlinear Optimization in Finite Dimensions,* Kluwer, Dordrecht, 2000.
16. H.Th. Jongen, J.-J. Rückmann, O. Stein, *Generalized semi-infinite optimization: a first order optimality condition and examples,* Mathematical Programming, Vol. 83 (1998), 145-158.
17. P.-J. Laurent, *Approximation et Optimisation,* Hermann, Paris, 1972.
18. V.H. Nguyen, J.J. Strodiot, *Computing a global optimal solution to a design centering problem,* Mathematical Programming, Vol. 53 (1992), 111-123.
19. O. Stein, *On parametric semi-infinite optimization,* Thesis, Shaker, Aachen, 1997.
20. O. Stein, *On level sets of marginal functions,* Optimization, Vol. 48 (2000), 43-67.
21. O. Stein, *Bi-level Strategies in Semi-infinite Programming,* Kluwer, Boston, 2003.
22. O. Stein, G. Still, *On generalized semi-infinite optimization and bilevel optimization,* European Journal of Operational Research, Vol. 142 (2002), 444-462.
23. O. Stein, G. Still, *Solving semi-infinite optimization problems with interior point techniques,* SIAM Journal on Control and Optimization, Vol. 42 (2003), 769-788.
24. W. Wetterling, *Definitheitsbedingungen für relative Extrema bei Optimierungs- und Approximationsaufgaben,* Numerische Mathematik, Vol. 15 (1970), 122-136.
25. G. Zwier, *Structural Analysis in Semi-Infinite Programming,* Thesis, University of Twente, 1987.

Part III

Set-Valued Optimization

Contraction mapping fixed point algorithms for solving multivalued mixed variational inequalities

Pham Ngoc Anh[1] and Le Dung Muu[2]

[1] Department of Mathematics, Posts and Telecommunications Institute of Technology, Hanoi, Vietnam Ngocanhncs@yahoo.com
[2] Optimization and Control Department, Institute of Mathematics, Hanoi, Vietnam ldmuu@math.ac.vn

Summary. We show how to choose regularization parameters such that the solution of a multivalued strongly monotone mixed variational inequality can be obtained by computing the fixed point of a certain multivalued mapping having a contraction selection. Moreover a solution of a multivalued cocoercive variational inequality can be computed by finding a fixed point of a certain mapping having nonexpansive selection. By the Banach contraction mapping principle it is easy to establish the convergence rate.

Keywords. Multivalued mixed variational inequality, cocoerciveness, contraction and nonexpansiveness, Banach fixed point method.

2001 Mathematics Subject Classification: 65K10, 90 C25

1 Introduction

The contraction and nonexpansive fixed point–methods for solving variational inequalities have been developed by several authors (see e.g. [1, 2, 6, 8, 15, 16] and the references therein). In our recent paper [1] we have used the auxiliary problem-method and the Banach contraction mapping fixed point principle to solve mixed variational inequalities involving single valued strongly monotone and cocoercive operators. Then in [2] we extended our method and combined it with the proximal point algorithm to solve mixed monotone variational inequalities.

In this paper we further extend the idea in [1, 2] to mixed multivalued variational inequalities involving strongly monotone and cocoercive cost operators with respect to the Hausdorff distance. Namely, we show that a necessary and sufficient condition for a point to be the solution of a multivalued strongly

S. Dempe and V. Kalashnikov (eds.), *Optimization with Multivalued Mappings*, pp. 231-249
©2006 Springer Science + Business Media, LLC

monotone mixed variational inequality is that it is the fixed point of a certain multivalued mapping having a contractive selection. For mixed variational inequalities involving multivalued cocoercive cost operators we show that their solutions can be computed by finding fixed points of corresponding multivalued mappings having a nonexpansive selection. These results allow that the Banach contraction mapping principle and its modifications can be applied to solve strongly monotone and cocoercive multivalued mixed variational inequalities. By the Banach contraction fixed point principle it is straightforward to obtain the convergence rate of the proposed algorithms.

2 Fixed Point Formulations

Let C be a nonempty, closed, convex subset of $I\!R^n$, let $F : I\!R^n \to 2^{I\!R^n}$ be a multivalued mapping. Throughout this paper we suppose that dom$F$ contains C and that $F(x)$ is closed, convex for every $x \in C$. We suppose further that we are given a convex, subdifferentiable function $\varphi : C \to I\!R$. We consider the following multivalued mixed variational inequality problem that we shall denote by (VIP) :

Find $x^* \in C$ and $w^* \in F(x^*)$ such that

$$\langle w^*, x - x^* \rangle + \varphi(x) - \varphi(x^*) \geq 0 \quad \forall x \in C. \tag{2.1}$$

This problem has been considered by some authors (see e.g., [4, 9, 12, 13, 14] and the references quoted therein). As usual in what follows we shall refer to F as cost operator and to C as constraint set.

As an example we consider an oligopolistic Cournot market model where there are n-firms producing a common homogeneous commodity. We assume that the price p_i of firm i depends on the total quantity of the commodity. Let h_i denote the cost of firm i when its production level is x_i. Suppose that the profit of firm i is given by

$$f_i(x_1, ..., x_n) = x_i p_i(\sum_{i=1}^{n} x_i) - h_i(x_i) \ (i = 1, ..., n).$$

Let U_i denote the strategy set of firm i and $U := U_1 \times ... \times U_n$ be the strategy set of the model. In the classical Cournot model the price and the cost functions for each firm are assumed to be affine of the forms

$$p_i(\sigma) = \alpha_i - \beta_i \sigma, \quad \alpha_i \geq 0, \beta_i > 0, \sigma = \sum_{i=1}^{n} x_i,$$

$$h_i(x_i) = \mu_i x_i + \xi_i, \ \mu_i > 0, \xi_i \geq 0 \ (i = 1, ..., n).$$

The problem is to find a point $x^* = (x_1^*, ..., x_n^*) \in U$ such that

$$f_i(x_1^*, ..., x_{i-1}^*, y_i, x_{i+1}^*, ..., x_n^*) \leq f_i(x^*) \ \forall y_i \in U_i, \forall i.$$

A vector $x^* \in U$ satisfying this inequality is called a Nash-equilibrium point of the model.

It is not hard to show (see also [10], and [9] for the case $\beta_i \equiv \beta$ for all i) that the problem of finding a Nash-equilibrium point can be formulated in the form (2.1) where

$$C = U := U_1 \times, ..., \times U_n, \ \varphi(x) := \sum_{i=1}^{n} \beta_i x_i^2 + \sum_{i=1}^{n} h_i(x_i), F(x) := Bx - \alpha$$

with

$$B := \begin{pmatrix} 0 & \beta_1 & \beta_1 & ... & \beta_1 \\ \beta_2 & 0 & \beta_2 & ... & \beta_2 \\ ... & ... & ... & ... & ... \\ \beta_n & \beta_n & \beta_n & ... & 0 \end{pmatrix}$$

and $\alpha := (\alpha_1, ..., \alpha_n)^T$. Some practical problems that can be formulated in a problem of form (2.1) can be found, for example, in [6, 9, 11].

For each fixed $x \in C$ and $w \in F(x)$, we denote by $h(x, w)$ the unique solution of the strongly convex program

$$\min\{\frac{1}{2}\langle y - x, G(y - x)\rangle + \langle w, y - x\rangle + \varphi(y) \ |y \in C\}, \tag{2.2}$$

where G is a symmetric, positive definite matrix. It is well known (see e.g., [5, 9, 11]) that $h(x, w)$ is the solution of (2.2) if and only if $h(x, w)$ is the solution of the variational inequality

$$\langle w + G(h(x, w) - x) + z, y - h(x, w)\rangle \geq 0 \ \ \forall y \in C, \tag{2.3}$$

for some $z \in \partial\varphi(h(x, w))$.

Now for each $x \in C$, we define the multivalued mapping

$$H(x) := \{h(x, w)|w \in F(x)\}.$$

Clearly, H is a mapping from $I\!R^n$ to C and, since $C \subseteq \text{dom}H$, we have $C \subseteq \text{dom } H \subseteq \text{dom}F$.

The next lemma shows that a point x^* is a solution to (VIP) if and only if it is a fixed point of H.

Lemma 2.1 *x^* is a solution to (VIP) if and only if $x^* \in H(x^*)$.*

Proof. Let x^* solve (VIP). It means that there exists $w^* \in F(x^*)$ such that (x^*, w^*) satisfies inequality (2.1). Let $h(x^*, w^*)$ be the unique solution of Problem (2.2) corresponding to x^*, w^* and some positive definite matrix G. We replace x by $h(x^*, w^*)$ in (2.1) to obtain

$$\langle w^*, h(x^*, w^*) - x^*\rangle + \varphi(h(x^*, w^*)) - \varphi(x^*) \geq 0. \tag{2.4}$$

From (2.3) it follows that there exists z^* in $\partial\varphi(h(x^*,w^*))$ such that

$$\langle w^* + G(h(x^*,w^*) - x^*) + z^*, y - h(x^*,w^*)\rangle \geq 0 \ \ \forall y \in C. \tag{2.5}$$

Replacing y by $x^* \in C$ in (2.5) we have

$$\langle w^* + G(h(x^*,w^*) - x^*) + z^*, x^* - h(x^*,w^*)\rangle \geq 0. \tag{2.6}$$

From inequalities (2.4) and (2.6) we obtain

$$\langle G(h(x^*,w^*) - x^*), x^* - h(x^*,w^*)\rangle + \langle z^*, x^* - h(x^*,w^*)\rangle$$

$$+\varphi(h(x^*,w^*)) - \varphi(x^*) \geq 0, \tag{2.7}$$

for some $z \in \partial\varphi(h(x,w))$. Since φ is convex on C, by the definition of subdifferential of a convex function, we have

$$\langle z^*, x^* - h(x^*,w^*)\rangle \leq \varphi(x^*) - \varphi(h(x^*,w^*)) \ \ \forall z^* \in \partial\varphi(h(x^*,w^*)).$$

Hence

$$\langle z^*, x^* - h(x^*,w^*)\rangle - \varphi(x^*) + \varphi(h(x^*,w^*)) \leq 0 \ \ \forall z^* \in \partial\varphi(h(x^*,w^*)). \tag{2.8}$$

From inequalities (2.7) and (2.8), it follows that

$$\langle G(h(x^*,w^*) - x^*), x^* - h(x^*,w^*)\rangle \geq 0.$$

Since G is symmetric, positive definite, the latter inequality implies that $h(x^*,w^*) = x^*$.
Now suppose $x^* \in H(x^*)$. Then there is w^* in $F(x^*)$ such that $x^* = h(x^*,w^*)$. But for every $x \in C, w \in F(x)$, we always have

$$\langle w + G(h(x,w) - x) + z, y - h(x,w)\rangle \geq 0 \ \ \forall y \in C, \tag{2.9}$$

for some $z \in \partial\varphi(h(x,w))$. Replacing x, w, z by $x^* = h(x^*,w^*), w^*, z^*$, respectively, in inequality (2.9) we obtain

$$\langle w^* + z^*, y - x^*\rangle \geq 0 \ \ \forall y \in C, \tag{2.10}$$

for some $z^* \in \partial\varphi(x^*)$. Using the definition of subdifferential of a convex function, we can write

$$\varphi(y) - \varphi(x^*) \geq \langle z^*, y - x^*\rangle \ \ \forall y \in C. \tag{2.11}$$

From inequalities (2.10) and (2.11) we have

$$\langle w^*, y - x^*\rangle + \varphi(y) - \varphi(x^*) \geq 0 \ \ \forall y \in C,$$

which means that x^* is a solution of Problem (VIP). □

Now we recall some well known definitions (see [3, 5]) about multivalued mappings that we need in the sequel.

• Let A, B be two nonempty subsets in $I\!R^n$. Let $\rho(A, B)$ denote the Hausdorff distance of A and B that is defined as

$$\rho(A, B) := \max\{d(A, B), d(B, A)\},$$

where

$$d(A, B) := \sup_{a \in A} \inf_{b \in B} ||a - b||, \quad d(B, A) := \sup_{b \in B} \inf_{a \in A} ||a - b||.$$

Let $\emptyset \neq M \subseteq I\!R^n$ and $K : I\!R^n \to I\!R^n$ be a multivalued mapping such that $M \subseteq \text{dom}K$.

• K is said to be closed at x if $x^k \to x$, $y^k \in K(x^k)$, $y^k \to y$ as $k \to +\infty$, then $y \in F(x)$. We say that K is closed on M if it is closed at every point of M.

• K is said to be upper semicontinuous at x if for every open set G containing $K(x)$ there exists an open neighborhood U of x such that $K(U) \subset G$. We say that K is upper semicontinuous on M if it is upper semicontinuous at every point of M.

• K is said to be Lipschitz with a constant L (briefly L-Lipschitz) on M if

$$\rho(K(x), K(y)) \leq L||x - y|| \quad \forall x, y \in M.$$

K is called a contractive mapping if $L < 1$ and K is said to be nonexpansive if $L = 1$.

• We say that K has a L-Lipschitz selection on M if for every $x, y \in M$ there exist $w(x) \in K(x)$ and $w(y) \in K(y)$ such that

$$||w(x) - w(y)|| \leq L||x - y||.$$

If $0 < L < 1$ (resp. $L = 1$) we say that K has a contractive (resp. nonexpansive) selection on M. It is easy to check that a multivalued Lipschitz mapping with compact, convex values has a Lipschitz selection. This is why in the sequel, for short, we shall call a mapping having a Lipschitz selection a quasi-Lipschitz mapping. Likewise, a mapping having a contractive (resp. nonexpansive) selection is called quasicontractive (resp. quasinonexpansive).

• K is said to be monotone on M if

$$\langle w - w', x - x' \rangle \geq 0 \quad \forall x, x' \in M, \ \forall w \in K(x), \ \forall w' \in K(x').$$

• K is said to be strongly monotone with modulus $\beta > 0$ (briefly β-strongly monotone) on M if

$$\langle w - w', x - x' \rangle \geq \beta||x - x'||^2 \quad \forall x, x' \in M, \ \forall w \in K(x), \ \forall w' \in K(x').$$

• K is said to be cocoercive with modulus $\delta > 0$ (briefly δ-cocoercive) on M if

$$\langle w - w', x - x' \rangle \geq \delta\rho^2(K(x), K(x')) \ \ \forall x, x' \in M, \ \forall w \in K(x), \ \forall w' \in K(x'),$$

where ρ stands for the Hausdorff distance.

Note that in an important case when $G = \alpha I$ with $\alpha > 0$ and I being the identity matrix, problem (2.2) becomes

$$\min\{\frac{\alpha}{2}||y - x||^2 + \langle w, y - x \rangle + \varphi(y) \ |y \in C\}.$$

In the sequel we shall restrict our attention to this case. The following theorem shows that with a suitable value of regularization parameter α, the mapping H defined above is quasicontractive on C .

In what follows we need the following lemma:

Lemma 2.2 *Suppose that $C \subseteq I\!R^n$ is nonempty, closed, convex and $F : I\!R^n \to 2^{I\!R^n}$ is L-Lipschitz on C such that $F(x)$ is closed, convex for every $x \in C$. Then for every $x, x' \in C$ and $w \in F(x)$, there exists $w' \in F(x')$, in particular $w' = P_{F(x')}(w)$, such that $||w - w'|| \leq L||x - x'||$.*

Here, $P_{F(x')}(w)$ denotes the projection of the point w on the set $F(x')$.

Proof. Since $w \in F(x)$, by the definition of the projection and the Hausdorff distance, we have

$$\begin{aligned} ||w - w'|| = \inf_{v' \in F(x')} ||w - v'|| &\leq \sup_{v \in F(x)} \inf_{v' \in F(x')} ||v - v'|| \\ &\leq \rho\Big(F(x), F(x')\Big) \leq L||x - x'||. \end{aligned}$$

□

Theorem 2.1 *Suppose that F is β- strongly monotone and L- Lipschitz on C, and that $F(x)$ is closed, convex for every $x \in C$. Then the mapping H is quasicontractive on C with constant $\delta := \sqrt{1 - \frac{2\beta}{\alpha} + \frac{L^2}{\alpha^2}}$ whenever $\alpha > \frac{L^2}{2\beta}$. Namely,*

$$||h(x, w(x)) - h(x', w(x')) \leq \delta||x - x'|| \ \forall x, x' \in C \ \forall w(x) \in F(x)$$

where $w(x')$ is the Euclidean projection of $w(x)$ onto $F(x')$.

Proof. Problem (2.2) with $G = \alpha I$ can be equivalently rewritten as

$$\min_y\{\frac{1}{2}\langle\alpha||y - x||^2\rangle + \langle w, y - x \rangle + \varphi(y) + \delta_C(y)\},$$

where δ_C is the indicator function of C. Let $h(x, w)$ be the unique solution of this unconstrained problem. Then we have

$$0 \in \alpha(h(x, w) - x) + w + N_C(h(x, w)) + \partial\varphi(h(x, w)),$$

where $N_C(h(x, w))$ is the normal cone to C at the point $h(x, w)$. Thus there are $z_1 \in N_C(h(x, w))$ and $z_2 \in \partial\varphi(h(x, w))$ such that

$$\alpha(h(x, w) - x) + w + z_1 + z_2 = 0.$$

Therefore

$$h(x, w) = x - \frac{1}{\alpha}w - \frac{1}{\alpha}z_1 - \frac{1}{\alpha}z_2. \tag{2.12}$$

Similarly for $x' \in C, w' \in F(x')$, we have

$$h(x', w') = x' - \frac{1}{\alpha}w' - \frac{1}{\alpha}z_1' - \frac{1}{\alpha}z_2', \tag{2.13}$$

where $z_1' \in N_C(h(x', w'))$ and $z_2' \in \partial\varphi(h(x', w'))$.
Since N_C is monotone, we have

$$\langle z_1 - z_1', h(x, w) - h(x', w')\rangle \geq 0. \tag{2.14}$$

Substituting z_1 from (2.12) and z_1' from (2.13) into (2.14) we obtain

$$\langle x - x' - \frac{1}{\alpha}(w - w') - \frac{1}{\alpha}(z_2 - z_2') - (h(x, w) - h(x', w')), h(x, w) - h(x', w')\rangle \geq 0,$$

which implies

$$||h(x, w) - h(x', w')||^2 \leq \langle x - x' - \frac{1}{\alpha}(w - w') - \frac{1}{\alpha}(z_2 - z_2'), h(x, w) - h(x', w')\rangle$$

$$= \langle x - x' - \frac{1}{\alpha}(w - w'), h(x, w) - h(x', w')\rangle - \frac{1}{\alpha}\langle z_2 - z_2', h(x, w) - h(x', w')\rangle. \tag{2.15}$$

Since $\partial\varphi$ is monotone on C, we have

$$\langle h(x, w) - h(x', w'), z_2 - z_2'\rangle \geq 0 \ \forall z_2 \in \partial\varphi(h(x, w)), z_2' \in \partial\varphi(h(x', w')). \tag{2.16}$$

From (2.15), (2.16) it follows that

$$||h(x, w) - h(x', w')||^2 \leq \langle x - x' - \frac{1}{\alpha}(w - w'), h(x, w) - h(x', w')\rangle$$

$$\leq ||x - x' - \frac{1}{\alpha}(w - w')|| \ ||h(x, w) - h(x', w')||.$$

Thus

$$||h(x, w) - h(x', w')||^2 \leq ||x - x' - \frac{1}{\alpha}(w - w')||^2 \tag{2.17}$$

$$= ||x - x'||^2 - \frac{2}{\alpha}\langle x - x', w - w'\rangle + \frac{1}{\alpha^2}||w - w'||^2.$$

Since F is L-Lipschitz on C and $F(x')$ is closed, for every $w(x) \in F(x)$, by Lemma 2.2, there exists $w(x') \in F(x')$ such that

$$||w(x) - w(x')|| \leq L||x - x'||$$

which together with strong monotonicity of F implies

$$||x - x' - \frac{1}{\alpha}(h(x, w(x)) - h(x', w(x'))||^2 \leq (1 - \frac{2\beta}{\alpha} + \frac{L^2}{\alpha^2})||x - x'||^2. \quad (2.18)$$

Finally, from (2.17) and (2.18) we have

$$||h(x, w(x)) - h(x', w(x'))|| \leq \sqrt{1 - \frac{2\beta}{\alpha} + \frac{L^2}{\alpha^2}}||x - x'||. \quad (2.19)$$

Let $\delta := \sqrt{1 - \frac{2\beta}{\alpha} + \frac{L^2}{\alpha^2}}$, then

$$||h(x, w(x)) - h(x', w(x'))|| \leq \delta||x - x'|| \;\; \forall x, x' \in C.$$

Note that if $\alpha > \frac{L^2}{2\beta}$ then $\delta \in (0, 1)$. Thus the multivalued mapping H has a contractive selection on C with constant δ. □

Remark 2.1 *From the definition of H and Theorem 2.1 it follows that when F is single-valued, the mapping H is contractive on C.*

Note that if φ is η-strongly convex and subdifferentiable on C, then its subdifferential is η- strongly monotone on C (see e.g., [5]). This means that

$$\langle z - z', x - x'\rangle \geq \eta||x - x'||^2 \;\; \forall x, x' \in C, z \in \partial\varphi(x), x' \in \partial\varphi(x').$$

In the following theorem the strong monotonicity of F is replaced by the strong convexity of φ.

Theorem 2.2 *Suppose that F is monotone and L- Lipschitz on C, that $F(x)$ is closed, convex for every $x \in C$ and that φ is η-strongly convex and subdifferentiable on C. Then the mapping H is quasicontractive on C with constant*

$$\delta := \frac{\sqrt{L^2 + \alpha^2}}{\alpha + \eta},$$

whenever $\alpha > \frac{L^2 - \eta^2}{2\eta}$.

Proof. By the same way as in the proof of Theorem 2.1 we obtain

$$||h(x, w) - h(x', w')||^2 \leq \langle x - x' - \frac{1}{\alpha}(w - w') - \frac{1}{\alpha}(z_2 - z_2'), h(x, w) - h(x', w')\rangle$$

$$= \langle x - x' - \frac{1}{\alpha}(w - w'), h(x, w) - h(x', w')\rangle - \frac{1}{\alpha}\langle z_2 - z_2', h(x, w) - h(x', w')\rangle$$

from which it follows that

$$||h(x,w) - h(x',w')||^2 \le \langle x - x' - \frac{1}{\alpha}(w - w'), h(x,w) - h(x',w')\rangle$$

$$-\frac{1}{\alpha}\langle z_2 - z_2', h(x,w) - h(x',w')\rangle, \qquad (2.20)$$

for all $x, x' \in C, w \in F(x), w' \in F(x')$, for some $z_2 \in \partial\varphi(h(x,w)), z_2' \in \partial\varphi(h(x',w'))$.
Since $\partial\varphi$ is strongly monotone with modulus η on C, we have

$$\langle z_2 - z_2', h(x,w) - h(x',w')\rangle \ge \eta||h(x,w) - h(x',w')||^2$$

$$\forall z_2 \in \partial\varphi(h(x,w)), z_2' \in \partial\varphi(h(x',w'))$$

$$\Leftrightarrow -\frac{1}{\alpha}\langle z_2 - z_2', h(x,w) - h(x',w')\rangle \le -\frac{\eta}{\alpha}||h(x,w) - h(x',w')||^2. \qquad (2.21)$$

Combining (2.20) and (2.21) yields

$$(1 + \frac{\eta}{\alpha})^2||h(x,w) - h(x',w')||^2 \le ||x - x' - \frac{1}{\alpha}(w - w')||^2$$

$$= ||x - x'||^2 - \frac{2}{\alpha}\langle x - x', w - w'\rangle + \frac{1}{\alpha^2}||w - w'||^2. \qquad (2.22)$$

Since F is Lipschitz with constant L on C and $F(x)$ is closed, convex, it follows that for every $x, x' \in C$, $w(x) \in F(x)$, there exists $w(x') \in F(x')$ satisfying

$$||w(x) - w(x')|| \le L||x - x'||.$$

Since F is monotone, we have

$$\langle w(x) - w(x'), x - x'\rangle \ge 0,$$

which together with (2.22) implies

$$(1 + \frac{\eta}{\alpha})^2||h(x,w(x)) - h(x',w(x'))||^2 \le (1 + \frac{L^2}{\alpha^2})||x - x'||^2$$

$$\Leftrightarrow ||h(x,w(x)) - h(x',w(x'))|| \le \delta||x - x'|| \ \ \forall x, x' \in C,$$

where $\delta := \frac{\sqrt{L^2+\alpha^2}}{\alpha+\eta}$. It is easy to verify that $\delta \in (0,1)$ when $\alpha > \frac{L^2-\eta^2}{2\eta}$. □

In the next theorem we weaken strong monotonicity of F by cocoercivity.

Theorem 2.3 *Suppose that F is γ-cocoercive on C, and that $F(x)$ is closed, convex for every $x \in C$. Then the mapping H is quasinonexpansive on C.*

Proof. By the same way as in the proof of Theorem 2.1, for every $x, x' \in C$, we have

$$||h(x,w) - h(x',w')||^2 \leq ||x - x' - \frac{1}{\alpha}(w - w')||^2 \ \forall w \in F(x), \ \forall w' \in F(x'). \tag{2.23}$$

From the cocoercivity of F on C with modulus γ, it follows that

$$\gamma\rho^2(F(x), F(x')) \leq \langle x - x', w - w' \rangle \ \ \forall x, x' \in C, w \in F(x), w' \in F(x').$$

Hence, for every $x, x' \in C$ and $w \in F(x)$, $w' \in F(x')$ we have

$$\begin{aligned} ||x - x' - \frac{1}{\alpha}(w - w')||^2 &= ||x - x'||^2 - \frac{2}{\alpha}\langle x - x', w - w' \rangle + \frac{1}{\alpha^2}||w - w'||^2 \\ &\leq ||x - x'||^2 - \frac{2\gamma}{\alpha}\rho^2(F(x), F(x')) + \frac{1}{\alpha^2}||w - w'||^2. \end{aligned}$$

Let $w(x) \in F(x), w(x') \in F(x')$, such that $\rho(F(x), F(x')) = ||w(x) - w(x')||$. Substituting $w(x)$ and $w(x')$ into the last inequality we obtain

$$||x - x' - \frac{1}{\alpha}(w(x) - w(x'))||^2 \leq ||x - x'||^2 - (\frac{2\gamma}{\alpha} - \frac{1}{\alpha^2})||w(x) - w(x')||^2.$$

Since $\alpha \geq \frac{1}{2\gamma}$, we have

$$||x - x' - \frac{1}{\alpha}(w(x) - w(x')||^2 \leq ||x - x'||^2 \ \ \forall x, x' \in C. \tag{2.24}$$

From (2.23) and (2.24) it follows that

$$||h(x, w(x)) - h(x', w(x'))|| \leq ||x - x'|| \ \ \forall x, x' \in C.$$

□

Remark 2.2 *From the proof we can see that the theorem remains true if we weaken the cocoercivity of F by the following one*

$$\forall x, x' \in C, \forall w \in F(x), \exists \pi'(w) \in F(x') : \gamma\rho^2\big(F(x), F(x')\big) \leq \langle w - \pi'(w), x - x' \rangle$$

Below is given a simple example for a multivalued mapping which is both monotone and Lipschitz.

Example 2.1 Let $C = \{(x,0)|x \geq 0\} \subseteq I\!R^2$, and $F : C \to 2^{I\!R^2}$ be given as

$$F(x,0) = \{(x,y)|0 \leq y \leq x\}.$$

It is easy to see that F is monotone and Lipschitz on C with constant $L = \sqrt{2}$. The mapping $G := I + F$ with I identity on $I\!R^2$ is strongly monotone with modulus $\beta = 1$ and Lipschitz on C with constant $L = \sqrt{2} + 1$.

Indeed, by definition of F, it is clear that F is monotone on C. Using the definition of the Hausdorff distance we have

$$\rho(F(x,y), F(x',y')) = \sqrt{2}||(x,y) - (x',y')|| \ \ \forall (x,y), (x',y') \in C.$$

Thus F is Lipschitz on C with constant $L = \sqrt{2}$. △

3 Algorithms

The results in the preceding section lead to algorithms for solving multivalued mixed variational inequalities by the Banach contraction mapping principle or its modifications. By Theorem 2.1 and 2.2, when either F is strongly monotone or φ is strongly convex, one can choose a suitable regularization parameter α such that the solution mapping H is quasi-contractive. In this case, by the Banach contraction principle the unique fixed point of H, thereby the unique solution of Problem (2.1) can be approximated by iterative procedures

$$x^{k+1} \in H(x^k), k = 0, 1...$$

where x^0 can be any point in C.
According to the definition of H, computing x^{k+1} amounts to solving a strongly convex mathematical program. In what follows by ε-solution of (VIP) we mean a point $x \in C$ such that $||x - x^*|| \leq \varepsilon$ where x^* is an exact solution of (VIP).

The algorithm then can be described in detail as follows:
Algorithm 3.1. Choose a tolerance $\varepsilon \geq 0$.
Choose $\alpha > \frac{L^2}{2\beta}$, when F is β-strongly monotone (and choose $\alpha > \frac{L^2-\eta^2}{2\eta}$, when φ is η-strongly convex), where L is the Lipschitz constant of F.
Seek $x^0 \in C, w^0 \in F(x^0)$.
Iteration k $(k = 0, 1, 2...)$
Solve the strongly convex program

$$P(x^k): \quad \min\{\frac{1}{2}\alpha||x - x^k||^2 + \langle w^k, x - x^k\rangle + \varphi(x) | x \in C\},$$

to obtain its unique solution x^{k+1}. Find $w^{k+1} \in F(x^{k+1})$ such that $||w^{k+1} - w^k|| \leq L||x^{k+1} - x^k||$, for example $w^{k+1} := P_{F(x^{k+1})}(w^k)$ (the projection of w^k onto $F(x^{k+1})$) .
If $||x^{k+1} - x^k|| \leq \varepsilon\frac{(1-\delta)}{\delta^k}$, then terminate: x^k is an ε-solution to Problem (2.1).
Otherwise, if $||x^{k+1} - x^k|| > \varepsilon\frac{(1-\delta)}{\delta^k}$, then increase k by 1 and go to iteration k.

By Theorems 2.1 and 2.2 and the Banach contraction principle it is easy to prove the following estimation:

$$||x^{k+1} - x^*|| \leq \frac{\delta^{k+1}}{1-\delta}||x^0 - x^1|| \ \forall k,$$

where $0 < \delta < 1$ is the quasicontractive constant of h. According to Theorem 2.1 $\delta = \sqrt{1 - \frac{2\beta}{\alpha} + \frac{L^2}{\alpha^2}}$, when F is β-strongly monotone, and according to Theorem 2.2 $\delta = \frac{\sqrt{L^2+\alpha^2}}{\alpha+\eta}$ when φ is η-strongly convex.

Theorem 3.1 *Under the assumptions of Theorem 2.1 (or Theorem 2.2), the sequence* $\{x^k\}$ *generated by Algorithm 3.1 satisfies*

$$||x^k - x^*|| \leq \frac{\delta^{k+1}}{1-\delta}||x^0 - x^1|| \;\; \forall k, \qquad (3.1)$$

where x^* is the solution of (VIP). If, in addition F is closed on C, then the sequence $\{w^k\}$ converges to $w^* \in F(x^*)$ with the rate

$$||w^k - w^*|| \leq \frac{L\delta^k}{1-\delta}||x^0 - x^1|| \;\; \forall k.$$

Proof. First we suppose that the assumptions of Theorem 2.1 are satisfied. Let x^* be the solution of (2.1). By Lemma 2.1,

$$x^* \in H(x^*) := \{h(x^*, w)|w \in F(x^*)\}.$$

Let $w^* \in F(x^*)$ such that $x^* = h(x^*, w^*) \in H(x^*)$. By the choice of w^{k+1} in the algorithm

$$||w^{k+1} - w^k|| \leq L||x^{k+1} - x^k|| \;\; \forall k.$$

Then as shown in Theorem 2.1 we have

$$||h(x^{k+1}, w^{k+1}) - h(x^k, w^k)|| \leq \delta||x^{k+1} - x^k|| \;\; \forall k,$$

Since $h(x^{k+1}, w^{k+1}) = x^{k+2}$, we have

$$||x^{k+2} - x^{k+1}|| \leq \delta||x^{k+1} - x^k|| \;\; \forall k,$$

from which, by the Banach contraction mapping fixed point principle, it follows that

$$||x^k - x^*|| \leq \frac{\delta^{k+1}}{1-\delta}||x^0 - x^1|| \;\; \forall k.$$

Thus $x^k \to x^*$ as $k \to +\infty$. Moreover using again the contraction property we have

$$||x^{p+k} - x^k|| \leq \delta^k \frac{(1-\delta^p)}{1-\delta}||x^{k+1} - x^k|| \; \forall k, p.$$

Letting $p \to +\infty$ we obtain

$$||x^k - x^*|| \leq \frac{\delta^k}{1-\delta}||x^{k+1} - x^k|| \; \forall k.$$

Thus if $||x^{k+1} - x^k|| \leq \varepsilon\frac{(1-\delta)}{\delta^k}$, then it follows that $||x^k - x^*|| \leq \varepsilon$ which means that x^k is an ε-solution to (VIP).

On the other hand, since

$$||w^{k+1} - w^k|| \leq L||x^{k+1} - x^k||$$

we have

$$||w^{k+p} - w^k|| \le ||w^{k+1} - w^k|| + ||w^{k+2} - w^{k+1}|| + ... + ||w^{k+p} - w^{k+p-1}||$$
$$\le L(||x^{k+1} - x^k|| + ||x^{k+2} - x^{k+1}|| + ... + ||x^{k+p} - x^{k+p-1}||)$$
$$\le L(\delta^k + \delta^{k+1} + ... + \delta^{k+p-1})||x^1 - x^0||.$$

Thus

$$||w^{k+p} - w^k|| < L\delta^k \frac{\delta^p - 1}{\delta - 1}||x^1 - x^0||, \tag{3.2}$$

which means that $\{w^k\}$ is a Cauchy sequence. Hence the sequence $\{w^k\}$ converges to some $w^* \in C$. Since F is closed, $w^* \in F(x^*)$. From (3.2) and letting $p \to +\infty$ we have

$$||w^k - w^*|| \le \frac{L\delta^k}{1-\delta}||x^1 - x^0|| \ \ \forall j.$$

The proof can be done similarly under the assumptions of Theorem 2.3. □

Remark 3.1 *From* $\delta := \sqrt{1 - \frac{2\beta}{\alpha} + \frac{L^2}{\alpha^2}}$ *(resp.* $\delta = \frac{\sqrt{L^2+\alpha^2}}{\alpha+\eta}$*) we see that the contraction coefficient* δ *is a function of the regularization parameter* α*. An elementary computation shows that* δ *takes its minimum when* $\alpha = \frac{L^2}{\beta}$ *(resp.* $\alpha = \frac{L^2-\eta^2}{\eta}$*). Therefore for the convergence, in Algorithm 3.1 the best way is to choose* $\alpha = \frac{L^2}{\beta}$ *(resp.* $\alpha = \frac{L^2-\eta^2}{\eta}$*).*

Remark 3.2 *In Algorithm 3.1, at each iteration k, it requires finding* $w^{k+1} \in F(x^{k+1})$ *such that* $|w^{k+1} - w^k|| \le L||x^{k+1} - x^k||$*, which can be done when* $F(x)$ *has a special structure, for example, box, ball , simplex or a convex set given explicitly. One may ask whether the algorithm remains convergent if it takes any point from* $F(x^{k+1})$*. To our opinion, there is less hope for a positive answer to this question except cases when the set* $F(x^{k+1})$ *can be represented by any of its elements.*

Now we consider a special case that often occurs in practice.

Let $\mu = \sup\{\text{diam } F(x) | x \in C\}$, $\tau = \text{diam } C$. It is well known that if C is compact and F is upper semicontinuous on C, then μ and τ are finite.

Algorithm 3.2. Choose a tolerance $\varepsilon > 0$, $\alpha > \frac{L_0^2}{2\beta}$ when F is β-strongly monotone (and choose $\alpha > \frac{L_0^2-\eta^2}{2\eta}$ when φ is η-strongly convex), where $L_0 \ge \frac{L\tau+\mu}{\varepsilon(1-\delta)}$ and $\delta := \sqrt{1 - \frac{2\beta}{\alpha} + \frac{L_0^2}{\alpha^2}}$ when F is β-strongly monotone ($\delta = \frac{\sqrt{L_0^2+\alpha^2}}{\alpha+\eta}$ when φ is η-strongly convex).

Seek $x^0 \in C, w^0 \in F(x^0)$.

Iteration k $(k = 0, 1, 2...)$

Solve the strongly convex program

$$P(x^k): \ \ \min\{\frac{1}{2}\alpha||x - x^k||^2 + \langle w^k, x - x^k\rangle + \varphi(x) | x \in C\},$$

to obtain its unique solution x^{k+1}. Choose $w^{k+1} \in F(x^{k+1})$.
If $||x^{k+1} - x^k|| \leq \varepsilon\frac{(1-\delta)}{\delta^k}$, then terminate: x^k is an ε-solution to Problem (2.1).

Otherwise, if $||x^{k+1} - x^k|| > \varepsilon\frac{(1-\delta)}{\delta^k}$, then increase k by 1 and go to iteration k.

Theorem 3.2 *Suppose that C is compact and F is upper semicontinuous on C. Then under the assumptions of Theorem 2.1 or Theorem 2.2, the sequence $\{x^k\}$ generated by Algorithm 3.2 satisfies*

$$||x^k - x^*|| \leq \frac{\delta^{k+1}}{1-\delta}||x^0 - x^1|| \quad \forall k,$$

where x^* is the solution of (2.1). Moreover

$$||w^k - w^*|| \leq \frac{L_0\delta^k}{1-\delta}||x^0 - x^1|| \quad \forall k.$$

Proof. By the same argument as in the proof of Theorem 3.1 we see that if $||x^{k+1} - x^k|| \leq \varepsilon\frac{(1-\delta)}{\delta^k}$, then indeed, x^k is an ε-solution.

Now suppose $||x^{k+1} - x^k|| > \varepsilon\frac{(1-\delta)}{\delta^k}$. For every $w^{k+1} \in F(x^{k+1})$, since $L_0 \geq \frac{L\tau+\mu}{\varepsilon(1-\delta)}$, we have

$$||w^{k+1} - w^k|| \leq d(w^k, F(x^{k+1})) + \text{diam}F(x^{k+1}) \leq L||x^{k+1} - x^k|| + \mu$$

$$\leq L\tau + \mu \leq L_0\varepsilon(1-\delta) < L_0||x^{k+1} - x^k||.$$

where the last inequality follows from $L_0 \geq \frac{L\tau+\mu}{\varepsilon(1-\delta)}$ and $\delta^k \leq 1$ for all k. Since $||x^{k+1} - x^k|| > \varepsilon\frac{(1-\delta)}{\delta^k}$, we have

$$||w^{k+1} - w^k|| \leq L_0||x^{k+1} - x^k|| \; \forall k.$$

Using this inequality we can prove the theorem by the same way as in the proof of Theorem 2.1 (or Theorem 2.2 when φ is strongly convex). □

Now we return to the case when F is cocoercive. Note that in this case Problem (VIP) is not necessarily uniquely solvable. By Theorem 2.3, a solution of (VIP) can be obtained by computing a fixed point of mapping H. Since H has a nonexpansive selection, its fixed point may be computed using the following theorem.

Theorem 3.3 *Let $C \subseteq I\!R^n$ be a nonempty, closed, convex set and $S : C \to 2^C$. Suppose that $S(x)$ is compact and that S has a nonexpansive selection on C. For $0 < \lambda < 1$ define*

$$S_\lambda := (1-\lambda)I + \lambda S.$$

Then the sequences $\{x^k\}$, $\{y^k\}$ defined by $x^{k+1} \in S_\lambda(x^k)$, i.e.,

$$x^{k+1} := (1-\lambda)x^k + \lambda y^k,$$

with $y^k \in S(x^k)$ *satisfy*

$$||y^{k+1} - y^k|| \leq ||x^{k+1} - x^k|| \quad \forall k = 0, 1, 2, ...$$

$$||x^k - y^k|| \to 0 \text{ as } k \to +\infty,$$

Moreover any cluster point of the sequence $\{x^k\}$ *is a fixed point of* S.

To prove this theorem we need the following lemma:

Lemma 3.1 *Under the assumptions of Theorem 3.3, for all* $i, m = 0, 1, ...$, *we have*

$$||y^{i+m} - x^i|| \geq (1-\lambda)^{-m}[||y^{i+m} - x^{i+m}|| - ||y^i - x^i||] + (1+\lambda m)||y^i - x^i||. \tag{3.3}$$

Proof. We proceed by induction on m, assuming that (3.3) holds for a given m and for all i. Clearly, (3.3) is trivial if $m = 0$. Replacing i with $i+1$ in (3.3) yields

$$||y^{i+m+1} - x^{i+1}|| \geq (1-\lambda)^{-m}[||y^{i+m+1} - x^{i+m+1}|| - ||y^{i+1} - x^{i+1}||]$$
$$+(1+\lambda m)||y^{i+1} - x^{i+1}||. \tag{3.4}$$

Since $x^{k+1} := (1-\lambda)x^k + \lambda y^k$ with $y^k \in S(x^k)$ that

$$||y^{i+m+1} - x^{i+1}|| = ||y^{i+m+1} - [(1-\lambda)x^i + \lambda y^i]||$$
$$\leq \lambda ||y^{i+m+1} - y^i|| + (1-\lambda)||y^{i+m+1} - x^i||$$
$$\leq (1-\lambda)||y^{i+m+1} - x^i|| + \lambda \sum_{k=0}^{m} ||x^{i+k+1} - x^{i+k}||. \tag{3.5}$$

Combining (3.4) and (3.5) we obtain

$$||y^{i+m+1} - x^i|| \geq (1-\lambda)^{-(m+1)}[||y^{i+m+1} - x^{i+m+1}|| - ||y^{i+1} - x^{i+1}||]$$
$$+(1-\lambda)^{-1}(1+\lambda m)||y^{i+1} - x^{i+1}|| - \lambda(1-\lambda)^{-1} \sum_{k=0}^{n} ||x^{i+k+1} - x^{i+k}||.$$

Since $||x^{i+k+1} - x^{i+k}|| = \lambda ||y^{k+i} - x^{k+i}||$ and since the sequence $\{||y^m - x^m||\}$ is decreasing, from

$$\lambda ||y^m - x^m|| = ||x^{m+1} - x^m|| = ||(1-\lambda)x^m + \lambda y^m - [(1-\lambda)x^{m-1} - \lambda y^{m-1}]||$$
$$\leq (1-\lambda)||x^m - x^{m-1}|| + \lambda ||y^m - y^{m-1}|| \leq ||x^m - x^{m-1}|| = \lambda ||y^{m-1} - x^{m-1}||$$

and $1 + m\lambda \leq (1-\lambda)^{-m}$, we have

$$||y^{i+m+1} - x^i|| \geq (1-\lambda)^{-(m+1)}[||y^{i+m+1} - x^{i+m+1}|| - ||y^{i+1} - x^{i+1}||]$$

$$+ (1-\lambda)^{-1}(1+\lambda m)||y^{i+1} - x^{i+1}|| - \lambda^2(1-\lambda)^{-1}(m+1)||y^i - x^i||$$

$$= (1-\lambda)^{-(m+1)}[||y^{i+m+1} - x^{i+m+1}|| - ||y^i - x^i||]$$

$$+ [(1-\lambda)^{-1}(1+\lambda m) - (1-\lambda)^{-(m+1)}]||y^{i+1} - x^{i+1}||$$

$$+ [(1-\lambda)^{-(m+1)} - \lambda^2(1-\lambda)^{-1}(m+1)]||y^i - x^i||$$

$$\geq (1-\lambda)^{-(m+1)}[||y^{i+m+1} - x^{i+m+1}|| - ||y^i - x^i||]$$

$$+ [(1-\lambda)^{-1}(1+\lambda m) - (1-\lambda)^{-(m+1)}]||y^i - x^i||$$

$$+ [(1-\lambda)^{-(m+1)} - \lambda^2(1-\lambda)^{-1}(m+1)]||y^i - x^i||$$

$$= (1-\lambda)^{-(m+1)}[||y^{i+m+1} - x^{i+m+1}|| - ||y^i - x^i||] + [1+\lambda(m+1)]||y^i - x^i||.$$

Thus (3.5) holds for $m+1$. □

Proof of Theorem 3.3. Let $d := \sup\{\text{diam } S(x) | x \in C\}$, and suppose that $\lim_{m\to\infty} ||y^m - x^m|| = r > 0$. Select $m \geq \frac{d}{r\lambda}$ and ε is a sufficiently small positive number such that $\varepsilon(1-\lambda)^{-m} < r$. Since $\{||y^m - x^m||\}$ is decreasing, there exists an integer i such that

$$0 \leq ||y^i - x^i|| - ||y^{m+i} - x^{m+i}|| \leq \varepsilon.$$

Therefore, using (3.3) we arrive at the contradiction

$$d + r \leq (1+m\lambda)r \leq (1+m\lambda)||y^i - x^i||$$

$$\leq ||y^{m+i} - x^i|| + (1-\lambda)^{-m}[||y^i - x^i|| - ||y^{m+i} - x^{m+i}||]$$

$$\leq ||y^{m+i} - x^i|| + (1-\lambda)^{-m}\varepsilon < d + r.$$

Consequently $r = 0$, thus $\lim_{m\to\infty} ||x^m - y^m|| = 0$. Since S is a bounded-valued mapping on C and S is closed, we have that any cluster point of convergent sequences $\{x^m\}$ is a fixed point of S. □

Now applying Theorem 3.3 to H we can solve Problem (2.1) with F being cocoercive on C by finding a fixed point of H.

Algorithm 3.3. *Step 0.* Choose a tolerance $\varepsilon \geq 0$ and $\lambda \in (0,1)$, $\alpha \geq \frac{1}{2\gamma}$ and seek $x^0 \in C, w^0 \in F(x^0)$. Let $k = 0$.

Step 1. Solve the strongly convex program

$$P(x^k): \quad \min\{\frac{1}{2}\alpha||y - x^k||^2 + \langle w^k, y - x^k\rangle + \varphi(y)|y \in C\}$$

to obtain its unique solution y^k.
If $||y^k - x^k|| \leq \varepsilon$, then the algorithm terminates.
Otherwise go to Step 2.

Step 2. Take

$$x^{k+1} := (1 - \lambda)x^k + \lambda y^k.$$

Find $w^{k+1} := P_{F(x^{k+1})}(w^k)$.
Let $k \leftarrow k + 1$ and return to Step 1.

Theorem 3.4 *In addition to the assumptions of Theorem 2.3, suppose that C is compact, and F is upper semicontinous on C. Then, if Algorithm 3.3 does not terminate, the sequence $\{x^k\}$ is bounded and any cluster point is a solution of Problem (VIP). In addition, it holds $d(x^k, H(x^k)) \to 0$ as $k \to \infty$.*

Proof. In Algorithm 3.3, we have $w^{k+1} := P_{F(x^{k+1})}(w^k)$ with $w^k \in F(x^k)$. From Lemma 2.2 and the definition of $\rho(F(x^k), F(x^{k+1}))$ it follows that

$$||w^{k+1} - w^k|| \leq \rho(F(x^k), F(x^{k+1})).$$

From the cocoercivity of F on C with modulus γ, we have

$$\gamma\rho^2(F(x^k), F(x^{k+1})) \leq \langle x^k - x^{k+1}, w^k - w^{k+1}\rangle.$$

Thus

$$\begin{aligned}
&||x^k - x^{k+1} - \frac{1}{\alpha}(w^k - w^{k+1})||^2 \\
&= ||x^k - x^{k+1}||^2 - \frac{2}{\alpha}\langle x^k - x^{k+1}, w^k - w^{k+1}\rangle + \frac{1}{\alpha^2}||w^k - w^{k+1}||^2 \\
&\leq ||x^k - x^{k+1}||^2 - \frac{2\gamma}{\alpha}||w^k - w^{k+1}||^2 + \frac{1}{\alpha^2}||w^k - w^{k+1}||^2 \\
&= ||x^k - x^{k+1}||^2 - (\frac{2\gamma}{\alpha} - \frac{1}{\alpha^2})||w^k - w^{k+1}||^2.
\end{aligned}$$

Since $\alpha > \frac{1}{2\gamma}$, we have

$$||x^k - x^{k+1} - \frac{1}{\alpha}(w^k - w^{k+1})||^2 \leq ||x^k - x^{k+1}||^2$$

which together with quasinonexpansiveness of H implies

$$||y^{k+1} - y^k|| \leq ||x^{k+1} - x^k||,$$

where

$$y^k = h(x^k, w^k) \in H(x^k), y^{k+1} = h(x^{k+1}, w^{k+1}) \in H(x^{k+1}).$$

By Theorem 3.3, every cluster point of the sequence $\{x^k\}$ is the fixed point x^* of H which is also a solution to Problem (2.1).

Furthermore, since C is compact and F is upper semicontinous on C, it follows from $w^k \in F(x^k)$ that the sequence $\{w^k\}$ is bounded. Thus, without loss of generality, we may assume that the sequence $\{w^k\}$ converges to some w^*. Since F is closed at x^*, we have $w^* \in F(x^*)$ and $x^* \in C$.

To prove $d(x^k, H(x^k)) \to 0$ we observe that $y^k \in H(x^k)$, and therefore

$$d(x^k, H(x^k)) \leq ||x^k - y^k|| \ \forall k.$$

By Theorem 3.3, we have $d(x^k, H(x^k)) \to 0$ as $k \to +\infty$. □

Acknowledgment. A part of this work was done while the second author was on leave at the University of Namur, Belgium under the form of FUNDP Research Scholarship. The authors wish to thank V.H. Nguyen and J.J. Strodiot for their helpful comments and remarks on the papers.

References

1. P.N. Anh P.N;, L.D. Muu, V.H. Nguyen and J. J. Strodiot., *On contraction and nonexpansiveness properties of the marginal mapping in generalized variational inequalities involving co-coercive operators*, in: Generalized Convexity and Generalized Monotonicity and Applications. Edited by A. Eberhard, N. Hadjisavvas and D. T. Luc, Springer **Chapter 5** (2005) 89-111.
2. P. N. Anh, L. D. Muu, V. H. Nguyen and J. J. Strodiot, *Using the Banach contraction principle to implement the proximal point method for multivalued monotone variational inequalities*, J. of Optimization Theory and Applications, **124** (2005) 285-306.
3. J.P. Aubin and I. Ekeland, *Applied Nonlinear Analysis*, John Wiley and Sons, New York, 1984.
4. T.Q. Bao and P.Q. Khanh, *A projection-type algorithm for pseudomonotone nonlipschitzian multivalued variational inequalities*, in: Generalized Convexity and Generalized Monotonicity and Applications. Edited by A. Eberhard, N. Hadjisavvas and D. T. Luc, Springer **Chapter 6** (2005) 113-129.
5. F. H. Clarke, *Optimization and Nonsmooth Analysis*, Wiley, New Yowk, 1983.
6. F. Facchinei and J. S. Pang, *Finite-Dimensional Variational Inequalities and Complementarity Problem.* Springer 2002.
7. K. Goebel and W. A. Kirk, *Topics in Metric Fixed Point Theory*, Cambridge University Press, 1990.
8. B.S. He, *A class of projection and for monotone variational inequalities*, Applied Mathematical Optimization, **35** (1997) 69-76.
9. I. Konnov, *Combined Relaxation Methods for Variational Inequalities*, Springer, 2001.
10. L. D. Muu, V. H. Nguyen, N. V. Quy, *On Nash-Cournot oligopolistic market equilibrium models with concave costs function* (to appear).
11. A. Narguney, *Network Economics: a Variational Inequality Approach*, Kluwer Academic Publishers, 1993.

12. M. A. Noor, *Iterative schemes for quasimonotone mixed variational inequalities*, J. of Optimization Theory and Applications, **50** (2001) 29-44.
13. G. Salmon, V. H. Nguyen and J. J. Strodiot, *Coupling the auxiliary problem principle and epiconvergence theory to solve general variational inequalities*, J. of Optimization Theory and Applications, **104** (2000) 629-657.
14. G. Salmon, V. H. Nguyen and J. J. Strodiot, *A bundle method for solving variational inequalities*, SIAM J. Optimization, **14** (2004) 869-893.
15. R. U. Verma, *A class of projection-contraction methods applied to monotone variational inequalities,* Applied Mathematics Letters, **13** (2000) 55-62
16. R. U. Verma, *A class of quasivariational inequalities involving cocoercive mappings,* Advanced Nonlinear Variational Inequalities, **2** (1999) 1-12.
17. D. Zhu and P. Marcotte, *A new classes of generalized monotonicity,* J. of Optimization Theory and Applications, **87** (1995) 457-471.

Optimality conditions for a d.c. set-valued problem via the extremal principle

N. Gadhi

Département de Mathématiques et Informatique, Faculté des Sciences Dhar-Mahraz, B.P. 1796 Atlas, Fès, Morocco `ngadhi@Math.net`

Summary. Set-valued optimization is known as a useful mathematical model for investigating some real world problems with conflicting objectives, arising from economics, engineering and human decision-making. Using an *extremal principle* introduced by Mordukhovich, we establish optimality conditions for D.C. (difference of convex) set-valued optimization problems. An application to vector fractional mathematical programming is also given.

Key Words: Extremal principle, Fréchet normal cone, Cone-convex set-valued mappings, Optimality conditions, Support function, Set-valued optimization.

2001 Mathematics Subject Classification. Primary 90C29, 90C26, 90C70; Secondary 49K99.

1 Introduction

In very recent years, the analysis and applications of D.C. mappings (difference of convex mappings) have been of considerable interest [4, 7, 8, 10, 12, 11, 13, 14, 23, 28]. Generally, nonconvex mappings that arise in nonsmooth optimization are often of this type. Recently, extensive work on the analysis and optimization of D.C. mappings has been carried out [5, 6, 22]. However, much work remains to be done. For instance, if the data of the objective function of a standard problem are not exactly known, we are entitled to replace the objective by a set-valued objective representing fuzzy outcomes.

In this paper, we are concerned with the set-valued optimization problem

$$(P) : \begin{cases} \min F(x) - G(x) \\ \text{subject to } x \in C, \end{cases}$$

where C is a nonempty subset of $\mathbb{R}^n$, $F : \mathbb{R}^n \rightrightarrows \mathbb{R}^p$ and $G : \mathbb{R}^n \rightrightarrows \mathbb{R}^p$ are $\mathbb{R}^p_+$-convex set-valued mappings.

S. Dempe and V. Kalashnikov (eds.), *Optimization with Multivalued Mappings*, pp. 251-264
©2006 Springer Science + Business Media, LLC

It is well known that the convex separation principle plays a fundamental role in many aspects of nonlinear analysis and optimization. The whole convex analysis revolves around the use of separation theorems; see Rockafellar [26]. In fact, many crucial results with their proofs are based on separation arguments which are applied to convex sets (see [25]). There is another approach initiated by Mordukhovich [15, 16], which does not involve any convex approximations and convex separation arguments. It is based on a different principle to study the extremality of set systems, which is called the *extremal principle* [18, 20].

The essence of the *extremal principle* is to provide necessary conditions for set extremality in terms of suitable normal cones in dual spaces that are not generated by tangent approximations in primal spaces and that may be non-convex. In the early work summarized in the books of Mordukhovich [17, 20], versions of the *extremal principle* were employed to derive necessary optimality conditions in various problems of optimization and optimal control and to obtain calculus rules for the nonconvex normal cones and subdifferentials generated in this approach.

Our approach consists of using a support function together with the *extremal principle* for the study of necessary and sufficient optimality conditions in set-valued optimization. This technique extends the results obtained in scalar case by Hiriart-Urruty [7] and in vector case by Taa [27]. In [3], Dien gave a characterization of a set-valued mapping by its support function. The advantage of this characterization is that it allows the theory of generalized derivative of single valued mappings to be used for set-valued mappings.

The rest of the paper is organized in this way: Section 2 contains basic definitions and preliminary material from nonsmooth variational analysis. Section 3 and Section 4 address main results (optimality conditions). Section 5 discusses an application to vector fractional mathematical programming.

Throughout this work, we use standard notations. The symbol $\|\cdot\|$ is used for denoting the Euclidean norm on the respective space, $\mathbb{B}_{\mathbb{R}^n}$ stand for the closed unit balls in the space.

2 Preliminaries

Given an extended real-valued function $\varphi : \mathbb{R}^n \to \mathbb{R} := (\infty, \infty]$ finite at $\bar{x}$ and a nonempty set $\Omega \subset \mathbb{R}^n$, let us recall some definitions.
When φ is taken lower semicontinuous around $\bar{x}$, the *Fréchet subdifferential* of φ at $\bar{x}$ is

$$\widehat{\partial}\varphi(\bar{x}) := \left\{ x^* \in \mathbb{R}^n : \liminf_{x \to \bar{x}} \frac{\varphi(x) - \varphi(\bar{x}) - \langle x^*, x - \bar{x}\rangle}{\|x - \bar{x}\|} \geq 0 \right\}.$$

Let $\Omega \subset \mathbb{R}^n$ be locally closed around $\bar{x} \in \Omega$. The *Fréchet normal cone* $\widehat{N}(\bar{x}; \Omega)$ to Ω at $\bar{x}$ is defined by

$$\widehat{N}(\bar{x};\Omega) := \left\{ x^* \in \mathbb{R}^n : \limsup_{x \xrightarrow{\Omega} \bar{x}} \frac{\langle x^*, x - \bar{x} \rangle}{\|x - \bar{x}\|} \leq 0 \right\}, \tag{1}$$

where $x \xrightarrow{\Omega} \bar{x}$ stands for $x \to \bar{x}$ with $x \in \Omega$. For more details, see [15] and [21]. Note that $\widehat{\partial}\varphi(\bar{x})$ reduces to the classical Fréchet derivative of φ at $\bar{x}$ if φ is Fréchet differentiable at this point. One clearly has

$$\widehat{N}(\bar{x};\Omega) = \widehat{\partial}\delta(\bar{x};\Omega),$$

where $\delta(\cdot;\Omega)$ is the indicator function of Ω.

Definition 2.1 *Let Ω_1 and Ω_2 be nonempty closed subsets of $\mathbb{R}^n$. We say that $\{\Omega_1, \Omega_2\}$ is an extremal system in $\mathbb{R}^n$ if these sets have at least one (locally)* extremal point $\bar{x} \in \Omega_1 \cap \Omega_2$; *that is, there exists a neighborhood U of $\bar{x}$ such that for every $\varepsilon > 0$ there is a vector $a \in \varepsilon\mathbb{B}_{\mathbb{R}^n}$ with*

$$(\Omega_1 + a) \cap \Omega_2 \cap U = \emptyset;$$

in this case, for any $\varepsilon > 0$ there are $x_1 \in \Omega_1 \cap (\overline{x} + \varepsilon\mathbb{B}_{\mathbb{R}^n})$, $x_2 \in \Omega_2 \cap (\overline{x} + \varepsilon\mathbb{B}_{\mathbb{R}^n})$ and $x^* \in \mathbb{R}^n$ such that $\|x^*\| = 1$

$$x^* \in \left(\widehat{N}(x_1;\Omega_1) + \varepsilon\mathbb{B}^*_{\mathbb{R}^n}\right) \cap \left(-\widehat{N}(x_2;\Omega_2) + \varepsilon\mathbb{B}^*_{\mathbb{R}^n}\right).$$

See [19, 20] for extremal systems of finitely many sets and more discussions.

Remark 2.1 *A common point of sets is locally extremal if these sets can be locally pushed apart by a (linear)* small translation *in such a way that the resulting sets have empty intersection.*

Since convexity plays an important role in the following investigations, we give the definition of cone-convex set-valued mappings.

Definition 2.2 *[2] Let $A \subset \mathbb{R}^n$ be a convex set. The set-valued mapping F from A into $\mathbb{R}^p$ is said to be $\mathbb{R}^p_+$-convex on A, if $\forall x_1,\ x_2 \in A$, $\forall \lambda \in [0,1]$*

$$\lambda F(x_1) + (1-\lambda) F(x_2) \subset F(\lambda x_1 + (1-\lambda) x_2) + \mathbb{R}^p_+,$$

where $\mathbb{R}^p_+ = \{x \in \mathbb{R}^p : x \geq 0\}$. Moreover, the polar cone of $\mathbb{R}^p_+$ is defined as

$$\left(\mathbb{R}^p_+\right)^\circ = \{y^* \in \mathbb{R}^p : \langle y^*, y \rangle \leq 0 \text{ for all } y \in \mathbb{R}^p_+\} = \mathbb{R}^p_-.$$

Proposition 2.1 *Let $A \subset \mathbb{R}^n$ be a convex set. Considering $\gamma^* \in \left(-\mathbb{R}^p_+\right)^\circ$, if F is $\mathbb{R}^p_+$-convex on A then $\varphi(\gamma^*, x) := \inf_{y \in F(x)} \langle \gamma^*, y \rangle$ is a convex function on A.*

Proof. Let $x_1,\ x_2 \in A,\ y_1 \in F(x_1),\ y_2 \in F(x_2),\ \gamma^* \in (-\mathbb{R}^p_+)^\circ$ and $\lambda \in [0,1]$. From the $\mathbb{R}^p_+$-convexity assumption of F, there exist $y_\lambda \in F(\lambda x_1 + (1-\lambda)x_2)$ and $p \in \mathbb{R}^p_+$ such that

$$\lambda y_1 + (1-\lambda) y_2 = y_\lambda + p.$$

Consequently,

$$\lambda \langle \gamma^*, y_1 \rangle + (1-\lambda) \langle \gamma^*, y_2 \rangle = \langle \gamma^*, y_\lambda \rangle + \langle \gamma^*, p \rangle .$$

Since $p \in \mathbb{R}^p_+$ and $\gamma^* \in (-\mathbb{R}^p_+)^\circ$, one has $\langle \gamma^*, p \rangle \geq 0$. It follows that

$$\lambda \langle \gamma^*, y_1 \rangle + (1-\lambda) \langle \gamma^*, y_2 \rangle \geq \langle \gamma^*, y_\lambda \rangle, \text{ for all } y_1 \in F(x_1) \text{ and } y_2 \in F(x_2).$$

Then

$$\lambda \varphi(\gamma^*, x_1) + (1-\lambda) \varphi(\gamma^*, x_2) \geq \langle \gamma^*, y_\lambda \rangle \geq \varphi(\gamma^*, \lambda x_1 + (1-\lambda) x_2),$$

which means that $\varphi(\gamma^*, \cdot)$ is convex on A. □

3 Approximate/fuzzy necessary optimality conditions

This section is completely devoted to applications of the extremal principle to problems of set-valued optimization.
Let $F : \mathbb{R}^n \rightrightarrows \mathbb{R}^p$ and $G : \mathbb{R}^n \rightrightarrows \mathbb{R}^p$ be two closed-graph set-valued mappings between spaces $\mathbb{R}^n$ and $\mathbb{R}^p$. For all the sequel, the set-valued mappings F and G are assumed to be $\mathbb{R}^p_+$-convex on a locally closed set C. The domain and the graph of F are denoted respectively by

$$\operatorname{dom}(F) := \{x \in \mathbb{R}^n : F(x) \neq \emptyset\},$$

$$\operatorname{gph}(F) := \{(x,y) \in \mathbb{R}^n \times \mathbb{R}^p : y \in F(x)\}.$$

If V is a nonempty subset of $\mathbb{R}^n$, then

$$F(V) = \bigcup_{x \in V} F(x).$$

We remind the reader that a point $(\overline{x}, \overline{y} - \overline{z}) \in \operatorname{gph}(F-G)$ with $\overline{x} \in C$ is said to be a weak local Pareto minimal point with respect to $\mathbb{R}^p_+$ of the problem (P) if there exists a neighborhood V of $\overline{x}$ such that

$$(F-G)(V \cap C) \subset \overline{y} - \overline{z} + \mathbb{R}^p \setminus (-\operatorname{int} \mathbb{R}^p_+). \tag{2}$$

Here, int denotes the topological interior. When $V = \mathbb{R}^n$, we say that $(\overline{x}, \overline{y} - \overline{z})$ is a weak Pareto minimal point with respect to $\mathbb{R}^p_+$ of the problem (P).

Lemma 3.1 *Assume that* $(\overline{x},\overline{y},\overline{z})$ *is a weak local Pareto minimal point of* (P). *Then,* $(\overline{x},\overline{y},\overline{z})$ *is a weak local Pareto minimal point of*

$$(P^{\triangleright}):\left\{\begin{array}{c}\min F(x)-z\\ \textit{subject to }\ (x,z)\in(C\times\mathbb{R}^{p})\cap\operatorname{gph}(G).\end{array}\right.$$

Theorem 3.1 *Assume that* $(\overline{x},\overline{y}-\overline{z})$ *is a weak local Pareto minimal point of* (P). *Then, for any* $\varepsilon>0$ *there are* $x_1,x_2\in\overline{x}+\varepsilon\mathbb{B}_{\mathbb{R}^n}$, $y_2\in\overline{y}+\varepsilon\mathbb{B}_{\mathbb{R}^p}$, $z_2\in\overline{z}+\varepsilon\mathbb{B}_{\mathbb{R}^p}$, $x_1\in C$, $y_2\in F(x_2)$, $z_2\in G(x_2)$ *and* $\gamma^*\in\left(-\mathbb{R}^p_+\right)^{\circ}\setminus\{0\}$, *such that*

$$\left\{\begin{array}{l}\widehat{\partial}\psi(\gamma^*,\cdot)(x_2)\subset\partial\varphi(\gamma^*,\cdot)(x_2)+\widehat{N}(x_1,C),\\ \varphi(\gamma^*,x_2)=\langle\gamma^*,y_2\rangle\ \textit{and}\ \psi(\gamma^*,x_2)=\langle\gamma^*,z_2\rangle,\end{array}\right.\tag{3}$$

where $\psi(\gamma^*,\cdot)(x):=\sup_{y\in F(x)}\langle\gamma^*,y\rangle$. *Here,* $\partial f(x)$ *stands for the subdifferential (of convex analysis) of* f *at* $\overline{x}$.

Proof. Since $(\overline{x},\overline{y}-\overline{z})$ is a weak local Pareto minimal point of (P), there exists a neighborhood V of $\overline{x}$ such that for all $x\in V\cap C$

$$F(x)-G(x)\subset\overline{y}-\overline{z}+\mathbb{R}^p\setminus\left(-\operatorname{int}\mathbb{R}^p_+\right).$$

The proof of this theorem consists of several steps.

- It is easy to see that $(\overline{x},\overline{y},\overline{z})$ is a weak local Pareto minimal point of

$$(P^{\triangleright}):\left\{\begin{array}{c}\min F(x)-z\\ \text{subject to }\ (x,z)\in(C\times\mathbb{R}^{p})\cap\operatorname{gph}(G).\end{array}\right.$$

- Setting $H(x,z)=F(x)-z$ for all $(x,z)\in\operatorname{gph}(G)$ and $\overline{h}=\overline{y}-\overline{z}$, let us start by relating $\left(\overline{x},\overline{z},\overline{h}\right)$ to an extremal point in the sense of Definition 2.1. Put

$$\Omega_1:=(C\times\mathbb{R}^p)\times\left(\overline{h}-\mathbb{R}^p_+\right)\ \text{and}\ \Omega_2:=\operatorname{gph}(H).$$

  Then $\left(\overline{x},\overline{z},\overline{h}\right)$ is an extremal point of the system (Ω_1,Ω_2). Indeed, suppose that it is not the case, i.e., for any neighborhood U of $\left(\overline{x},\overline{z},\overline{h}\right)$ there is $(x,z,h)\in\operatorname{gph}(H)\cap U$ close to $\left(\overline{x},\overline{z},\overline{h}\right)$ with $(x,z,h)\in(C\times\mathbb{R}^p)\times\left(\overline{h}-\operatorname{int}\mathbb{R}^p_+\right)$. Hence $h\in\overline{h}-\operatorname{int}\mathbb{R}^p_+$, which contradicts the fact that $\left(\overline{x},\overline{z},\overline{h}\right)$ is a weak local Pareto minimal point of $(P^{\triangleright})$. Finally, $(\overline{x},\overline{y}-\overline{z})$ is not a weak local Pareto minimal point of (P).

Given $1/4\geq\varepsilon>0$, we employ the extremal principle from Definition 2.1. This gives $x_1,x_2\in\overline{x}+\varepsilon\mathbb{B}_{\mathbb{R}^n}$, $z_1,z_2\in\overline{z}+\varepsilon\mathbb{B}_{\mathbb{R}^p}$, $h_1,h_2\in\overline{h}+\varepsilon\mathbb{B}_{\mathbb{R}^p}$, $x_1\in C$, $z_1\in\mathbb{R}^p$, $h_1\in\overline{h}-\mathbb{R}^p_+$, $h_2\in F(x_2)-z_2$, $z_2\in G(x_2)$ and $(x^*,z^*,h^*)\in\mathbb{R}^n\times\mathbb{R}^p\times\mathbb{R}^p$ such that

$$\|(x^*,z^*,h^*)\|=1,(x^*,z^*,h^*)\in-\widehat{N}((x_2,z_2,h_2);\Omega_2)+\varepsilon\mathbb{B}_{\mathbb{R}^n\times\mathbb{R}^p\times\mathbb{R}^p},\tag{4}$$

and

$$(x^*, z^*, h^*) \in \widehat{N}((x_1, z_1, h_1); \Omega_1) + \varepsilon \mathbb{B}_{\mathbb{R}^n \times \mathbb{R}^p \times \mathbb{R}^p}. \tag{5}$$

There exist $(a_2^*, b_2^*, c_2^*) \in \mathbb{B}_{\mathbb{R}^{n+2p}}$ and $(u^*, v^*, w^*) \in \widehat{N}((x_2, z_2, h_2); \Omega_2)$ satisfying

$$(u^*, v^*, w^*) = -(x^*, z^*, h^*) + \varepsilon (a_2^*, b_2^*, c_2^*).$$

Using (5), there exist

$$(a_1^*, b_1^*, c_1^*) \in \mathbb{B}_{\mathbb{R}^{n+2p}}$$

and

$$(\alpha_1^*, \beta_1^*, \gamma_1^*) \in \widehat{N}((x_1, z_1, h_1); \Omega_1)$$

such that

$$(u^*, v^*, w^*) = -(\alpha_1^*, \beta_1^*, \gamma_1^*) + \varepsilon (a_2^*, b_2^*, c_2^*) - \varepsilon (a_1^*, b_1^*, c_1^*). \tag{6}$$

- By the definition of Fréchet normals (1), (4) implies that

  $$0 \geq \langle u^*, x - x_2 \rangle + \langle v^*, z - z_2 \rangle + \langle w^*, h - h_2 \rangle - \varepsilon \|(x - x_2, z - z_2, h - h_2)\|$$

  for $(x, z, h) \in \operatorname{gph}(H)$ sufficiently close to (x_2, z_2, h_2) and $(x, z) \in \operatorname{gph}(G)$. Consequently, there exists $y_2 \in F(x_2)$ such that $h_2 = y_2 - z_2$ and

  $$\begin{aligned} 0 \geq\ & \langle -\alpha_1^*, x - x_2 \rangle + \langle -\beta_1^*, z - z_2 \rangle + \langle -\gamma_1^*, y - z - y_2 + z_2 \rangle \\ & + \varepsilon \langle a_2^* - a_1^*, x - x_2 \rangle + \varepsilon \langle b_2^* - b_1^*, z - z_2 \rangle + \varepsilon \langle c_2^* - c_1^*, y - z - y_2 + z_2 \rangle \\ & - \varepsilon \|(x - x_2, z - z_2, y - z - y_2 + z_2)\| \end{aligned}$$

  for $(x, z) \in \operatorname{gph}(G)$ sufficiently close to (x_2, z_2) and $(x, y) \in \operatorname{gph}(F)$ sufficiently close to (x_2, y_2).
- By the definition of Fréchet normals (1), (5) implies that

  $$0 \geq \langle \alpha_1^*, x - x_1 \rangle + \langle \beta_1^*, z - z_1 \rangle + \langle \gamma_1^*, h - h_1 \rangle - \varepsilon \|(x - x_1, z - z_1, h - h_1)\| \tag{7}$$

  for (x, z, h) sufficiently close to (x_1, z_1, h_1) such that $x \in C$ and $h \in \overline{h} - \mathbb{R}_+^p$. Taking $x = x_1$ and $h = h_1$, one gets

  $$0 \geq \langle \beta_1^*, z - z_1 \rangle - \varepsilon \|z - z_1\| \text{ for all } z \in \mathbb{R}^p.$$

  Thus, $\beta_1^* = 0$.

Then,

$$\begin{aligned} & \langle \alpha_1^*, x - x_2 \rangle + \langle \gamma_1^*, y - z - y_2 + z_2 \rangle \geq \\ & \varepsilon \langle (a_2^* - a_1^*, b_2^* - b_1^*, c_2^* - c_1^*), (x - x_2, z - z_2, y - z - y_2 + z_2) \rangle - \\ & \qquad - \varepsilon \|(x - x_2, z - z_2, y - z - y_2 + z_2)\|, \end{aligned}$$

for $(x, z) \in \operatorname{gph}(G)$ sufficiently close to (x_2, z_2) and $(x, y) \in \operatorname{gph}(F)$ sufficiently close to (x_2, y_2). Consequently,

$$\langle \alpha_1^*, x - x_2 \rangle + \langle \gamma_1^*, y - z - y_2 + z_2 \rangle \geq -3\varepsilon \| (x - x_2, z - z_2, y - z - y_2 + z_2) \|,$$

for all $(x, z) \in \text{gph}\,(G)$ sufficiently close to (x_2, z_2) and $(x, y) \in \text{gph}\,(F)$ sufficiently close to (x_2, y_2).
Then,

$$\begin{aligned} &\langle \alpha_1^*, x - x_2 \rangle + \langle \gamma_1^*, y \rangle + \langle \gamma_1^*, z_2 \rangle \\ \geq &\langle \gamma_1^*, z \rangle + \langle \gamma_1^*, y_2 \rangle - 3\varepsilon \| (x - x_2, z - z_2, y - z - y_2 + z_2) \|, \end{aligned}$$

for all $(x, y) \in \text{gph}\,(F)$ and $(x, z) \in \text{gph}\,(G)$ sufficiently close to (x_2, y_2) and (x_2, z_2) respectively.
As a special case, for $x = x_2$, $y = u$ and $z = w$, one has

$$\langle \gamma_1^*, u \rangle - \langle \gamma_1^*, w \rangle \geq \langle \gamma_1^*, y_2 \rangle - \langle \gamma_1^*, z_2 \rangle - 3\varepsilon \| (w - z_2, u - w - y_2 + z_2) \|, \quad (8)$$

for all $(x, u) \in \text{gph}\,(F)$ and $(x, w) \in \text{gph}\,(G)$ sufficiently close to (x_2, y_2) and (x_2, z_2) respectively.
Setting $\overline{F}\,(x) := F\,(x)$ and $\overline{G}\,(x) := G\,(x)$ over $\overline{x} + \varepsilon \mathbb{B}_{\mathbb{R}^n}$,

$$\begin{aligned} &\langle \alpha_1^*, x - x_2 \rangle + \langle \gamma_1^*, y \rangle + \langle \gamma_1^*, z_2 \rangle \\ \geq &\langle \gamma_1^*, z \rangle + \langle \gamma_1^*, y_2 \rangle - 3\varepsilon \| (x - x_2, z - z_2, y - z - y_2 + z_2) \|, \end{aligned}$$

for all $(x, y) \in \text{gph}\,(F)$ and $(x, z) \in \text{gph}\,(G)$ sufficiently close to (x_2, y_2) and (x_2, z_2) respectively. Then,

$$\begin{aligned} &\langle \alpha_1^*, x - x_2 \rangle + \inf_{y \in \overline{F}(x)} \langle \gamma_1^*, y \rangle + \sup_{z_2 \in \overline{G}(x_2)} \langle \gamma_1^*, z_2 \rangle \\ \geq &\sup_{z \in \overline{G}(x)} \langle \gamma_1^*, z \rangle + \inf_{y_2 \in \overline{F}(x_2)} \langle \gamma_1^*, y_2 \rangle - 3\varepsilon \| (x - x_2, z - z_2, y - z - y_2 + z_2) \|. \end{aligned}$$

Remark that $\| (x - x_2, z - z_2, y - z - y_2 + z_2) \| \leq \| x - x_2 \| \| z - z_2 \| \| y - z - y_2 + z_2 \| \leq \| x - x_2 \|$, since $(x, y) \in \text{gph}\,(F)$ and $(x, z) \in \text{gph}\,(G)$ sufficiently close to (x_2, y_2) and (x_2, z_2).
Now, we have

$$\langle \alpha_1^*, x - x_2 \rangle + \overline{\varphi}\,(\gamma_1^*, x) + \overline{\psi}\,(\gamma_1^*, x_2) \geq \overline{\psi}\,(\gamma_1^*, x) + \overline{\varphi}\,(\gamma_1^*, x_2) - 3\varepsilon \| x - x_2 \|. \quad (9)$$

In addition, by (8),

$$\begin{cases} \overline{\varphi}\,(\gamma_1^*, x_2) \geq \langle \gamma_1^*, y_2 \rangle, \\ \overline{\psi}\,(\gamma_1^*, x_2) \leq \langle \gamma_1^*, z_2 \rangle. \end{cases}$$

Here, $\overline{\varphi}\,(\gamma_1^*, x) := \inf_{y \in \overline{F}(x)} \langle \gamma_1^*, y \rangle$ and $\overline{\psi}\,(\gamma_1^*, x) := \sup_{z \in \overline{G}(x)} \langle \gamma_1^*, z \rangle$ for all x sufficiently close to x_2. Then,

$$\begin{cases} \langle \alpha_1^*, x - x_2 \rangle + \overline{\varphi}\,(\gamma_1^*, x) + \overline{\psi}\,(\gamma_1^*, x_2) \geq \overline{\psi}\,(\gamma_1^*, x) + \overline{\varphi}\,(\gamma_1^*, x_2) - 3\varepsilon \| x - x_2 \|, \\ \overline{\varphi}\,(\gamma_1^*, x_2) = \langle \gamma_1^*, y_2 \rangle, \\ \overline{\psi}\,(\gamma_1^*, x_2) = \langle \gamma_1^*, z_2 \rangle. \end{cases}$$

- $\|\gamma_1^*\| > 0$. By contrast, suppose that $\|\gamma_1^*\| = 0$.
  On the one hand, since $\|(\alpha_1^*, \beta_1^*, \gamma_1^*) + \varepsilon (a_1^*, b_1^*, c_1^*)\| = 1$ and $\beta_1^* = 0,\ \gamma_1^* = 0$ one has $1 \leq \|\alpha_1^*\| + \varepsilon \|(a_1^*, b_1^*, c_1^*)\|$. Due to $(a_1^*, b_1^*, c_1^*) \in \mathbb{B}_{\mathbb{R}^{n+2p}}$, one gets

$$\|\alpha_1^*\| \geq 1 - \varepsilon.$$

  On the other hand, from (7),

$$0 \geq \langle \alpha_1^*, x - x_1 \rangle - \varepsilon \|x - x_1\|$$

  for $x \neq x_1 \in C$ sufficiently close to x_1. Thus,

$$\varepsilon \geq \langle \alpha_1^*, \frac{x - x_1}{\|x - x_1\|} \rangle$$

  and then $\|\alpha_1^*\| \leq \varepsilon$; which contradicts the fact that $\varepsilon \leq 1/4$.
- $\gamma_1^* \in \left(-\mathbb{R}_+^p\right)^\circ$. Indeed, since $\mathbb{R}_+^p$ is convex and $\gamma_1^* \in \widehat{N}\left(h_1, \overline{h} - \mathbb{R}_+^p\right)$, one has

$$\gamma_1^* \in \widehat{N}\left(h_1, \overline{h} - \mathbb{R}_+^p\right).$$

  It follows that, (due to the convexity of $\overline{h} - \mathbb{R}_+^p$)

$$\langle \gamma_1^*, h - h_1 \rangle \leq 0 \text{ for all } h \in \overline{h} - \mathbb{R}_+^p; \tag{10}$$

  thus,

$$0 \leq \left\langle \gamma_1^*, h_1 - \overline{h} \right\rangle. \tag{11}$$

  Since $h_1 - \overline{h} \in -\mathbb{R}_+^p$, one has $2\left(h_1 - \overline{h}\right) \in -\mathbb{R}_+^p$; and from (10), we deduce

$$0 \geq \left\langle \gamma_1^*, h_1 - \overline{h} \right\rangle. \tag{12}$$

  Combining (11) and (12),

$$0 = \left\langle \gamma_1^*, h_1 - \overline{h} \right\rangle. \tag{13}$$

  Let $r \in -\mathbb{R}_+^p$. We have $\overline{h} + r \in \overline{h} - \mathbb{R}_+^p$ and then,

$$\left\langle \gamma_1^*, \overline{h} + r - h_1 \right\rangle \leq 0.$$

  Consequently,

$$\langle \gamma_1^*, r \rangle \leq 0.$$

  Because r is arbitrarily chosen, we have $\gamma_1^* \in \left(-\mathbb{R}_+^p\right)^\circ \setminus \{0\}$.

Let $T^* \in \widehat{\partial} \psi \left(\gamma_1^*, \cdot\right)(x_2)$ and let $\nu > 0$. Then, by definition,

$$\psi\left(\gamma_1^*, x\right) - \psi\left(\gamma_1^*, x_2\right) + \nu \|x - x_2\| \geq \langle T^*, x - x_2 \rangle,$$

for x sufficiently close to x_2. Consequently,

$$\overline{\psi}(\gamma_1^*, x) - \overline{\psi}(\gamma_1^*, x_2) + \nu \|x - x_2\| \geq \langle T^*, x - x_2 \rangle,$$

for x sufficiently close to x_2. From (9), we have

$$\overline{\varphi}(\gamma_1^*, x) - \overline{\varphi}(\gamma_1^*, x_2) + \langle \alpha_1^*, x - x_2 \rangle \geq \langle T^*, x - x_2 \rangle - (\nu + \varepsilon) \|x - x_2\|,$$

for x sufficiently close to x_2. Then,

$$\varphi(\gamma_1^*, x) - \varphi(\gamma_1^*, x_2) + \langle \alpha_1^*, x - x_2 \rangle \geq \langle T^*, x - x_2 \rangle - (\nu + \varepsilon) \|x - x_2\|,$$

for x sufficiently close to x_2. Thus,

$$T^* - \alpha_1^* \in \widehat{\partial}\varphi(\gamma_1^*, x_2). \tag{14}$$

Finally,

$$\widehat{\partial}\psi(\gamma_1^*, \cdot)(x_2) \subset \widehat{\partial}\varphi(\gamma_1^*, \cdot)(x_2) + \widehat{N}(x_1, C).$$

Observing that $\varphi(\gamma_1^*, \cdot)$ is a convex function (due to Proposition 2.1), we get

$$\widehat{\partial}\psi(\gamma_1^*, \cdot)(x_2) \subset \partial\varphi(\gamma_1^*, \cdot)(x_2) + \widehat{N}(x_1, C)$$

and the proof is finished. □

With the following example, we illustrate the usefulness of the necessary conditions in Theorem 3.1.

Example 3.1 Let f and $g : \mathbb{R}^n \to \mathbb{R}^+$ be given functionals. Then, we consider the set valued mappings $F : \mathbb{R}^n \rightrightarrows \mathbb{R}$ and $G : \mathbb{R}^n \rightrightarrows \mathbb{R}$ with

$$F(x) := \{y \in \mathbb{R} : f(x) \leq y\} \text{ and } G(x) := \{z \in \mathbb{R} : g(x) \leq z\}.$$

Under these assumptions, we investigate the optimization problem

$$(P^\diamond) : \begin{cases} \min F(x) - G(x) \\ \text{subject to } x \in C, \end{cases}$$

where C is a nonempty closed subset of $\mathbb{R}^n$.
This is a special of the general type (P). In this example, the values of the objective may vary between the values of two known functions.
Next, assume that $(\overline{x}, f(\overline{x}) - g(\overline{x}))$ is a weak local Pareto minimal point of $(P^\diamond)$, and that f and g are continuous at $\overline{x}$ and convex. Consequently, F and G are $\mathbb{R}^+$-convex with closed graphs and

$$\varphi(\gamma^*, x) = \gamma^* f(x) \text{ and } \psi(\gamma^*, x) = \gamma^* g(x).$$

Then,

$$\partial g(\overline{x}) \subset \partial f(\overline{x}) + \widehat{N}(\overline{x}, C).$$

△

Consider the following set-valued optimization problem

$$(P_1): \quad \begin{cases} \mathbb{R}^p_+ - \text{Minimize } f(x) - g(x) \\ \qquad \text{subject to } x \in C, \end{cases}$$

where $f := (f_1, \ldots, f_p)$ and $g := (g_1, \ldots, g_p)$ and the functions f_i and g_i are convex and lower semicontinuous for $i = 1, \ldots, p$.

Corollary 3.1 *[27]Let $\overline{x}$ be a weak local Pareto minimal point of (P_1), where C is a locally closed set. Then for any $\varepsilon > 0$ there are $x_1, x_2 \in \overline{x} + \varepsilon \mathbb{B}_{\mathbb{R}^n}$, $f(x_2) \in f(\overline{x}) + \varepsilon \mathbb{B}_{\mathbb{R}^p}$, $g(x_2) \in g(\overline{x}) + \varepsilon \mathbb{B}_{\mathbb{R}^p}$, $x_1 \in C$ and $(\gamma_1^*, \ldots, \gamma_p^*) \in (-\mathbb{R}^p_+)^\circ \setminus \{0\}$ such that*

$$\partial \left(\sum_{i=1}^{p} \gamma_i^* g_i \right)(x_2) \subset \widehat{N}(x_1, C) + \partial \left(\sum_{i=1}^{p} \gamma_i^* f_i \right)(x_2).$$

Remark 3.1 *When $p = 1$, we get the well known necessary optimality conditions established in scalar case by Hiriart-Urruty [7]. Sufficient optimality conditions are also obtained (see Corollary 4.1).*

4 Sufficient optimality conditions

Let $\varepsilon > 0$ and let g be a function from $\mathbb{R}^n$ into $\mathbb{R}$. Recall that the ε-subdifferential of g is defined by

$$\partial_\varepsilon g(\overline{x}) = \{x^* \in \mathbb{R}^n : g(x) - g(\bar{x}) \geq \langle x^*, x - \overline{x} \rangle - \varepsilon \text{ for all } x \in \mathbb{R}^n\}.$$

The following lemma will be needed to prove Theorem 4.1. For the rest of this section, we assume that F and G are $\mathbb{R}^p_+$-convex on a nonempty closed convex subset C of $\mathbb{R}^n$.

Lemma 4.1 *Let $\varepsilon > 0$ and $\gamma^* \in (-\mathbb{R}^p_+)^\circ \setminus \{0\}$ such that $\psi(\gamma^*, \overline{x}) = \langle \gamma^*, \overline{z} \rangle$. Then,*

$$\partial_\varepsilon \psi(\gamma^*, \cdot)(\overline{x})$$
$$= \{t^* \in \mathbb{R}^p : \langle y^*, z \rangle - \langle y^*, \overline{z} \rangle \geq \langle t^*, x - \bar{x} \rangle - \varepsilon \textit{ for all } x \in \mathbb{R}^n \textit{ and } z \in G(x)\}.$$

Proof. The proof is evident. □

The proof of the following theorem uses some ideas of [4].

Theorem 4.1 *Let $\bar{x} \in C$. Assume that there exists $\gamma^* \in (-\mathbb{R}^p_+)^\circ \setminus \{0\}$ such that*

$$\psi(\gamma^*, \overline{x}) = \langle \gamma^*, \overline{z} \rangle, \quad \varphi(\gamma^*, \overline{x}) = \langle \gamma^*, \overline{y} \rangle,$$

and

$$\partial_\varepsilon \psi(\gamma^*, \cdot)(\overline{x}) \subset \partial_\varepsilon (\varphi(\gamma^*, \cdot) + \delta_C)(\overline{x}), \textit{ for all } \varepsilon \in \mathbb{R}^*_+. \tag{15}$$

Then $\bar{x}$ is a weak Pareto minimal point of (P).

Proof. Suppose that $\overline{x}$ is not a weak Pareto minimal solution of (P); there exist $x_0,\ y_0 \in F(x_0)$ and $z_0 \in G(x_0)$ such that

$$x_0 \in C,\ y_0 - z_0 - \overline{y} + \overline{z} \in -\operatorname{int}\mathbb{R}^p_+.$$

By the fact that $\gamma^* \in \left(-\mathbb{R}^p_+\right)^\circ \setminus \{0\}$, it follows that

$$\langle \gamma^*, y_0 - z_0 - \overline{y} + \overline{z} \rangle < 0. \tag{16}$$

Now, let $\varepsilon > 0$ and $t^* \in \mathbb{R}^p$. Consider the following inequalities.

$$\langle \gamma^*, z \rangle - \langle \gamma^*, \overline{z} \rangle \geq \langle t^*, x - \bar{x} \rangle - \varepsilon \text{ for all } x \in \mathbb{R}^n \text{ and } z \in G(x). \tag{17}$$

$$\langle \gamma^*, y \rangle + \delta_C(x) - \langle \gamma^*, \overline{y} \rangle - \delta_C(\overline{x}) \geq \langle t^*, x - \bar{x} \rangle - \varepsilon \text{ for all } x \in \mathbb{R}^n \text{ and } y \in F(x). \tag{18}$$

Using Lemma 4.1, (15) is equivalent to

$$(17) \Longrightarrow (18), \text{ for all } \varepsilon > 0.$$

Then,

$$\langle \gamma^*, y \rangle + \delta_C(x) - \langle \gamma^*, \overline{y} \rangle - \delta_C(\overline{x}) \geq \langle \gamma^*, z \rangle - \langle \gamma^*, \overline{z} \rangle$$

for all $y \in F(x)$ and $z \in G(x)$. Particularly, for $x = x_0,\ y = y_0$ and $z = z_0$, we have

$$\langle \gamma^*, y_0 \rangle - \langle \gamma^*, \overline{y} \rangle \geq \langle \gamma^*, z_0 \rangle - \langle \gamma^*, \overline{z} \rangle;$$

which is contradiction with (16). □

Considering the optimization problem (P_1) and applying Theorem 4.1, we get one of the results of [6].

Corollary 4.1 *[6] Let $\bar{x} \in C$. Assume that there exists $\gamma^* \in \left(-\mathbb{R}^p_+\right)^\circ \setminus \{0\}$ such that*

$$\partial_\varepsilon(\gamma^* \circ g)(\bar{x}) \subset \partial_\varepsilon(\gamma^* \circ f + \delta_C)(\overline{x}), \textit{ for all } \varepsilon \in \mathbb{R}^*_+.$$

Then $\bar{x}$ is a weak Pareto minimal point of (P_1).

5 Application

In this section, we give an application to vector fractional mathematical programming. Let $f_1, \ldots, f_p, g_1, \ldots, g_p : \mathbb{R}^n \to \mathbb{R}$ be convex and lower semi-continuous functions such that

$$f_i(x) \geq 0 \text{ and } g_i(x) > 0 \text{ for all } i = 1, \ldots, p.$$

We denote by φ the mapping defined as follows

$$\varphi(x) := \frac{f(x)}{g(x)} = \left(\frac{f_1(x)}{g_1(x)}, \dots, \frac{f_p(x)}{g_p(x)}\right).$$

We consider the set valued mapping $H : \mathbb{R}^n \rightrightarrows \mathbb{R}^p$ with

$$H(x) := \left\{y \in \mathbb{R}^p : y \in \varphi(x) + \mathbb{R}^p_+\right\}.$$

Under these assumptions, we investigate the optimization problem

$$(P^*) : \begin{cases} \mathbb{R}^p_+ - \text{Minimize } H(x) \\ \text{subject to} : x \in C \end{cases}$$

Where C is a nonempty locally closed subset of $\mathbb{R}^n$.
We will need the following lemma.

Lemma 5.1 *Let $\bar{x}$ be a feasible point of problem (P^*). $\bar{x}$ is a local weak Pareto minimal point of (P^*) if and only if $\bar{x}$ is a local weak Pareto minimal point of the following problem*

$$\left(P''\right) : \begin{cases} \mathbb{R}^p_+ - \textit{Minimize } H_1(x) - H_2(x) \\ \textit{subject to} : x \in C \end{cases}$$

where

$$\varphi_i(\bar{x}) = \frac{f_i(\bar{x})}{g_i(\bar{x})},\ H_1(x) := (f_1(x), \dots, f_p(x)),$$
$$H_2(x) := (\varphi_1(\bar{x})\, g_1(x), \dots, \varphi_p(\bar{x})\, g_p(x))$$

Proof. Let $\bar{x}$ be a local weak Pareto minimal point of (P^*). If there exists $x_1 \in \bar{x} + \mathbb{B}_{\mathbb{R}^n}$ such that $x_1 \in C$ and

$$(f_i(x_1) - \varphi_i(\bar{x})\, g_i(x_1)) - (f_i(\bar{x}) - \varphi_i(\bar{x})\, g_i(\bar{x})) \in -\operatorname{int}\left(\mathbb{R}^p_+\right).$$

Since $f_i(\bar{x}) - \varphi_i(\bar{x})\, g_i(\bar{x}) = 0$, one has

$$\frac{f_i(x_1)}{g_i(x_1)} - \frac{f_i(\bar{x})}{g_i(\bar{x})} \in -\operatorname{int}\left(\mathbb{R}^p_+\right),$$

which contradicts the fact that $\bar{x}$ is a local weak Pareto minimal point of (P^*). So $\bar{x}$ is a local weak Pareto minimal point of $\left(P''\right)$. The converse implication can be proved in the similar way. The proof is thus completed. □

Applying Corollary 3.1 and Corollary 4.1 to $\left(P''\right)$, we deduce necessary and sufficient optimality conditions for the vector fractional mathematical programming problem (P^*).

Theorem 5.1 ***Necessary optimality conditions.*** *If $\bar{x}$ is a local weak Pareto minimal point of (P^*) then for any $\varepsilon > 0$ there are $x_1, x_2 \in \overline{x} + \varepsilon\mathbb{B}_{\mathbb{R}^n}$, $f(x_2) \in f(\overline{x}) + \varepsilon\mathbb{B}_{\mathbb{R}^p}$, $g(x_2) \in g(\overline{x}) + \varepsilon\mathbb{B}_{\mathbb{R}^p}$, $x_1 \in C$ and $(\gamma_1^*, \ldots, \gamma_p^*) \in \left(-\mathbb{R}_+^p\right)^\circ \setminus \{0\}$ such that*

$$\sum_{i=1}^{p} \gamma_i^* \varphi_i(\overline{x})\, \partial g_i(x_2) \subset \widehat{N}(x_1, C) + \partial\left(\sum_{i=1}^{p} \gamma_i^* f_i\right)(x_2).$$

Theorem 5.2 ***Sufficient optimality conditions.*** *Let $\bar{x} \in C$. Assume that C is convex and that there exists $\alpha = (\alpha_1, \ldots, \alpha_p) \in \left(-\mathbb{R}_+^p\right)^\circ \setminus \{0\}$ such that*

$$\partial_\varepsilon\left(\sum_{i=1}^{p} \alpha_i \varphi_i(\overline{x})\, g_i\right)(\overline{x}) \subset \partial_\varepsilon\left(\sum_{i=1}^{p} \alpha_i f_i + \delta_C\right)(\overline{x}), \text{ for all } \varepsilon \in \mathbb{R}_+^*.$$

Then $\bar{x}$ is a weak Pareto minimal point of (P^*).

Acknowledgment. This work has been supported by the Alexander-von-Humboldt foundation. I would like to acknowledge the contribution of the following professors :
Professor Boris Mordukhovich for sending many of his insightful articles;
Professor Stephan Dempe for encouragements, remarks and suggestions which improved the original version of this work.

References

1. Clarke, F. H. : Optimization and Nonsmooth Nnalysis, Wiley, New York (1983) .
2. Corley, H. W. : Optimality conditions for maximization of set- valued functions, Journal of Optimization Theory and Application, 58 (1988) , 1-10.
3. Dien, P. H., Locally Lipschitzian set-valued maps and general extremal problems with inclusion constraints. Acta Math Vietnamica. Vol 1, pp. 109-122, 1983.
4. Dinh Tao, P. and Hoai An, L. T. : Convex analysis approach to D. C. programming : theory, algorithms and applications, Acta Mathematica Vietnamica, 22 (1997) 289-355.
5. Flores-Bazan, F. and Oettli, W. : Simplified optimality conditions for minimizing the difference of vector-valued functions, Journal of Optimization Theory and Application, 108 (2001) , 571-586.
6. Gadhi, N. and Metrane, A. : Sufficient Optimality Condition for Vector Optimization Problems Under D.C. Data, To appear in Journal of Global Optimization.
7. Hiriart-Urruty, J. B. : From Convex Optimization to Nonconvex Optimization. Nonsmooth Optimization and Related Topics. Editors F.H.Clarke, V.F. Demyanov, F. Giannessi, Plenum Press, (1989), 219-239.
8. Horst, R. and Tuy, H., Global Optimization (Deterministic Approach), 3rd ed., Springer-Verlag, New York, 1996.
9. Kruger, A. Y. : Properties of generalized differentials, Siberian Mathematical Journal, 26 (1985) 822-832.

10. Lemaire, B. : Subdifferential of a convex composite functional. Application to optimal control in variational inequalities in nondifferentiable optimization, Proceeding Sopron, Sept. 1984 lecture notes in economics and mayhematical systems springer-verlag, pp. 103-117, 1985.
11. Lemaire, B. : Duality in reverse convex optimization, Siam J. Optim, Vol. 8, pp. 1029-1037, 1998.
12. B. Lemaire and M. Volle, Duality in D.C. programing, in Nonconvex Optimization and Its Applications, 27 (1998) 331-345.
13. J-E. Martínez-Legaz, and M. Volle, Duality in D.C. programming : the case of several D.C. constraints, J. Math. Anal. Appl., 237 (1999) 657–671.
14. Martinez-Legaz, J. E. and Seeger, A. : A formula on the approximata subdifferential of the difference of convex functions, Bulletin of the Australian Mathematical Society, 45 (1992) 37-41.
15. Mordukhovich, B. S. : Maximum principle in problems of time optimal control with nonsmooth constraints, J. Appl. Math. Mech., 40 (1976) 960-969.
16. Mordukhovich, B. S. : Metric approximations and necessary optimality conditions for general classes of nonsmooth extremal problems, Soviet. Math. Dokl., 22 (1980) 526-530.
17. Mordukhovich, B. S. : Approximation methods in problems of optimization and control, Nauka, Moscow; 2nd English edition (to appear).
18. Mordukhovich, B. S. : Generalized differential calculus for nonsmooth and set-valued mappings, J. Math. Anal. Appl., 183 (1994) 250-288.
19. Mordukhovich, B. S. : The extremal principle and its applications to optimization and economics, in Optimization and Related Topics, A. Rubinov and B. Glover, eds., Applied Optimization Volumes 47, Kluwer, Dordrecht, The Netherlands (2001) 343–369.
20. Mordukhovich, B. S. : Variational Analysis and Generalized Differentiation, I: Basic Theory, II: Applications, Grundlehren Series (Fundamental Principles of Mathematical Sciences), 330 and 331, Springer, Berlin, 2005.
21. Mordukhovich, B. S. and Shao, Y. : Nonsmooth sequential analysis in Asplund spaces, Transactions of the American Mathematical Society, 348 (1996) 1235-1280.
22. Oettli, W. : Kolmogorov conditions for minimizing vectorial optimization problems, OR Spektrum, 17 (1995) 227-229.
23. Penot, J-P. : Duality for anticonvex programs, Journal of Global optimization, 19 (2001) 163-182.
24. Phelps, R. R. : Convex Functions, Monotone Operators and Differentiability, 2nd edition, Springer, Berlin (1993) .
25. Pontryagin, L. S., Boltyanskii, V. G., Gamkrelidze, R. V. and Mishchenko, E. F. : The mathematical theory of optimal processe, Wiley, New York (1962) .
26. Rockafellar, R. T. : Convex analysis. Priceton Univ. Press, Princeton, New Jersey (1970) .
27. A. Taa, Optimality conditions for vector optimization problems of a difference of convex mappings, To appear in Journal of Global optimization.
28. Tuy, H. and Oettly, W. : On necessary and sufficient conditions for global optimality, Revista de Mathématicas Aplicadas, Vol. 5, pp. 39-41, 1994.

First and second order optimality conditions in set optimization

V. Kalashnikov[1], B. Jadamba[2], and A.A. Khan[3]

[1] Departamento de Ingenieria Industrial y de Sistemas, Centro de Calidad ITESM, Campus Monterrey Ave. Eugenio Garza Sada 2501 Sur Monterrey, N.L. Mexico, 64849. kalash@itesm.mx
[2] Department of Mathematical Sciences, Michigan Technological University, 1400 Townsend Drive, Houghton, MI 49931-1295, USA. jadamba@mtu.edu
[3] Department of Mathematics, University of Wisconsin, 1800 College Drive, Rice Lake, WI-54868, USA. akhan@uwc.edu

Summary. By using the second order asymptotic cone two epiderivatives for set-valued maps are proposed and employed to obtain second order necessary optimality conditions in set optimization. These conditions extend some known results in optimization.

Key words. Set optimization, weak minimality, second order epiderivatives, second order asymptotic cone, optimality conditions.

2000 Mathematics Subject Classification: 90C26, 90C29, 90C30.

1 Introduction

In this paper we are concerned with the optimization problems of the following type:

$$(*) \qquad \text{minimize}_{\mathcal{C}}\, \Phi(x) \qquad \text{subject to} \quad x \in \Omega.$$

Here $\mathcal{X}$ and $\mathcal{Y}$ are real normed spaces (unless stated otherwise all the spaces here will be real), $\Omega \subset \mathcal{X}$, $\mathcal{C} \subset \mathcal{Y}$ is a proper pointed convex cone, the map $\Phi : \mathcal{X} \hookrightarrow \mathcal{Y}$ is set-valued and the minimum may be taken in the sense that we seek $(\bar{x}, \bar{y}) \in \mathcal{X} \times \mathcal{Y}$ such that $\bar{y} \in \Phi(\bar{x})$ and $(\cup_{x\in\Omega}\Phi(x)) \cap (\{\bar{y}\} - \mathcal{C}) = \{\bar{y}\}$. Other possibilities are of finding weak minimizers or proper minimizers (see [13]). We recall that given a pointed convex cone $\mathcal{C} \subset \mathcal{Y}$, the set of minimal points of some $A \subset \mathcal{Y}$, henceforth denoted by $\mathrm{Min}(A, \mathcal{C})$, is defined as

$$\mathrm{Min}(A, \mathcal{C}) := \{x \in A |\; A \cap (\{x\} - \mathcal{C}) = \{x\}\}.$$

S. Dempe and V. Kalashnikov (eds.), *Optimization with Multivalued Mappings*, pp. 265-276
©2006 Springer Science + Business Media, LLC

Being equipped with this terminology, a minimizer to $(*)$ is a point $(\bar{x}, \bar{y}) \in \mathcal{X} \times \mathcal{Y}$ such that $\bar{y} \in \Phi(\bar{x}) \cap \mathrm{Min}(\Phi(\Omega), \mathcal{C})$ where $\Phi(\Omega) := \cup_{x \in \Omega} \Phi(x)$.

Notice that if the map Φ is single-valued then $(*)$ collapses to the known vector optimization problem. Additionally, if $\mathcal{Y} = R$ and $\mathcal{C} = R_+ := \{t \in R | \ t \geq 0\}$ then we recover the framework of classical optimization problems.

The problems of the above type belong to the realm of set optimization, a subject that has attracted a great deal of attention in recent years. In general, set optimization represents the optimization problems with set-valued objective and/or set-valued constraints. Besides the intrinsic interests that these problems bring forth an important generalization and unification of the scalar and the vector optimization problems, there are many research domains which lead directly to the problems of the above kind. Two useful examples for depicting the appearance of set optimization problems are the duality principles in vector optimization and the gap functions for vector variational inequalities (cf. [9]).

In recent years a great deal of attention has been given to the set optimization problems. Starting point for this interesting research domain was the influential paper by Corley [7] where the contingent derivatives and the circatangent derivative were employed to give general optimality conditions. His results were substantially improved by Luc-Malivert [18]. In these works the derivative notion revolves around the graphs of the involved set-valued maps. Another useful approach based on employing the epigraphs of the involved set-valued maps was initiated by Jahn-Rauh [17] which was further pursued in [10, 14, 15, 20], among others. Another interesting approach for set optimization problems that is based on the notion of Mordukhovich's coderivatives, is given in [8]. Although there are now a great variety of results available for first-order optimality conditions, the issue of second order optimality conditions in set optimization as well as in nonsmooth vector optimization is still in need to be exploited (cf. [21]). In [16], using the second-order tangent sets two epiderivatives were introduced and employed to give second-order necessary and sufficient optimality conditions. Although, in some respect the approach of defining the second order epiderivatives is similar to that earlier used for the first order epiderivatives, one sharp contrast is that the epigraph of the second-order epiderivatives is only a closed set whereas the epigraph of the first order epiderivatives is always a cone. On the other hand, while following the recent developments one notices an important concept of second order asymptotic tangent cone due to Penot [23] and Cambini et.al. [4].

In this short paper, we intend to employ the asymptotic cone to introduce two new notions of epiderivative and give new optimality conditions in set-optimization.

This paper is divided into three sections. In Section 2, after recalling some basic definitions, we give two epiderivatives and prove the existence of one of them. Section 3 is devoted to the optimality condition for the local weak minimizers. This section contains two main results, the first one (Theorem 3.1) modifies slightly results of [16] and the second one (Theorem 3.2), gives a different kind of second order optimality condition. Several particular cases are given.

2 Preliminaries

We begin with recalling the notion of the second order contingent set and second-order asymptotic tangent cone (see [1, 22] for details).

Definition 2.1 *Let Ξ be a real normed space, let $\mathcal{S} \subset \Xi$ be nonempty and let $w \in \Xi$.*

1. *The second order contingent set $\mathcal{T}^2(\mathcal{S}, \bar{z}, w)$ of $\mathcal{S}$ at $\bar{z} \in \mathrm{cl}(\mathcal{S})$ (closure of $\mathcal{S}$) in the direction $w \in \Xi$ is the set of all $z \in \Xi$ such that there are a sequence $(z_n) \subset \Xi$ with $z_n \to z$ and a sequence $(\lambda_n) \subset P := \{t \in R \mid t > 0\}$ with $\lambda_n \searrow 0$ so that $\bar{z} + \lambda_n w + \frac{1}{2}\lambda_n^2 z_n \in \mathcal{S}$.*
2. *The second order asymptotic tangent cone $\widetilde{\mathcal{T}}^2(\mathcal{S}, \bar{z}, w)$ of $\mathcal{S}$ at $\bar{z} \in \mathrm{cl}(\mathcal{S})$ in the direction $w \in \Xi$ is the set of all $z \in \Xi$ such that there are a sequence $(z_n) \subset \Xi$ with $z_n \to z$ and a sequence $(\lambda_n, \delta_n) \subset P \times P$ with $(\lambda_n, \delta_n) \searrow (0, 0)$ so that $\bar{z} + \lambda_n w + \frac{1}{2}\lambda_n \delta_n z_n \in \mathcal{S}$.*
3. *The contingent cone $\mathcal{T}(\mathcal{S}, \bar{z})$ of $\mathcal{S}$ at $\bar{z} \in cl(\mathcal{S})$ is the set of all $z \in \Xi$ such that there are a sequence $(z_n) \subset \mathcal{S}$ with $z_n \to \bar{z}$ and a sequence $(\lambda_n) \subset P$ with $\lambda_n \to \infty$ so that $\lambda_n(z_n - \bar{z}) \to z$.*

Remark 2.1 It is known that the contingent cone $\mathcal{T}(\mathcal{S}, \bar{z})$ is a nonempty closed cone (cf. [1]). However, $\mathcal{T}^2(\mathcal{S}, \bar{z}, w)$ is only a closed set (possibly empty), non-connected in general, and it may be nonempty only if $w \in \mathcal{T}(\mathcal{S}, \bar{z})$. On the other hand, $\widetilde{\mathcal{T}}^2(\mathcal{S}, \bar{z}, w)$ is a closed cone (possibly empty) which may be nonempty only if $w \in \mathcal{T}(\mathcal{S}, \bar{z})$. Moreover, if $\mathcal{S}$ is convex and $\widetilde{\mathcal{T}}^2(\mathcal{S}, \bar{z}, w) \neq \emptyset$ then $\mathcal{T}^2(\mathcal{S}, \bar{z}, w) \subset \widetilde{\mathcal{T}}^2(\mathcal{S}, \bar{z}, w)$. Some details and examples of these cone are given in [1, 4, 19, 22, 23, 24].

Let $\mathcal{X}$ and $\mathcal{Y}$ be real normed spaces and let $\Phi : \mathcal{X} \hookrightarrow \mathcal{Y}$ be a set-valued map. The effective domain and the graph of Φ are given by $\mathrm{dom}(\Phi) := \{x \in \mathcal{X} | \quad \Phi(x) \neq \emptyset\}$ and $\mathrm{graph}(\Phi) := \{(x, y) \in \mathcal{X} \times \mathcal{Y} | \quad y \in \Phi(x)\}$, respectively. Given a proper convex cone $\mathcal{C} \subset \mathcal{Y}$, the so-called profile map $\Phi_+ : \mathcal{X} \hookrightarrow \mathcal{Y}$ is defined by: $\Phi_+(x) := \Phi(x) + \mathcal{C}$, for every $x \in \mathrm{dom}(\Phi)$. Moreover the epigraph of Φ is the graph of Φ_+, that is, $\mathrm{epi}(\Phi) = \mathrm{graph}(\Phi_+)$.

Definition 2.2 *Let $\Phi : \mathcal{X} \hookrightarrow \mathcal{Y}$ be a set-valued map, let $(\bar{x}, \bar{y}) \in \mathrm{graph}(\Phi)$, and let $(\bar{u}, \bar{v}) \in \mathcal{X} \times \mathcal{Y}$.*

(i) *A set-valued map $\widetilde{\mathcal{D}}_a^2\Phi(\bar{x}, \bar{y}, \bar{u}, \bar{v}) : \mathcal{X} \hookrightarrow \mathcal{Y}$ defined by*

$$\widetilde{\mathcal{D}}_a^2\Phi(\bar{x}, \bar{y}, \bar{u}, \bar{v})(x) := \left\{ y \in \mathcal{Y} | \quad (x, y) \in \widetilde{\mathcal{T}}^2(\mathrm{graph}(\Phi), (\bar{x}, \bar{y}), (\bar{u}, \bar{v})) \right\}$$

is called second order asymptotic derivative of Φ at $(\bar{x},\bar{y})$ in the direction $(\bar{u},\bar{v})$.

(ii) *A set-valued map $\mathcal{D}_c^2\Phi(\bar{x},\bar{y},\bar{u},\bar{v}) : \mathcal{X} \hookrightarrow \mathcal{Y}$ defined by*

$$\mathcal{D}_c^2\Phi(\bar{x},\bar{y},\bar{u},\bar{v})(x) := \{y \in \mathcal{Y} | \quad (x,y) \in \mathcal{T}^2(\text{graph}(\Phi),(\bar{x},\bar{y}),(\bar{u},\bar{v}))\}$$

is called second order contingent derivative of Φ at $(\bar{x},\bar{y})$ in the direction $(\bar{u},\bar{v})$.

Details of the second order contingent derivative are available in [1] and the second order asymptotic derivative is based on the ideas in [22].

It is clear that if $(\bar{u},\bar{v}) = (0_{\mathcal{X}},0_{\mathcal{Y}})$ in the above definition, where $0_{\mathcal{X}}$ and $0_{\mathcal{Y}}$ are the zero elements in $\mathcal{X}$ and $\mathcal{Y}$, we recover the contingent derivative $\mathcal{D}_c\Phi(\bar{x},\bar{y})$ of Φ at $(\bar{x},\bar{y})$ (cf. [1]). In particular, if $\Phi : \mathcal{X} \to \mathcal{Y}$ is a single valued map which is twice continuously Fréchet differentiable around $\bar{x} \in \Omega \subset \mathcal{X}$, then the second order contingent derivative of the restriction Φ_Ω of Φ to Ω at $\bar{x}$ in a direction $\bar{u}$ is given by the formula (see [1, p. 215]):

$$\mathcal{D}_c^2\Phi_\Omega(\bar{x},\Phi(\bar{x}),\bar{u},\Phi'(\bar{x})(\bar{u}))(x) = \Phi'(\bar{x})(x) + \Phi''(\bar{x})(\bar{u},\bar{u}) \text{ for } x \in \mathcal{T}^2(\Omega,\bar{x},\bar{u}). \tag{1}$$

It is empty when $x \notin \mathcal{T}^2(\Omega,\bar{x},\bar{u})$.

Now we are in a position to introduce two new second order epiderivatives.

Definition 2.3 *Let $\mathcal{X}$ and $\mathcal{Y}$ be real normed spaces and let $\mathcal{C} \subset \mathcal{Y}$ be a pointed convex cone. Let $\Phi : \mathcal{X} \hookrightarrow \mathcal{Y}$ be a set-valued map, let $(\bar{x},\bar{y}) \in \text{graph}(\Phi)$ and let $(\bar{u},\bar{v}) \in \mathcal{X} \times \mathcal{Y}$.*

(a) *A single-valued map $\widetilde{\mathcal{D}}^2\Phi(\bar{x},\bar{y},\bar{u},\bar{v}) : \mathcal{X} \to \mathcal{Y}$ defined by*

$$\text{epi}(\widetilde{\mathcal{D}}^2\Phi(\bar{x},\bar{y},\bar{u},\bar{v})) = \widetilde{\mathcal{T}}^2(\text{epi}(\Phi),(\bar{x},\bar{y}),(\bar{u},\bar{v})) \tag{2}$$

is called second order asymptotic epiderivative of Φ at $(\bar{x},\bar{y})$ in direction $(\bar{u},\bar{v})$.

(b) *A set-valued map $\widetilde{\mathcal{D}}_g^2\Phi(\bar{x},\bar{y},\bar{u},\bar{v}) : \mathcal{X} \hookrightarrow \mathcal{Y}$ defined by*

$$\widetilde{\mathcal{D}}_g^2\Phi(\bar{x},\bar{y},\bar{u},\bar{v})(x) = \text{Min}(\widetilde{\mathcal{D}}_a^2\Phi_+(\bar{x},\bar{y},\bar{u},\bar{v})(x),\mathcal{C}),$$

$$x \in \text{dom}(\widetilde{\mathcal{D}}_a^2\Phi_+(\bar{x},\bar{y},\bar{u},\bar{v}))$$

is called generalized second order asymptotic epiderivative of Φ at $(\bar{x},\bar{y})$ in direction $(\bar{u},\bar{v})$.

If the cone $\widetilde{\mathcal{T}}^2$ in (a) is replaced by the set $\mathcal{T}^2$, then we get the second order contingent epiderivative (see [16]) which we will denote by $\mathcal{D}^2\Phi(\bar{x},\bar{y},\bar{u},\bar{v})$. If in (b) we replace $\widetilde{\mathcal{D}}_a^2\Phi_+(\bar{x},\bar{y},\bar{u},\bar{v})(x)$ by $\mathcal{D}_c^2\Phi_+(\bar{x},\bar{y},\bar{u},\bar{v})(x)$, then the generalized second-order epiderivative is obtained, which we will denote by $\mathcal{D}_g^2\Phi(\bar{x},\bar{y},\bar{u},\bar{v})$ (see [16]). In this case, if $\mathcal{D}_c^2\Phi_+(\bar{x},\bar{y},\bar{u},\bar{v})(x) = \emptyset$, we set $\mathcal{D}_g^2\Phi(\bar{x},\bar{y},\bar{u},\bar{v})(x) = \emptyset$. Moreover, if in the above $(\bar{u},\bar{v}) = (0_{\mathcal{X}},0_{\mathcal{Y}})$, then we recover the contingent epiderivative $\mathcal{D}\Phi(\bar{x}, \bar{y})$ (cf. [17]) and the generalized contingent epiderivative $\mathcal{D}_g\Phi(\bar{x}, \bar{y})$ (cf. [6]) of Φ at $(\bar{x}, \bar{y})$, respectively.

Since, in general, the epigraph of Φ has nicer properties than the graph of Φ, it is advantageous to employ the epiderivatives in set optimization. For example, it is a less stringent requirement that the epi(Φ) is convex than that the graph(Φ) is convex.

Now we give some auxiliary results for the second order derivative and the epiderivatives.

Proposition 2.1 *For every* $x \in \mathrm{dom}(\widetilde{\mathcal{D}}^2_a\Phi(\bar{x}, \bar{y}, \bar{u}, \bar{v}))$, *the following relation holds:*

$$\widetilde{\mathcal{D}}^2_a\Phi(\bar{x}, \bar{y}, \bar{u}, \bar{v})(x) + \mathcal{C} \subseteq \widetilde{\mathcal{D}}^2_a\Phi_+(\bar{x}, \bar{y}, \bar{u}, \bar{v})(x).$$

Proof. Let $y \in \widetilde{\mathcal{D}}^2_a\Phi(\bar{x}, \bar{y}, \bar{u}, \bar{v})(x)$ and let $c \in \mathcal{C}$ be arbitrarily chosen. Because $(x, y) \in \mathrm{graph}(\widetilde{\mathcal{D}}^2_a\Phi(\bar{x}, \bar{y}, \bar{u}, \bar{v}))$ we have $(x, y) \in \widetilde{\mathcal{T}}^2(\mathrm{graph}(\Phi), (\bar{x}, \bar{y}), (\bar{u}, \bar{v}))$. Therefore there exist sequences $(\lambda_n, \delta_n) \subset P \times P$, $((x_n, y_n)) \subset \mathcal{X} \times \mathcal{Y}$ such that $(\lambda_n, \delta_n) \searrow (0, 0)$, $(x_n, y_n) \to (x, y)$ and $\bar{y} + \lambda_n\bar{v} + \frac{1}{2}\lambda_n\delta_n y_n \in \Phi(\bar{x} + \lambda_n\bar{u} + \frac{1}{2}\lambda_n\delta_n x_n)$. By setting $\bar{y}_n := y_n + c$, we notice that

$$\begin{aligned}\bar{y} + \lambda_n\bar{v} + \frac{1}{2}\lambda_n\delta_n\bar{y}_n &= \bar{y} + \lambda_n\bar{v} + \frac{1}{2}\lambda_n\delta_n y_n + \frac{1}{2}\lambda_n\delta_n c\\ &\in \Phi(\bar{x} + \lambda_n\bar{u} + \frac{1}{2}\lambda_n\delta_n x_n) + \mathcal{C},\end{aligned}$$

and this implies that $(\bar{x}, \bar{y}) + \lambda_n(\bar{u}, \bar{v}) + \frac{1}{2}\lambda_n\delta_n(x_n, \bar{y}_n) \in \mathrm{epi}(\Phi)$. Since $\bar{y}_n$ converges to $y + c$, we deduce that $y + c \in \mathcal{D}^2_a\Phi_+(\bar{x}, \bar{y}, \bar{u}, \bar{v})(x)$. The proof is complete. □

Notice that, in view of the fact that $\Phi_+(\cdot) + \mathcal{C} = \Phi_+(\cdot)$ the above result implies that for every $x \in \mathrm{dom}(\mathcal{D}^2_a\Phi_+(\bar{x}, \bar{y}, \bar{u}, \bar{v}))$, the following relation holds:

$$\widetilde{\mathcal{D}}^2_a\Phi_+(\bar{x}, \bar{y}, \bar{u}, \bar{v})(x) + \mathcal{C} = \widetilde{\mathcal{D}}^2_a\Phi_+(\bar{x}, \bar{y}, \bar{u}, \bar{v})(x).$$

Recall that a convex cone $\mathcal{C}$ is called regular (cf. [11]), if each $\mathcal{C}$-decreasing and $\mathcal{C}$-lower bounded (see [9, 11] for details) sequence converges to an element of $\mathcal{C}$. We also need to recall the following

Lemma 2.1 [11] *Let* $\mathcal{Y}$ *be a real normed space and let* $\mathcal{C} \subset \mathcal{Y}$ *be a convex regular cone. Let* $\mathcal{D} \subset \mathcal{Y}$ *be closed and* $\mathcal{C}$*-lower bounded. Then* $\mathrm{Min}(\mathcal{D}, \mathcal{C}) \neq \emptyset$ *and* $\mathcal{D} \subseteq \mathrm{Min}(\mathcal{D}, \mathcal{C}) + \mathcal{C}$.

The following is an existence theorem.

Theorem 2.1 *Let* $\mathcal{X}$ *and* $\mathcal{Y}$ *be real normed spaces and let* $\mathcal{C} \subset \mathcal{Y}$ *be a regular convex cone. Let* $\Phi : \mathcal{X} \hookrightarrow \mathcal{Y}$ *be a set-valued map, let* $(\bar{x}, \bar{y}) \in \mathrm{graph}(\Phi)$ *and let* $(\bar{u}, \bar{v}) \in \mathcal{X} \times \mathcal{Y}$. *Let for every* $x \in A := \mathrm{dom}(\widetilde{\mathcal{D}}^2_a\Phi_+(\bar{x}, \bar{y}, \bar{u}, \bar{v}))$, *the set*

$\widetilde{\mathcal{D}}_a\Phi_+(\bar{x},\bar{y},\bar{u},\bar{v})(x)$ *have a* $\mathcal{C}$*-lower bound. Then* $\widetilde{\mathcal{D}}_g^2\Phi(\bar{x},\bar{y},\bar{u},\bar{v})(x) \neq \emptyset$ *for every* $x \in A$. *Moreover the following relations hold:*

$$\widetilde{\mathcal{D}}_a^2\Phi_+(\bar{x},\bar{y},\bar{u},\bar{v})(x) \subseteq \widetilde{\mathcal{D}}_g^2\Phi(\bar{x},\bar{y},\bar{u},\bar{v})(x) + \mathcal{C} \quad \text{for every} \quad x \in A. \tag{3}$$

$$\text{epi}(\widetilde{\mathcal{D}}_g^2\Phi(\bar{x},\bar{y},\bar{u},\bar{v})) = \widetilde{\mathcal{T}}^2(\text{epi}(F),(\bar{x},\bar{y}),(\bar{u},\bar{v})). \tag{4}$$

Proof. The proof is very similar to that of Theorem 2.1 in [16] where the existence of the generalized second order contingent epiderivative was given. □

We conclude this section with a remark that several auxiliary results given in [16] for the generalized second order contingent epiderivative can be extended to the present setting.

3 Main results

Consider the following set optimization problem:

$$\text{(P)} \qquad \text{minimize}_{\mathcal{C}}\ \Phi(x) \qquad \text{subject to} \quad x \in \Omega.$$

Here for a nonempty set $\Omega \subset \mathcal{X}$, the map $\Phi : \Omega \hookrightarrow \mathcal{Y}$ is set-valued (if $\Omega \subset \text{dom}(\Phi)$ then one can work with the restriction of Φ on Ω), $\mathcal{X}$ and $\mathcal{Y}$ are real normed spaces, where the space $\mathcal{Y}$ is partially ordered by a nontrivial pointed closed convex cone $\mathcal{C} \subset \mathcal{Y}$ with nonempty topological interior and $\partial\mathcal{C}$ as its boundary. We are interested in local weak minimizers of (P). Recall that a pair $(\bar{x},\bar{y}) \in \text{graph}(\Phi)$ is called *local weak minimizer* of (P) if there exists a neighborhood U of $\bar{x}$ such that

$$\Phi(\Omega \cap U) \cap (\{\bar{y}\} - \text{int}(\mathcal{C})) = \emptyset, \quad \text{where} \quad \Phi(\Omega \cap U) := \bigcup_{x \in \Omega \cap U} \Phi(x).$$

If this property holds for $U = \mathcal{X}$, then $(\bar{x},\bar{y})$ is called a *weak minimizer* of (P).

Throughout the rest of the paper we assume that the cones and sets involved either in defining the derivatives and epiderivatives or used otherwise are nonempty.

3.1 Optimality via second order asymptotic derivative

The following second order necessary optimality condition is the main result of this section.

Theorem 3.1 *Let* $(\bar{x},\bar{y}) \in \text{graph}(\Phi)$ *be a local weak minimizer of* (P) *and let* $\bar{u} \in \mathcal{D}_0 := \text{dom}(\mathcal{D}\Phi_+(\bar{x},\bar{y}))$ *be arbitrary. Then:*

(i) *For every* $\bar{v} \in \mathcal{D}_c\Phi_+(\bar{x},\bar{y})(\bar{u}) \cap (-\partial\mathcal{C})$ *and*

$$x \in \mathcal{D}_1 := \mathrm{dom}(\mathcal{D}_c^2\Phi_+(\bar{x},\bar{y},\bar{u},\bar{v})),$$

we have

$$\mathcal{D}_c^2\Phi_+(\bar{x},\bar{y},\bar{u},\bar{v})(x) \cap (-\mathrm{int}(\mathcal{C}) - \{\bar{v}\}) = \emptyset. \tag{5}$$

(ii) *For every* $\bar{v} \in \mathcal{D}_c\Phi_+(\bar{x},\bar{y})(\bar{u}) \cap (-\partial\mathcal{C})$ *and*

$$x \in \widetilde{\mathcal{D}}_1 := \mathrm{dom}(\widetilde{\mathcal{D}}_a^2\Phi_+(\bar{x},\bar{y},\bar{u},\bar{v})),$$

we have

$$\widetilde{\mathcal{D}}_a^2\Phi_+(\bar{x},\bar{y},\bar{u},\bar{v})(x) \cap (-\mathrm{int}(\mathcal{C}) - \{\bar{v}\}) = \emptyset. \tag{6}$$

Proof. We will only prove (6) as the proof of (5) is quite similar and is available in [16] with an equivalent definition of the second order contingent set. Since $(\bar{x},\bar{y}) \in \mathrm{graph}(\Phi)$ is a local weak minimizer of (P), there is a neighborhood U of $\bar{x}$ such that

$$\Phi(\Omega \cap U) \cap (\{\bar{y}\} - \mathrm{int}(\mathcal{C})) = \emptyset. \tag{7}$$

Assume that for some $x \in \widetilde{\mathcal{D}}_1$ there exists

$$y \in \widetilde{\mathcal{D}}_a^2\Phi_+(\bar{x},\bar{y},\bar{u},\bar{v})(x) \cap (-\mathrm{int}(\mathcal{C}) - \{\bar{v}\}).$$

Since the above containment implies that

$$(x,y) \in \mathrm{graph}(\widetilde{\mathcal{D}}_a^2\Phi_+(\bar{x},\bar{y},\bar{u},\bar{v})) = \widetilde{\mathcal{T}}^2(\mathrm{epi}(\Phi),(\bar{x},\bar{y}),(\bar{u},\bar{v})),$$

there are sequences $(\lambda_n,\delta_n) \subset P \times P$, and $((x_n,y_n)) \subset \mathrm{epi}(\Phi)$ such that

$$(\lambda_n,\delta_n) \searrow (0,0) \tag{8}$$

$$(x_n,y_n) \to (x,y) \tag{9}$$

$$(\bar{x},\bar{y}) + \lambda_n(\bar{u},\bar{v}) + \frac{1}{2}\lambda_n\,\delta_n(x_n,y_n) \in \mathrm{epi}(\Phi). \tag{10}$$

From $y \in -\mathrm{int}(\mathcal{C}) - \{\bar{v}\}$, $y_n \to y$ and $\delta_n \searrow 0$, there exists $n_1 \in N$ such that

$$\bar{v} + \frac{1}{2}\delta_n y_n \in -\mathrm{int}(\mathcal{C}) \quad \text{for every} \quad n \geq n_1.$$

Since $\lambda_n \in P$, we get

$$\lambda_n\bar{v} + \frac{1}{2}\lambda_n\delta_n y_n \in -\mathrm{int}(\mathcal{C}) \quad \text{for every} \quad n \geq n_1.$$

The above inequality further implies that for every $n \geq n_1$ we have

$$\bar{y} + \lambda_n\bar{v} + \frac{1}{2}\lambda_n\delta_n y_n \in \bar{y} - \mathrm{int}(\mathcal{C}). \tag{11}$$

However, we also have

$$\bar{y} + \lambda_n \bar{v} + \frac{1}{2}\lambda_n \delta_n y_n \in \Phi(\bar{x} + \lambda_n \bar{u} + \frac{1}{2}\lambda_n \delta_n x_n) + \mathcal{C}.$$

Set

$$a_n := \bar{x} + \lambda_n \bar{u} + \frac{1}{2}\lambda_n \delta_n x_n$$
$$b_n := \bar{y} + \lambda_n \bar{v} + \frac{1}{2}\lambda_n \delta_n y_n.$$

In view of the inclusion $b_n \in \Phi(a_n) + \mathcal{C}$ there exists some $w_n \in \Phi(a_n)$ with $b_n \in w_n + \mathcal{C}$. From this we deduce that $w_n \in b_n - \mathcal{C}$. In view of the containments $\mathrm{int}(\mathcal{C}) + C \subseteq \mathrm{int}(\mathcal{C})$ and $b_n \in \bar{y} - \mathrm{int}(\mathcal{C})$, we conclude that

$$w_n \in \bar{y} - \mathrm{int}(\mathcal{C}) \quad \text{for every} \quad n \geq n_1.$$

Since $a_n \in \Omega$ and $a_n \to \bar{x}$, we can find $n_2 \in N$ such that $a_n \in \Omega \cap U$ for every $n \geq n_2$.

Therefore, we have shown that for sufficiently large $n \in N$ there exists $a_n \in \Omega \cap U$ such that $\Phi(a_n) \cap (\bar{y} - \mathrm{int}(\mathcal{C})) \neq \emptyset$. This, however, is a contradiction to the assumption that $(\bar{x}, \bar{y})$ is a local weak minimizer. The proof is complete. □

Remark 3.1 Notice that by setting $(\bar{u}, \bar{v}) = (0_{\mathcal{X}}, 0_{\mathcal{Y}})$ in (6), we obtain the following first order optimality condition: For every $x \in \mathrm{dom}(\mathcal{D}_c \Phi_+(\bar{x}, \bar{y}))$ we have

$$\mathcal{D}_c \Phi_+(\bar{x}, \bar{y})(x) \cap (-\mathrm{int}(\mathcal{C})) = \emptyset. \tag{12}$$

Remark 3.2 It is evident that (12) and (6) hold as first and second order necessary optimality conditions if $\mathcal{D}_c \Phi_+(\bar{x}, \bar{y})$ is replaced by $\mathcal{D}_g \Phi(\bar{x}, \bar{y})$ or $\mathcal{D}\Phi(\bar{x}, \bar{y})$ and $\widetilde{\mathcal{D}}_c^2 \Phi_+(\bar{x}, \bar{y}, \bar{u}, \bar{v})$ is replaced by $\widetilde{\mathcal{D}}_g^2 \Phi(\bar{x}, \bar{y}, \bar{u}, \bar{v})$ or $\widetilde{\mathcal{D}}^2 \Phi(\bar{x}, \bar{y}, \bar{u}, \bar{v})$, respectively.

3.2 Optimality via second order lower Dini derivative

Notice that Theorem 3.1 shows that the local weak minimality can be characterized as a disjunction in the image space of Φ. Before we extract some useful particular cases from Theorem 3.1, we would prefer to give another form of optimality conditions. For brevity we will employ a notion of second-order lower Dini derivative which is inspired by the lower Dini derivative of Penot [22].

Definition 3.1 *Let $\Phi : \mathcal{X} \hookrightarrow \mathcal{Y}$ be a set-valued map and $(\bar{x}, \bar{y}) \in \mathrm{graph}(\Phi)$. The second order lower Dini derivative of Φ at $(\bar{x}, \bar{y})$ in the direction $(\bar{u}, \bar{v}) \in \mathcal{X} \times \mathcal{Y}$ is the set-valued map such that $(x, y) \in \mathrm{graph}(\mathcal{D}_l^2 \Phi(\bar{x}, \bar{y}, \bar{u}, \bar{v}))$ if and only if for every $(\lambda_n) \subset P$ and for every $(x_n) \subset \mathcal{X}$ with $\lambda_n \downarrow 0$ and $x_n \to x$ there are a sequence $(y_n) \subset \mathcal{Y}$ with $y_n \to y$ and an integer $m \in N$ such that $\bar{y} + \lambda_n \bar{v} + \frac{1}{2}\lambda_n^2 y_n \in \Phi(\bar{x} + \lambda_n \bar{u} + \frac{1}{2}\lambda_n^2 x_n)$ for every $n \geq m$.*

We also need to recall the following notion of the second-order interiorly adjacent set.

Definition 3.2 ([25]) *The second-order interiorly adjacent set* $\mathcal{IT}^2(\mathcal{S},\bar{x},\bar{u})$ *of* $\mathcal{S}$ *at* $\bar{x}$ *in the direction* $\bar{u}$ *is the set of all* $x \in \mathcal{X}$ *such that for all sequences* $\{t_n\} \subset P$ *with* $t_n \searrow 0$, *and for all sequences* $(x_n) \subset X$ *with* $x_n \to x$, *we have* $\bar{x} + t_n\bar{u} + \frac{1}{2}t_n^2 x_n \in \mathcal{S}$, *for* n *large enough.*

Given a set-valued map $\Phi : \mathcal{X} \hookrightarrow \mathcal{Y}$, the *weak-inverse image* $\Phi[\mathcal{S}]^-$ of Φ with respect to a set $\mathcal{S} \subset \mathcal{Y}$ is

$$\Phi[\mathcal{S}]^- := \{x \in \mathcal{X} | \quad \Phi(x) \cap \mathcal{S} \neq \emptyset\}.$$

The following is the main result of this subsection.

Theorem 3.2 *Let* $(\bar{x},\bar{y}) \in \mathrm{graph}(\Phi)$ *be a local weak minimizer of* (P) *and let* $\bar{u} \in \mathcal{D}_0 := \mathrm{dom}(\mathcal{D}\Phi_+(\bar{x},\bar{y}))$ *be arbitrary. Then for every* $\bar{v} \in \mathcal{D}_c\Phi_+(\bar{x},\bar{y})(\bar{u}) \cap (-\partial\mathcal{C})$ *we have*

$$\mathcal{T}^2(\Omega,\bar{x},\bar{u}) \cap \mathcal{D}_l^2\Phi(\bar{x},\bar{y},\bar{u},\bar{v})[-\mathrm{int}(\mathcal{C}) - \bar{v}]^- = \emptyset. \tag{13}$$

Proof. To prove (13), we begin by claiming that

$$\mathcal{T}^2(\Omega,\bar{x},\bar{u}) \cap \mathcal{IT}^2(\Phi[\bar{y} - \mathrm{int}(\mathcal{C})]^-,\bar{x},\bar{u}) = \emptyset. \tag{14}$$

Assume that the acclaimed disjunction does not hold. Let $x \in \mathcal{X}$ be such that

$$x \in \mathcal{T}^2(\Omega,\bar{x},\bar{u}) \cap \mathcal{IT}^2(\Phi[\bar{y} - \mathrm{int}(\mathcal{C})]^-,\bar{x},\bar{u}).$$

Then, in view of the containment $x \in \mathcal{T}^2(\Omega,\bar{x},\bar{u})$ there are sequences $(\lambda_n) \subset P$ and $(x_n) \subset \mathcal{X}$ such that $x_n \to x$, $\lambda_n \searrow 0$ and $s_n := \bar{x} + \lambda_n\bar{u} + \frac{1}{2}\lambda_n^2 x_n \in \Omega$. Moreover, since $\lambda_n \searrow 0$ and $x_n \to x$, it follows from $x \in \mathcal{IT}^2(\Phi[\bar{y} - \mathrm{int}(\mathcal{C})]^-,\bar{x},\bar{u})$ that there is an integer $n_1 \in N$ such that $s_n = \bar{x} + \lambda_n\bar{u} + \frac{1}{2}\lambda_n^2 x_n \in \Phi[\bar{y} - \mathrm{int}(\mathcal{C})]^-$. This, however, implies that

$$\Phi(s_n) \cap \{\bar{y} - \mathrm{int}(\mathcal{C})\} \neq \emptyset \quad \text{for all } n \geq n_1.$$

This, however, is a contradiction the local weak minimality.

For (13) it suffices to show that

$$\mathcal{D}_l^2\Phi(\bar{x},\bar{y},\bar{u},\bar{v})[-\mathrm{int}(\mathcal{C}) - \bar{v}]^- \subseteq \mathcal{IT}^2(\Phi[\bar{y} - \mathrm{int}(\mathcal{C})]^-,\bar{x},\bar{u}). \tag{15}$$

Let $x \in \mathcal{D}_l^2\Phi(\bar{x},\bar{y},\bar{u},\bar{v})[-\mathrm{int}(\mathcal{C}) - \bar{v}]^-$. Therefore there exists

$$y \in \mathcal{D}_l^2\Phi(\bar{x},\bar{y},\bar{u},\bar{v})(x) \cap \{-\mathrm{int}(\mathcal{C}) - \bar{v}\}.$$

Let $(x_n) \subset \mathcal{X}$ and $(\lambda_n) \subset P$ be arbitrary sequences such that $x_n \to x$ and $\lambda_n \searrow 0$. It suffices to show that there exists $m \in N$ such that

$$\bar{x} + \lambda_n \bar{u} + \frac{1}{2}\lambda_n^2 x_n \in \Phi[\bar{y} - \text{int}(\mathcal{C})]^-$$

for every $n \geq m$.

By the definition of $\mathcal{D}_l^2\Phi(\bar{x}, \bar{y}, \bar{u}, \bar{v})$, there are a sequence $(y_n) \subset \mathcal{Y}$ with $y_n \to y$ and $n_1 \in N$ such that

$$\bar{y} + \lambda_n \bar{v} + \frac{1}{2}\lambda_n y_n \in \Phi(\bar{x} + \lambda_n \bar{u} + \frac{1}{2}\lambda_n^2 x_n)$$

for every $n \geq n_1$. Since $y \in (-\text{int}(\mathcal{C}) - \bar{v})$ and $y_n \to y$, there exists $n_2 \in N$ such that $\bar{v} + \frac{1}{2}\lambda_n y_n \in -\text{int}(\mathcal{C})$ for every $n \geq n_2$. This implies that

$$\bar{y} + \lambda_n \bar{v} + \frac{1}{2}\lambda_n^2 y_n \in \Phi(\bar{x} + \lambda_n \bar{u} + \frac{1}{2}\lambda_n^2 x_n) \cap (\bar{y} - \text{int}(\mathcal{C}))$$

for $n \geq m := \max\{n_1, n_2\}$. Hence for the sequences (x_n) and (λ_n) we have

$$\bar{x} + \lambda_n \bar{u} + \frac{1}{2}\lambda_n^2 x_n \in \Phi[\bar{y} - \text{int}(\Omega)]^-$$

for $n \geq m$. This, however, implies that $x \in \mathcal{IT}^2(\Phi[\bar{y} - \text{int}(\mathcal{C})]^-, \bar{x}, \bar{u})$. The proof is complete. □

3.3 Special cases

Theorem 3.1 and Theorem 3.2 extend the second order theory known from nonlinear programming to a general set optimization (and nonsmooth vector optimization) problem. For instance, the following result which generalizes [5, Theorem 5.3], is a direct consequence in Theorem 3.1 and Remark 3.1.

Corollary 3.1 *Let $\mathcal{Y} = R$ and let $\mathcal{C} = R_+ := \{t \in R|\ t \geq 0\}$ in Theorem 3.1. Then:*

$$\mathcal{D}_c\Phi_+(\bar{x}, \bar{y})(x) \subseteq R_+ \quad \text{for every } x \in \mathcal{D}_0 := \text{dom}(\mathcal{D}_c\Phi_+(\bar{x}, \bar{y})).$$

Furthermore, for $\bar{u} \in \mathcal{D}_0$ with $0 \in \mathcal{D}_c\Phi_+(\bar{x}, \bar{y})(\bar{u})$, we have

$$\mathcal{D}_c^2\Phi_+(\bar{x}, \bar{y}, \bar{u}, 0)(x) \subseteq R_+ \quad \text{for every} \quad x \in \text{dom}(\mathcal{D}_c^2\Phi_+(\bar{x}, \bar{y}, \bar{u}, 0)).$$

In the above results we have assumed that the effective domain of Φ is Ω. In fact if the effective domain of Φ contains the set Ω, then the above results hold for the restriction Φ_Ω of Φ to Ω. In view of this remark, equation (1) and Theorem 3.1, we obtain the following

Corollary 3.2 *Let $\Phi : \mathcal{X} \to \mathcal{Y}$ be a single-valued map being twice continuously Fréchet differentiable around a point $\bar{x} \in \Omega$ being assumed to be a local weak minimizer of* (P). *Then:*

$$\Phi'_{\Omega}(\bar{x})(x) \notin -\mathrm{int}(\mathcal{C}) \quad \text{for every} \quad x \in \mathcal{T}(\Omega, \bar{x}).$$

Furthermore, for every $\bar{u} \in \mathcal{T}(\Omega, \bar{x})$ *such that* $\bar{v} := \Phi'_{\Omega}(\bar{x})(\bar{u}) \in (-\partial\mathcal{C})$ *we have*

$$\Phi'_{\Omega}(\bar{x})(x) + \Phi''_{\Omega}(\bar{x})(\bar{u}, \bar{u}) \notin -\mathrm{int}(\mathcal{C}) - \{\bar{v}\} \quad \text{for every} \quad x \in \mathcal{T}^2(\Omega, \bar{x}, \bar{u}).$$

As a further specialization of Corollary 3.2 we obtain the following necessary optimality condition in finite dimensional mathematical programming. This result is well comparable to the similar results obtained in [2].

Corollary 3.3 *Assume that in Corollary 3.2, we have* $\mathcal{Y} = R^n$ *and* $\mathcal{C} = R^n_+ := \{x \in R^n |\ x_i \geq 0,$ *for every* $i \in I := \{1, 2, \ldots, n\}\}$. *For simplicity set* $\Phi_{\Omega} = \Phi$ *and define* $I(x) := \{i \in I |\ x_i = 0\}$. *Then:*

$$\Phi'(\bar{x})(x) \notin -\mathrm{int}(R^n_+) \quad \text{for every} \quad x \in \mathcal{T}(\Omega, \bar{x}). \tag{16}$$

Furthermore, for every $\bar{u} \in \mathcal{T}(\Omega, \bar{x})$ *such that* $\bar{v} := \Phi'(\bar{x})(\bar{u}) \in (-\partial R^n_+)$, *we have*

$$\Phi'(\bar{x})(x) + \Phi''(\bar{x})(\bar{u}, \bar{u}) \notin -\mathrm{int}(R^n_+) - \{\bar{v}\} \quad \text{for every} \quad x \in \mathcal{T}^2(\Omega, \bar{x}, \bar{u}). \tag{17}$$

Remark 3.3 Notice that (16) implies that there is no $x \in \mathcal{T}(\Omega, \bar{x})$ with $\Phi'_i(\bar{x})(x) < 0$ for every $i \in I$. Moreover, if for every $\bar{u} \in \mathcal{T}(\Omega, \bar{x})$ such that $\Phi_i(\bar{x})(\bar{u}) \leq 0$ for all $i \in I$ and $I(\bar{v}) \neq \emptyset$, we have the incompatibility of the system
(i) $x \in \mathcal{T}^2(\Omega, \bar{x}, \bar{u})$
(ii) $\Phi'_i(\bar{x})(x) + \Phi''_i(\bar{x})(\bar{u}, \bar{u}) < 0$ whenever $i \in I(\bar{v})$,
then this implies the condition (17).

The following corollary also extends results given in [3], [4] and in [25] for special cases.

Corollary 3.4 *Let* $\Phi : \mathcal{X} \to R$ *be a single-valued map being twice continuously Fréchet differentiable around a point* $\bar{x} \in \Omega$ *which is assumed to be a local weak minimizer of* (P). *Then:*

$$\Phi'_{\Omega}(\bar{x})(x) \geq 0 \quad \text{for every} \quad x \in \mathcal{T}(\Omega, \bar{x}).$$

Furthermore, for every $\bar{u} \in \mathcal{T}(\Omega, \bar{x})$ *such that* $\Phi'_{\Omega}(\bar{x})(\bar{u}) = 0$ *we have*

$$\Phi'_{\Omega}(\bar{x})(x) + \Phi''_{\Omega}(\bar{x})(\bar{u}, \bar{u}) \geq 0 \quad \text{for every} \quad x \in \mathcal{T}^2(\Omega, \bar{x}, \bar{u}).$$

Acknowledgments: The work of the first author was supported by the project SEP-2004-C01-45786 (Mexico).

The work of the third author was partly carried out during the author's stay at the Institute of Applied Mathematics, University of Erlangen-Nürnberg, Martensstr. 3, 91058 Erlangen, Germany. The third author is grateful to Prof. Dr. Johannes Jahn for the stimulating discussion and helpful remarks on this topic.

The authors are grateful to the referee for the careful reading and helpful remarks.

References

1. Aubin, J.P. and H. Frankowska, *Set valued analysis*, Birkhäuser, Boston (1990).
2. Bigi, G. and Castellani, M: *Second order optimality conditions for differentiable multiobjective problems*, RAIRO Oper. Res., 34, 411-426 (2000).
3. Cambini, A. and Martein, L.: *First and second order optimality conditions in vector optimization*, Preprint, Dept. Stat. Appl. Math., Univ. Pisa, 2000.
4. Cambini, A., Martein, L. and Vlach, M.: *Second order tangent sets and optimality conditions,* Math. Japonica, 49, 451-461 (1999).
5. Castellani, M. and Pappalardo, M.: *Local second-order approximations and applications in optimization,* Optimization, 37, 305-321 (1996).
6. Chen, G.Y. and Jahn, J.: *Optimality conditions for set-valued optimization problems*, Math. Meth. Oper. Res., 48, 187-200 (1998).
7. Corley, H.W.: *Optimality conditions for maximization of set-valued functions,* J. Optim. Theory Appl., 58, 1-10 (1988).
8. El Abdouni, B. and Thibault, L.: *Optimality conditions for problems with set-valued objectives*, J. Appl. Anal., 2, 183-201 (1996).
9. Giannessi, F.: *Vector variational inequalities and vector equilibria,* Kluwer Academic Publishers, Dordrecht, 2000.
10. Götz, A. and Jahn, J.: *The Lagrange multiplier rule in set-valued optimization,* SIAM J. Optim., 10, 331-344 (1999).
11. Isac, G.: *Sur l'existence de l'optimum de Pareto*, Riv. Mat. Univ. Parma (4) 9, 303–325 (1984).
12. Isac, G. and Khan, A.A.: *Dubovitski-Milutin approach in set-valued optimization.* Submitted to: SIAM J. Cont. Optim. (2004).
13. Jahn, J.: *Vector optimization. Theory, applications, and extensions.* Springer-Verlag, Berlin, 2004.
14. Jahn, J. and Khan, A.A.: *Generalized contingent epiderivatives in set-valued optimization*, Numer. Func. Anal. Optim., 27, 807-831 (2002).
15. Jahn, J. and Khan, A.A.: *Existence theorems and characterizations of generalized contingent epiderivatives*, J. Nonl. Convex Anal., 3, 315–330 (2002).
16. Jahn, J; Khan, A.A and Zeilinger, P.: *Second order optimality conditions in set-valued optimization,* J. Optim. Theory Appl., 125, 331–347 (2005).
17. Jahn, J. and Rauh, R.: *Contingent epiderivatives and set-valued optimization,* Math. Meth. Oper. Res., 46, 193-211 (1997).
18. Luc, D.T. and Malivert, C.: *Invex optimization problems*, Bull. Austral. Math. Soc., 46, 47–66 (1992).
19. Jiménez, B. and Novo, V: *Second order necessary conditions in set constrained differentiable vector optimization*, Math. Meth. Oper. Res., 58, 299-317 (2003).
20. Khan, A.A. and Raciti, F.: *A multiplier rule in set-valued optimization,* Bull. Austral. Math. Soc., 68, 93-100 (2003).
21. Mordukhovich, B.S. and Outrata, J.V.: *On second-order subdifferentials and their applications,* SIAM J. Optim., 12, 139-169 (2001).
22. Penot, J.P.: *Differentiability of relations and differential stability of perturbed optimization problems,* SIAM J. Cont. Optim., 22, 529-551 (1984).
23. Penot, J.P.: *Second-order conditions for optimization problems with constraints,* SIAM J. Cont. Optim., 37, 303-318 (1999).
24. Rockafellar, R.T. and Wets, J.B.: *Variational analysis*, Springer, Berlin (1997).
25. Ward, D.: *Calculus for parabolic second-order derivatives,* Set-Valued Anal., 1, 213-246 (1993).